1 MONTH OF
FREE
READING

at

www.ForgottenBooks.com

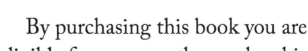

By purchasing this book you are eligible for one month membership to ForgottenBooks.com, giving you unlimited access to our entire collection of over 700,000 titles via our web site and mobile apps.

To claim your free month visit: www.forgottenbooks.com/free581206

ISBN 978-0-483-80601-6
PIBN 10581206

CONTENTS

No. 1. JULY

No. 2. AUGUST

No. 3. OCTOBER

No. 4. NOVEMBER

PROMPT PUBLICATION

The Author can greatly assist the Publishers of this Journal in attaining prompt publication of his paper by following these four suggestions:

1. *Abstract.* Send with the manuscript an Abstract containing not more than 250 words, in the precise form of The Bibliographic Service Card, so that the paper when accepted can be scheduled for a definite issue as soon as received by the Publisher from the Editor.

2. *Manuscript.* Send the Manuscript to the Editor prepared as described in the Notice to Contributors, to conform to the style of the Journal (see third page of cover).

3. *Illustrations.* Send the Illustrations in complete and finished form for engraving, drawings and photographs being protected from bending or breaking when shipped by mail or express.

4. *Proofs.* Send the Publisher early notice of any change in your address, to obviate delay. Carefully correct and mail proofs to the Editor as soon as possible after their arrival.

By assuming and meeting these responsibilities, the author avoids loss of time, correspondence that may be required to get the Abstract, Manuscript and Illustrations in proper form, and does all in his power to obtain prompt publication.

Resumen por el autor, James W. Buchanan.

La regulación de la formación de la cabeza en Planaria por
medio de los anestésicos.

Mediante experimentos en masa el autor demuestra que la
frecuencia de la cabeza en pedazos de Planaria puede regularse
sometiendo dichos pedazos, durante cortos periodos después de
cortarlos, a la acción de concentraciones apropiadas de cloretona,
cloroformo, hidrato de cloral, éter y alcohol etílico; en tales
concentraciones de estos agentes el aumento en la asimilación
de oxígeno después de la sección no se lleva a cabo.   Las pruebas
acumuladas demuestran que los factores que regulan la formación de la cabeza no son específicos y presta apoyo a las
conclusiones de Child respecto a la naturaleza de estos factores,
esto es, que la formación de la cabeza está determinada por las
actividades relativas de dos factores antagónicos: 1) La tendencia de las células en la vecindad de la superficie anterior del
corte a desdiferenciarse y a desarrollarse en la cabeza de un
nuevo individuo; 2) La tendencia del conjunto del pedazo, con
exclusión de las células destinadas al desarrollo de la nueva
cabeza, a mantener la diferenciación del antiguo individuo,
ejerciendo un cierto grado de regulación sobre estas células en
la proximidad de la superficie anterior del corte, y tendiendo a
impedir la formación de una nueva cabeza.   Los hechos encon-
trados por el autor indican que los anestésicos alteran la frecuencia de la cabeza: 1) Por inhibición directa de los procesos
del desarrollo de las células destinadas a la formación de una
nueva cabeza, produciendo disminuciones en la frecuencia de
la cabeza; 2) Mediante inhibicion del aumento de la actividad
metabólica del conjunto del pedazo a raíz de la sección; en
pedazos de ciertas regiones este efecto supera al efecto directo
de las células destinadas a la formación de la nueva cabeza sobre
los procesos del desarrollo, y en tales piezas la frecuencia de la
formación de la cabeza aumenta.

Translation by José F. Nonidez
Cornell Medical College, New York

THE CONTROL OF HEAD FORMATION IN PLANARIA BY MEANS OF ANESTHETICS

J. WILLIAM BUCHANAN

Hull Zoological Laboratory, University of Chicago

THREE FIGURES

Regeneration in Planaria has attracted the attention of investigators for more than a century, at least since 1791 (Randolph, '95). Record of mass experiments is absent from the literature, however, until the closing years of the nineteenth century (Morgan, '98). The present paper presents the results of an attempt to analyze with the aid of certain narcotics the conditions controlling the completeness of regeneration, particularly as regards the head, in pieces of Planaria. The data consist of mass experiments which have involved the cutting of more than forty thousand worms and the handling of more than one hundred and twenty thousand pieces.

THE PROBLEM

Cross-sections from the body of Planaria do not always reconstitute new individuals with anterior ends like those in nature. Abnormalities have been frequently produced experimentally and described in many species and their occurrence explained by diverse theories. In Planaria dorotocephala Child ('11 a) has for convenience distinguished five classes of anterior ends in regenerated pieces, each class continuous into the next. Figure 1 shows the external appearance of the five types: normal (*A*), in which the head is the usual form of those in nature; teratophthalmic (*B, C, D*), in which the shape of the head is normal, but the eyes show some degree of abnormality ranging from slight reductions in size and approximation to the median line, to a single median eye; teratomorphic (*E*), in which the shape of the head is abnormal, reduced in size, and the cephalic

lobes are approximated to the median line and the eye is single
and median; anophthalmic (F), in which there is more or less
anterior regeneration, but no eye; headless (G), in which there
is no appreciable anterior regeneration, merely a healed wound.
These external characteristics are indices of the degree of development of the cephalic ganglia (Child and McKie, '11).   The
term 'head frequency' has been applied to the frequency with
which heads of these types appear in a given number of pieces.
    An examination by Child of the nature of regeneration and
head frequency in pieces from different regions of the body of
several species of Planaria, particularly Planaria dorotocephala,
has led him to the conclusion that the conditions controlling

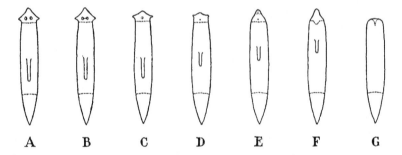

Fig. 1 A, Normal form.   B, C, D, three types of teratophthalmic forms. E,
teratomorphic form.   F, anophthalmic form.   G, acephalic, or headless form.

the degree of regeneration are primarily physiological and
quantitative, not morphological and specific.   He supports this
conclusion with data showing that controlled changes in the
physiological conditions of the animals by feeding, starvation,
temperature, mechanical stimulation, and the action of certain
chemicals bring about controlled changes in head frequency,
either increases or decreases, in pieces from any region of
the body.
    Furthermore, it is evident from the data that changes in the
physiological conditions in the animals which bring about changes
in head frequency are quantitative changes, and not specific.
For example, the rate of oxidative reactions, whether determined by differences in age, or nutrition, or by chemical agents

such as KNC, is a fundamental factor in determining head frequency (Child, '11 a, '19 a, '20 a).

By investigation of the regeneration of pieces of the same length from different regions of the body of animals of the same size, Child showed that there is a definite gradation in head frequency in such pieces, decreasing from the head posteriorly to the region at which fission usually occurs, indicating there the posterior limit of the first zooid. Then it rises abruptly to decrease again toward the end of the second zooid (Child, '11 a, '11 b).

These two facts, that conditions controlling regeneration are physiological and quantitative and that there is an anteroposterior gradation in head frequency in each zooid, when jointly considered lead to the inference that there must be some sort of quantitative physiological gradient along the axis of the animal. By means of certain poisons in solutions strong enough to kill slowly, Child ('11 b) was able to demonstrate a well-marked anteroposterior gradient in survival time. The reagent most extensively used for this purpose was KNC, which is known to interfere with the oxidative reactions. In appropriate solutions of this reagent, as well as others, the anterior end of the animal dies first, indicating that there the rate of oxidative reactions is most rapid, since the protoplasm is most susceptible there.[1] The death process then progresses posteriorly in each zooid, indicating the decreasing rate of metabolism from anterior to posterior (Child, '11 b, '13 a, '13 b).

Child's first conclusion, based on his results on regeneration and this differential susceptibility to poisons, that the gradient made evident by the data is a rough index of the relative rate

[1] The differential susceptibility to external agents of regions of different rates of metabolism has been itself the subject of extended investigation and discussion (Child, '13 a '14 b, '15 a, '20 b, and other papers; Bellamy, '19; Hyman, '19 a, '21 a). The findings have been that regions of high rate of metabolism are more affected by certain ranges of concentration or intensity of action of external agents than regions of lower rate. To certain ranges of lower concentrations or lower intensities of action, however, the regions of higher rate of metabolism are better able to acclimate or recover from the effects of the agent or condition and consequently show eventually less effect of such treatment than regions of lower rate.

of the metabolic processes, particularly the oxidative, in different regions of the animal, has received ample proof. This proof is afforded indirectly by subsequent work showing that the progress of death due to lack of oxygen follows in general the same course as in KNC (Child, '19 b); furthermore, as far as the work has been carried out, the rate of carbon-dioxide production measured both by indicator and the Tashiro methods furnishes direct proof that confirms the conclusions drawn from the susceptibility data (Child, '11 b, '15; Tashiro, '17). And last, but perhaps more important, Doctor Hyman has shown that the relative rate of oxygen consumption of different regions agrees in every case with the susceptibility data (Hyman, '21 b). The value of the susceptibility methods as an indicator of metabolic conditions has thus been sustained both by direct and indirect proof. To recapitulate, the existence of an anteroposterior gradient in rate of metabolic processes, particularly those involved in oxidative reactions, in Planaria dorotocephala may be considered to have been adequately demonstrated. This gradient is known to involve the following physiological processes: the rate of oxygen consumption, the rate of carbon-dioxide production, the rate of death in several types of poisons, and the rate of death due to lack of oxygen. The nature of this gradient and its existence in other forms and rôle in morphogenesis in general have been discussed in a number of papers by Child, Hyman, Bellamy, and others. A general review of the evidence for the existence of such gradients in axiate organisms and their relation to the problem of pattern is given in a recent paper (Child, '20 b).

The isolation of a piece by section introduces certain quantitative changes in the metabolic processes of the isolated part, which are to a greater or less extent dependent on its former level in the gradient. The first change to be considered and probably the first in chronological order is the stimulation produced in immediately adjacent cells by the injury of cutting. The stimulation of injury is a very general phenomenon and requires little discussion here. Tashiro ('17) considers its occurrence to be sufficient evidence that a part is alive. In pieces of Planaria it is indicated by a greatly increased susceptibility of the region to KNC (Child, '13 b). This method indicates that the region of the wound has been stimulated to a higher rate of metabolism than any other region of the piece, but that the rate is probably lower if the cut is through a lower level of the gradient (Child, '14 a, '11 b). We may therefore picture the first effect of section as establishing a region of very high rate of metabolism immediately adjacent to the cut surfaces of the piece, the rate being slightly lower in pieces from the lower levels of the gradient.

The new relations of the cells near the anterior cut surface are somewhat different from the relations of the cells of the corresponding region near the posterior cut surface of the piece. In the former case, since the path of nervous correlations and integrative control is chiefly in the anteroposterior direction, the cells near the anterior cut surface are to a considerable extent freed from such integrative and differentiative controls, especially from those arising in regions more anterior (Child, '11 c, '14 b). The cells near the posterior cut surface are, however, still largely under control of the piece. Attention is confined in the present paper to the conditions controlling the cells near the anterior cut surface, since they produce the new head.

The second result of isolation to be considered is the immediate stimulation of the remainder of the piece. This has been shown by susceptibility methods (Child, '14 a) and by direct measurement of carbon-dioxide production (Robbins and Child, '20) and oxygen consumption (Hyman, '19 b). This stimulation may be considered to arise from the severing of the nerve cords and is probably of nervous origin in all regions of the piece. The degree of stimulation is inverse to the length of the piece and inverse to its position in the gradient of the intact animal; short pieces are more stimulated than long pieces and pieces from posterior lower levels of the zooid are more stimulated than those from the anterior higher levels. The stimulation lasts for a number of hours, decreasing gradually, until within a day or two it completely disappears. In an isolated piece we may therefore picture the second effect of isolation as a tempo-

rary stimulation of the piece in all regions not directly injured by
the cuts, the degree and duration of stimulation depending on
the length of the piece and its original position in the gradient.

For convenience in discussion the cells near the anterior cut
surface that are directly involved in the wound stimulation and
concerned in head formation may be designated as the X cells
(fig. 2) and their rate of metabolism expressed as Rate X.   The
cells of the remainder of the piece that are affected by the nervous stimulation may be designated as the Y cells and their rate
of metabolism expressed as Rate Y.   The relation to each other,
during this period of stimulation, of these two new conditions

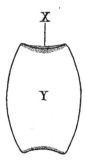

Fig. 2   An isolated piece from the body of Planaria dorotocephala. X, the
region adjacent to the anterior cut surface directly affected by the injury of
cutting; Y, the remainder of the piece, exclusive of the region adjacent to the
posterior cut surface.

set up by cutting and isolation has been shown by Child to be
the factor determining whether or not a head shall develop.   In
the first place, it has been shown that head determination, i.e.,
the establishment of the conditions which determine whether or
not a head shall develop, occurs very soon after isolation and
during the period of stimulation of the Y region by section.   Second, head frequency decreases and the degree of stimulation of
the Y region increases with decrease in length of the piece with
anterior end at a given level.   Third, it has been shown that
in pieces of equal length from different levels the degree of stimulation of the Y region increases and the head frequency decreases
as the level from which the piece is taken becomes more posterior.

Fourth, in young animals the rate of metabolism is higher than in old animals and the difference between Rate X and Rate Y is therefore not as great as in old animals and the head frequency in corresponding fractions of body length is less in young than in old animals. Fifth, the stimulation of the Y region can be eliminated by exposure for a few hours after section to dilute KNC and in pieces in which this stimulation is sufficient to inhibit head development, such treatment increases head frequency (Child, '14 a, '14 b, '16, '20 a). The facts indicate very clearly that in some way physiological conditions in Y tend to inhibit head formation in X and that this inhibiting action increases as the metabolic activity of Y increases in relation to that of X and vice versa.

A brief consideration of the conditions resulting from the isolation of a piece will serve to throw some light on this relation between X and Y. The result of section is not simply the localization of a cut surface at X with its effect on adjoining cells, but also the isolation of the cells at X from the influence of all physiological correlative factors originating in regions anterior to the level of the piece. Since these factors played a large part in determining the differentiation at X, their absence must tend to bring about dedifferentiation in these cells. But the region X is physiologically continuous with the region Y which is less directly affected by isolation of the piece and retains more or less completely its differentiation. Consequently, any correlative factors originating in Y and affecting X must tend to keep the X cells differentiated to some degree, i.e., to retard or inhibit their dedifferentiation. The greater the physiological activity of Y, the more effective are these factors and therefore the greater the degree of inhibition of head development and vice versa.

Since the experimental data already at hand indicate that the factors in this relation between X and Y which are concerned in head development are essentially quantitative and closely associated with the rate of metabolism, particularly of oxidations, in the two regions, we may say that head frequency varies in general with the ratio of Rate X to Rate Y or with the value of the expression $\dfrac{\text{Rate X}}{\text{Rate Y}}$. This is merely a brief statement of the fact that the cells at X give rise to a head if their metabolic rate is high enough in relation to that of Y and that the region Y may retard or prevent head formation in X if its rate of metabolism is high enough in relation to that of X.

It was noted above that head frequency can be either increased or decreased experimentally. This has been accomplished by altering the value $\dfrac{\text{Rate X}}{\text{Rate Y}}$ by a differential change of Rate X and Rate Y and thus altering the normal head frequency. Using KNC, which is known to reduce the rate of oxidative reactions in many organisms (Child, '19 a; Hyman, '19 a), Child was able by the use of appropriate concentrations to control head frequency to some extent in pieces of a certain length (Child, '16). In pieces of the same length from the extreme anterior regions of a given number of worms of the same size, e.g., A in figure 3, subjected immediately after section during the period of stimulation of Rate Y to certain concentrations of KNC, the head frequency was decreased. The result is attributed to the direct effect of the KNC on the cells of the X region; the Y region is known to be not greatly stimulated by section in such anterior pieces, and this stimulation has been found insufficient to inhibit head formation to any marked extent (Child, '13 a; Child and Robbins, '20; Hyman, '19 a). Consequently, head frequency in these pieces depends chiefly on Rate X. Therefore, in such experiments the value $\dfrac{\text{Rate X}}{\text{Rate Y}}$ is decreased by KNC in A pieces by reason of its direct inhibition of Rate X. In pieces of the same length from more posterior regions, e.g., C in figure 3, subjected to the same concentrations for the same length of time after section, the head frequency was increased. This result is attributed to the reduction or prevention of stimulation of section in the Y region by the depressing action of the KNC. The X region thus obtains the degree of independence necessary for the formation of a more or less complete head. As stated previously, the stimulation of the Y region is greatest at lower

levels of the gradient.   In fact, the data show that such posterior
pieces as C are stimulated to a rate of metabolism equal to or
greater than that of anterior pieces (Hyman, '21 b).   The depressing effect of KNC on Rate Y in such posterior pieces with
short time of exposure is sufficient to obscure in the end result
any direct effect on Rate X.   Later Hyman showed that the

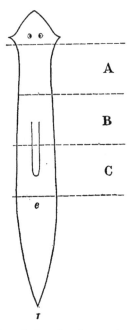

Fig. 3 Planaria dorotocephala, showing regions from which A, B, and C
pieces were taken. *E* to *F*, posterior zooid.

rate of oxygen consumption of Planaria is actually much reduced in such concentrations of KNC and Child showed that
carbon-dioxide production was similarly reduced (Hyman, '19 a;
Child, '19 a).

The data also show that head frequency may be shifted downward by KNC in pieces posterior to the A region (e.g., *B* and *C*
in fig. 3) by increasing the concentration or extending the time
of exposure. The decrease is greatest in A pieces, however.
In pieces from lower levels of the gradient the shift upward in

head frequency occurs, but the number of normals is decreased. The latter effect is obviously the result of the direct effect of the KNC on the X cells of these pieces.   The tendency to increased head frequency in such posterior pieces is therefore somewhat interfered with by the direct effect of the KNC on Rate X.

If such alterations of head frequency are due to the prevention of the stimulation of section and a differential depressing effect of KNC on Rate X and Rate Y, the results ought to be similar for all depressants under comparable experimental conditions. In the present paper I purpose to present the results on head frequency and on oxygen consumption of subjecting pieces from the anterior zooid of Planaria dorotocephala to various concentrations of a number of narcotics.   The work was begun in the autumn of 1915.   I desire here to acknowledge my deep obligations to Prof. C. M. Child, under whose direction the work was done, for his stimulating interest and helpful suggestions and criticism.   I am also indebted to Dr. L. H. Hyman for valuable suggestions and advice.

MATERIAL AND METHODS

A single species, Planaria dorotocephala, was used in this study.   Since it has been shown that the physiological condition of the animals controls the head frequency of the pieces to a considerable extent, consistent and uniform results can be expected from standardized material only.   The environment of the animals in nature varies with respect to light, temperature, food supply, and other conditions.   For this work, therefore, the worms after collection were kept in the laboratory for some time before being used for experiment—at least two weeks and in most cases several months—in order to bring the animals as nearly as possible to the same physiological condition.   Each collection was kept in separate pans under approximately the same conditions of light and temperature.   The city tap-water, because of its free chlorine content, was unfit for use, and the water used in the pans was drawn from a well.   The stocks were fed on beef liver three times a week.   In stocks so controlled, size is the best index of physiological condition, and for experiment the required number of worms of as nearly as possible the same size were selected at the same time from a single stock pan. Tests show that under such conditions two groups of animals so selected do not differ appreciably in rate of oxygen consumption. Nutritive conditions were controlled as the experiments required by withholding the food from the pan from which the selection was to be made for the designated period of time.

The narcotics used were a number of common anesthetics, chloroform, ether, chloretone, chloral hydrate, ethyl alcohol, and, in a few cases, methyl, normal propyl, and iso-butyl alcohols. When obtainable, Kahlbaum's and Squibb's products were used. All solutions were made up freshly just before using. The concentrations mentioned in the data are only approximate; although corrected for the specific gravity of the anesthetic and made up as carefully as possible, no attempt was made to determine the molecular concentrations more accurately.

The nature of the experiments divides them into two groups: those concerned with head frequency after short periods of exposure of the freshly cut pieces to relatively strong solutions of the anesthetics and with the rate of oxygen consumption during such exposure, and those concerned with the head frequency of pieces kept in more dilute solutions of the anesthetics during the entire period of regeneration and the effects of such solutions on the rate of oxygen consumption of the pieces during this period of time.

In general, for the head-frequency experiments one hundred or more worms of as nearly as possible identical size were selected from the same stock pan at the same time. Fifty of them served as controls and the anterior zooids were cut into A, B, and C pieces (fig. 3), the heads and posterior zooids being discarded. The pieces as soon as cut were dropped with a pipette into separate Erlenmeyer flasks of 500 cc. capacity filled with well-aerated water, A pieces in one flask, B in another, and C in another, each flask thus receiving fifty pieces from the same region of the animals. The flasks were then stoppered; a small air space was left under the stopper to allow for the changes in volume of the contents with temperature changes.

The remaining worms were cut in the same manner in groups of fifty, but the pieces were dropped as soon as cut into the appro priate concentration of the anesthetic. As soon as the cutting was completed, the solution was replaced by fresh and the flasks stoppered as the control. One series of A, B, and C pieces regenerating in water thus served as a control for one or more series subjected to the anesthetic. In experiments designed to test the effect on head frequency of short-period exposures to the anesthetics the solutions were poured off after a number of hours, the pieces washed several times and set aside to regenerate in water. In experiments designed to test the effect on head frequency of more dilute solutions of the anesthetics applied throughout the period of regeneration, the anesthetic was not replaced by water. In all experiments the water or solutions were replaced by fresh every forty-eight hours. Both control and experimental series were kept on the same shelf at approximately the same temperature and out of direct light. At the end of fourteen days the regenerated animals were removed from the flasks and examined under a dissecting microscope. The number of each type of anterior end was tabulated in per cent, and the result for each flask is considered the head frequency under the conditions of the experiment of the level from which the pieces were cut.

Slight differences in length of the pieces undoubtedly introduce an unavoidable error. Another source of error lies in slight differences in the physiological conditions of the animals. The total error is certainly not greater than 10 per cent, but a difference between the control and experimental pieces from the same level of less than 10 per cent that repeats itself throughout a series of comparable experiments must be considered as evidence of the effect of the anesthetic on head frequency. To reduce the subjective error of cutting pieces of unequal length as much as possible, in some of the earlier experiments the control and experimental sets were cut in alternate groups of ten. However, no differences in results could be noted between such series and those cut in two consecutive groups of fifty and the method was discontinued as time consuming.

Direct measurement of the oxygen consumed in such head-frequency experiments is not practicable; first, because fifty pieces do not afford sufficient bulk of living material to produce in a convenient period of time marked changes in the oxygen content of 500 cc. of liquid, which is the smallest amount from which it is practicable to obtain a fair sample of 125 cc. for analysis; second, the operation of weighing, necessary in calculating the rate of oxygen consumption for purposes of comparison, undoubtedly injures the pieces to some extent, particularly the delicate tissues of the new head, and may result in the loss of a few pieces. Either of these events vitiates the results of a head-frequency experiment. It was necessary, therefore, to measure the oxygen consumption in a separate series of experiments that are comparable to those on head frequency. The conditions of an oxygen-consumption experiment were of necessity somewhat different from those of a head-frequency experiment. It was not always practicable to use the same concentrations of the anesthetics employed in head-frequency experiments nor animals of the same size. More than fifty worms were used in most cases, and all three pieces, A, B, and C, were dropped into the same flask. Thus after the pieces were cut there were two or more flasks, one, the control, containing A, B, and C pieces in water and one or more others containing A, B, and C pieces in the proper solution of the anesthetic. The solution and water were then poured off and fresh introduced at the bottom of the flasks by means of inlet tubes passing through the rubber stoppers. The stoppers also held outlet siphons and holes to allow for overflow during the process of filling. The solution and water were allowed to overflow through the stoppers for some time, until all the liquid that had been exposed to air had been replaced. All air bubbles were thus forced out of the flasks. The holes were then plugged, the siphons and inlet tubes clamped, and the flasks thus sealed were placed in a water-bath in which the temperature did not vary one degree during the time of the experiment. Fair samples were taken from the original stock solution and water and after a certain time interval fair samples were taken from the flasks by means of the outlet siphons. After the withdrawal of the samples from the flasks, the remaining liquids were poured off and replaced by fresh. The above operations of filling and taking samples were then repeated as often as necessary. In experiments requiring a record of the rate of oxygen consumption over a considerable period of time, the required samples were taken after the first ten to twenty hours after cutting. The flasks were then set aside at room temperature, the solutions changed frequently, and on designated days the rate of oxygen consumption over a certain number of hours was taken in the same manner as that of freshly cut pieces.

The oxygen content of each sample was measured by the Winkler method. This method has been so widely used that no extended discussion or account of it is necessary here. Due precautions were exercised to exclude casual air or other sources of oxygen from the system. With slight modifications, the methods of analysis described by Birge and Juday ('11) and modified by Hyman ('19 a) were followed. Data presented by Hyman show that with apparatus for obtaining samples very like my own the error introduced is not greater than two hundredths of a cubic centimeter when well-aerated water is used, and since in all cases in this work well aerated water or solutions were used, the error is considered to be plus or minus two hundredths. When the rate of oxygen consumption per milligram of weight of the animals is calculated, this error becomes very small indeed, and variations in results must be considered to be due to variations in the physiological condition of the animals from day to day, due to temperature changes in the laboratory and other possible sources. The titrations were not affected by the presence of the anesthetics, so far as can be determined. As Hyman has shown ('19 a), the planarians add something to the water which decreases its iodine-absorbing power, and the oxygen content of water in which the planarians have been living measured by this method is not really exact. Whether or not the pieces add more of this substance or less when subjected to an anesthetic, I do not attempt to say. Certainly, the error thus introduced must be very small when compared with the relatively great differences of the oxygen content of samples from the flasks containing water and those containing solutions of anesthetics.

The measurement of the oxygen consumption of pieces from all three levels in bulk obviously does not afford any comparison of the effect of the anesthetics on the stimulation by section of pieces from different levels of the gradient. But it is known that anterior pieces are less stimulated than posterior and the measurement of the effect of the anesthetics on the stimulation of section in all pieces is quite satisfactory for the purpose of analyzing the factors concerned in head determination.

After the data on the oxygen content of the original stock solutions and water and of the liquids in the flasks had been secured, the pieces were weighed. The method of weighing was as follows: The pieces were poured into a funnel lined with hard-surfaced filter-paper. The liquid was then poured off and the filter-paper removed and drawn over a dry glass plate until it adhered firmly. The pieces were then picked up with a thin spatula and placed in a Paar weighing tube, the cover of the tube replaced, and the tube with its contents weighed, the weight of the tube being known. It is admittedly unavoidable not to weigh some of the liquid by this method, but it is more important to avoid excessive drying in experiments involving tests of oxygen consumption at intervals throughout the period of regeneration. Even with this method there was no doubt some slight stimulation resulting from partial drying and handling. The results were consistent, however. In comparable experiments pieces lost weight at the same rate during the period of regeneration. Worms of the same physiological condition showed the same rate of oxygen consumption per unit of weight. It seems, therefore, that the error introduced by weighing a small amount of the liquid with the pieces was fairly constant.

EFFECT OF STRONG SOLUTIONS ACTING FOR SHORT PERIODS AFTER SECTION

Chloretone

Head frequency. The data obtained on the head frequency of pieces subjected for short periods after section to various concentrations of chloretone are given complete in table 1A. These

J. WILLIAM BUCHANAN

TABLE 1A

Head frequency—chloretone. In per cent

SERIES	TIME	PART	NORM	T. OPHTH	T MOR	ANOPH	RDL	DEAD
3/4, '16· 10-mm. worms. Fed two days before cutting								
	hours							
Control		A	78	16				6
		B		16	10	72		2
		C	2	12		64	22	0
1. M. 1/175......	3	A	68	14				18
		B	2	26	16	46		10
		C	4	14	2	62	18	0
2. M. 1/450......	12	A	66	26				8
		B		14	6	70		10
		C		2	4	68	26	0
3/14, '16· 15-mm. worms. Fed two days before cutting								
Controls.........		A	90	10				0
		B		12	6	78		4
		C	10	12	2	60	16	0
3. M. 1/175......	3	A	86	10				4
		B		12	4	70	10	4
		C	20	16	6	40	18	0
4. M. 1/450......	3	A	78	22				0
		B	2	18	6	74		0
		C	12	18	6	36	28	0
4/22, '20· 16- to 18-mm. worms. Fed two days before cutting								
Control..........		A	80	20				0
		B		68	2	30		0
		C	2	30		60	8	0
5. M. 1/450......	12	A	64	36				0
		B	4	44		52		0
		C	4	48		48		0

data show very clearly that the head frequency in A pieces is
decreased and in C pieces increased by reason of exposure of these
regions of the same animals to the concentration of chloretone
for the same period of time.   This striking difference in results

TABLE 1A—*Continued*

SERIES	TIME	PART	NORM.	T OPHTH	T MOR.	ANOPH	HDL	DEAD
12/18, '20·	16- to 18-mm. worms. Fed three days before cutting							
	hours							
Control..........		A	76	22				2
		B		54	4	36		6
		C	2	36		32	28	2
6. M. 1/350......	5	A	52	38				10
		B	2	76		12		10
		C	12	44	2	32		10
7. M. 1/350......	4	A	62	32				6
		B	4	74		16		6
		C	8	36		50	6	0
2/15, '21·	16- to 18-mm. worms. Fed three days before cutting							
Control..........		A	84	16				0
		B	2	58		36	4	0
		C		26		48	26	0
8. M. 1/400.....	4	A	62	38				0
		B	2	58		36	2	0
		C	4	28		52	16	2
9. M. 1/400......	5	A	44	46				0
		B		58	4	34		4
		C	4	58	2	26	10	0

between the A and C pieces indicates some sort of difference in the effect of the chloretone on the factors controlling head formation in pieces from these regions.

If we accept Child's conclusions regarding the effects of KNC on head frequency (p. 8) in A pieces where the stimulation of section is least, the agent may be expected to produce a decrease in head frequency because of its direct inhibition of the X cells. In table 1A this result shows quite distinctly in all experiments except series 1 and 3; in both of these the number of deaths in the A pieces completely obscures any effect of the agent on head frequency.

In C pieces the data show a marked increase in head frequency in all experiments except one. On the basis of Child's conclusions, this effect is interpreted as the result of the elimination by the anesthetic of stimulation in the Y region. In C pieces in water this stimulation inhibits the developmental processes of the X region. Its elimination in chloretone removes to a greater or less extent this inhibitory factor. The effect of the removal of this inhibition is sufficient to obscure any direct effect of the chloretone on Rate X. In the apparent exception, series 2, the animals used were quite small, and consequently their susceptibility to external agents was higher and the stimulation of Y by section less than in older animals. Under such conditions, a more striking direct effect on Rate X is to be expected.

In B pieces the results may be regarded as intermediate between those in the A and C pieces. Changes in the head frequency in the B pieces appear to depend in these data on the concentration employed, the duration of the time of exposure, and the size of the animals. Very high concentrations or long periods of exposure bring about decreases as in series 3 and series 5. Shorter periods of exposure and lower concentrations bring about increases in head frequency as in series 1, 6, and 7, or no change as in series 8 and 9.

Oxygen consumption. The similarity of the head-frequency results given in table 1A to Child's results with KNC indicates that the effect of chloretone on the stimulation of section is similar to that of KNC. To test by actual direct measurements whether or not chloretone does prevent the stimulation of the pieces by section, the oxygen consumption of the pieces immediately after section in water and in appropriate concentrations of chloretone was measured. Table 1B gives the complete results of all experiments attempted.

Pieces subjected to relatively strong solutions of chloretone increase in weight during the period of exposure, and in order to arrive at a more nearly correct measurement of their rate of oxygen consumption per unit of weight it was found necessary to weigh them before the tests. In no other of the anesthetics

TABLE 1B

Oxygen consumption—chloretone

SERIES	WEIGHT BEFORE TEST	WEIGHT AFTER TEST	OXYGEN CONSUMPTION PER TWO HOURS PER MILLIGRAM

10/26, '20· 15- to 18-mm. worms. Fed two days before cutting. 3 hr. 20 deg.

	mgm.	*mgm*	*cc*
Control....................	381	374	0.00052
M. 1/350....................	411	441	0.00018
M. 1/350....................	399	435	0.00019
Whole animals..............	629	593	0.00032

10/30, '20· 16- to 18-mm. worms. Fed 24 hours before cutting. 4 hr. 20 deg.

Control....................	445	424	0.00041
M. 1/500....................	385	421	0.00026
M. 1/500....................	436	444	0.00024
Whole animals..............	642	636	0.00032

10/26, '20· 18- to 20-mm. worms. Fed seven days before cutting. 3 hr. 20 deg.

Control....................		390	0.00044
M. 1/175....................		523	0.00005
M. 1/175....................		544	0.00005
Whole animals..............		555	0.00030

10/17, '20· 19- to 21-mm. worms. Fed two days before cutting. 4 hr. 21 deg.

		First two hours	Second two hours
		cc	*cc*
Control....................	520	0.00050	0.00049
M. 1/350..............	564	0.00030	0.00037
M. 1/350	565	0.00032	0.00031
Whole animals	609	0.00033	0.00031

10/23, '20· 18- to 20-mm. worms. Fed four days before cutting. 4 hr. 22 deg.

Control....................	463	0.00048	0.00052
M. 1/500....................	459	0.00044	0.00040
M. 1/500....................	456	0.00028	0.00039
Whole animals..............	656	0.00037	0.00038

10/19, '20· 21- to 23-mm. worms. Fed four days before cutting. 4 hr. 21 deg.

Control....................		529	0.00048	0.00052
M. 1/350....................		583	0.00030	0.00025
M. 1/350....................		578	0.00017	0.00005
Whole animals..............		470	0.00032	0.00039

used in this work did the pieces increase in weight. Whether this increase in chloretone solutions is due to the imbibition of water by the tissues or to imbibition of water by the adherent mucus which they extrude is not possible to say at present. The data on the increases in weight are given in full in table 1B. Consideration of the data on oxygen consumption will be confined to those experiments in which the rate is calculated on the weight of the pieces before the test, since that is the more nearly correct weight of living material. The weighing before the tests of oxygen consumption might possibly increase the stimulation of the pieces, which should be evident in a higher rate of the control and to some extent in the chloretone series as well. But no increase of rate due to weighing is evident in the controls if we compare them with the controls in which the weighing was done after the test. And without exception the pieces in the chloretone had a much lower rate of oxygen consumption than that of their controls. In fact, there are indications that not only is stimulation largely or entirely prevented, but that there is also some degree of depression below the rate of oxygen consumption of whole animals. It must be pointed out that any depression below the rate of whole animals, since it is not merely the prevention of stimulation, but a general protoplasmic effect, must strongly affect the X cells. This depression, therefore, throws some light on the decreases in head frequency, so general in A pieces, and occasional in B pieces, and in one case, series 2, occurring in C pieces.

Chlordform

Head frequency. In general, the results obtained by treating pieces with various concentrations of chloroform for short periods immediately after section were similar to those obtained with chloretone. Table 2A gives in detail the least and greatest changes in head frequency obtained. Seven experiments were carried out in which the number of deaths, aways high in chloroform, did not obscure the basis for comparison with the control. The increases in head frequency in both B and C pieces were noticeably greater in the chloroform experiments than in those with chloretone. In exposures up to thirty-two hours there were no decreases in head frequency in B pieces, even when the concentration and time of exposure were very nearly lethal. These results indicate the greater effect of the chloroform on the stimulation of the Y region as compared with its effect on the cells of the X region.

TABLE 2A

Head frequency—chloroform. In per cent

SERIES	TIME	PART	NORM	T. OPHTH	T MOR.	ANOPH.	HDL.	DEAD
4/11, '16· 12-mm. worms. Fed six days before cutting								
	hours							
Control..........		A	80	20				0
		B		24	8	68		0
		C	2	8	2	52	36	0
1. M. 1/400......	8	A	72	22				6
		B	50	26				24
		C	26	30	2	20		22
2/5, '21· 14- to 16-mm. worms. Fed two weeks before cutting								
Control..........		A	68	32				0
		B		64		36		0
		C	2	42		42	14	0
2. M. 1/300......	5	A	50	44				6
		B	2	72	2	8		16
		C	12	72		2		14

Oxygen consumption. Table 2B gives the high and low extremes of results obtained in measuring the rate of oxygen consumption of pieces placed in suitable concentrations of chloroform immediately after section. The usual rate of oxygen consumption of whole animals of the size and condition of those used in table 2B is 0.00040 to 0.00045 cc. per milligram per two hours. It will be be noted that the pieces subjected to the chloroform have a rate very closely approximating that of intact animals. This was true in all the nine experiments carried out.

Pieces as small as one-third the anterior zooid do not move about to any great extent, consequently the higher rate of the controls cannot be accounted for by their greater motor activity. On the contrary, it is probable that motor activity is greater in the pieces subjected to the chloroform. An extended series of preliminary experiments with whole worms showed that chloroform increases the oxygen consumption of intact animals in the concentrations used during the first two to four hours of exposure; this increase is undoubtedly due in large measure to

TABLE 2B

Oxygen consumption—chloroform

SERIES	OXYGEN CONSUMPTION PER MILLIGRAM	
	First two hours	Second two hours
3/24, '20· 18- to 20-mm. worms. Fed three days before cutting. 20 deg.		
	cc.	*cc.*
Control..............................	0.00080	0.00066
M. 1/750..............................	0.00036	0.00031
3/17, '20· 18- to 20-mm. worms. Fed 24 hrs. before cutting		
Control..............................	0.00070	0.00057
M. 1/750..............................	0.00050	0.00046

Usual rate, whole worms of similar condition, 0.00040–45 cc. per two hr.

increased motor activity, for the animals wriggle constantly during that time. In another series of five experiments with halves of the first zooid, the rate of oxygen consumption was also slightly increased, and here, too, the increase is attributed to increased motor activity. But in pieces as small as one-third the first zooid muscular coordination is much reduced and the stimulation of section greater, especially in the C pieces; there is little wriggling, and motor activity, while probably greater than in the control, is not sufficient to obscure the fact that the chloroform prevents to a greater or less extent the nervous stimulation of section. Because of the complication introduced by the increased muscular activity actuated by the chloroform, the data in table 2B must be regarded as only indicative of the inhibition of stimulation and not necessarily a true measure of the degree of inhibition.

Ether

Head frequency. The least and greatest effects on head frequeney obtained by subjecting the pieces to various concentrations of ether for short periods after section are given in table 3A.

TABLE 3A

Head frequency—ether. In per cent

SERIES	TIME	PART	NORM.	T OPHTH	T MOR	ANOPH	HDL	DEAD
8/24, '20· 16- to 18-mm. worms. Fed 24 hrs. before cutting								
	hours							
Control..........		A	96	4				0
		B	4	94	2			0
		C	40	40		16	4	0
1. M. 1/40.......	16	A	94	2				4
		B	22	74		4		0
		C	54	38		8		0
2/3, '21· 14- to 16-mm. worms. Fed two weeks before cutting								
Control..........		A	78	22				0
		B		60		40		0
		C	2	60		30	8	0
2. M. 1/20.......	6	A	78	22				0
		B	2	86		12		0
		C	10	76		12	2	0

All other results fall between these two extremes. Eight experiments were carried out, and the results in general did not differ greatly from those obtained with chloretone and chloroform. They require no extended comment. The increases in head frequency in the B pieces in almost all cases and the failure to obtain marked decreases in the A pieces and any decreases whatever in the B pieces indicate, as in chloroform, that the effect of the ether is relatively greater on the nervous stimulation of the Y region than its effect on the general protoplasmic activity of the X region.

Oxygen consumption. Eight experiments were carried out on the effect of ether solutions on the rate of oxygen consumption of the pieces for some hours after section. The high and low extreme results are given in table 3B. All other results fall between these two extremes, and no comment is necessary except to point out that such solutions of ether obviously prevent to a greater or less degree the stimulation of section.

TABLE 3B

Oxygen consumption—ether

SERIES	OXYGEN CONSUMPTION PER MILLIGRAM	
	First two hours	Second two hours
10/9, '20· 15- to 17-mm. worms. Fed 5 days before cutting. 20 deg.		
	cc.	*cc*
Control.................................	0.00051	0.00054
M. 1/40.................................	0.00044	0.00047
Whole animals in water.................		0.00038
9/17, '20· 18- to 20-mm. worms. Fed 24 hrs. before cutting		
Control.................................	0.00056	0.00053
M. 1/20.................................	0.00033	0.00027
Whole animals in water.................		0.00049

Chloral hydrate

Head frequency. Table 4A gives the extreme high and low results obtained with chloral hydrate. Five experiments were done. The number of experiments is too small to warrant any conclusions other than that the head frequency is increased in C pieces and slightly decreased in A pieces by exposure to certain concentrations of chloral hydrate—conclusions which serve to extend to chloral hydrate the capacity to affect head frequency in the same directions as chloretone, chloroform, and ether.

Oxygen consumption. Table 4B gives the extreme high and low results obtained in measuring the rate of oxygen consumption of pieces exposed to such concentrations of chloral hydrate. Fifteen experiments were carried out. It is necessary only to state that in every case the chloral-hydrate solutions prevented to a considerable extent the increase of the rate of oxygen consumption attendant upon section, and in some cases there were indications that the rate of oxygen consumption of the pieces in chloral hydrate had been depressed below the rate of whole animals in water.

TABLE 4A

Head frequency—chloral hydrate. In per cent

SERIES	TIME	PART	NORM	T OPHTH	T MOR	ANOPH	HDL	DEAD
9/8, '20· 19- to 21-mm. worms. Fed two days before cutting								
	hours							
Control..........		A	98	2				0
		B	12	88				0
		C	46	40		14		0
1. M. 1/165......	6	A	100					0
		B	48	50		2		0
		C	78	18		4		0
11/20, '16· 15-mm. worms. Fed two days before cutting								
Control..........		A	94	6				0
		B	54	36	2	8		0
		C	24	26	4	34	12	0
2. M. 1/165......	24	A	100					0
		B	94	6				0
		C	86	10	2	2		0

TABLE 4B

Oxygen consumption—chloral hydrate

SERIES	OXYGEN CONSUMPTION PER MILLIGRAM PER TWO HOURS		
	First two hours	Second two hours	After 24 hours
4/13, '20· 16- to 18-mm. worms. Fed 24 hrs. before cutting. 20 deg.			
	cc.	*cc.*	*cc.*
Control........................	0.00080	0.00079 ·	0.00080
M. 1/165........................	0.00073	0.00061	0.00067
4/3, '20· 17- to 19-mm. worms. Fed three days before cutting. 20 deg.			
Control........................	0.00075	0.00075	
M. 1/82........................	0.00036	0.00039	

Ethyl alcohol

Head frequency. The results on head frequency of subjecting pieces to various concentrations of ethyl alcohol for short periods after section are similar in many cases to those obtained with the other anesthetics. Eleven experiments were performed with concentrations ranging from mol. 1/4 to mol. 1/20. Six of the eleven resulted in slight decreases in head frequency in the A pieces and increases in the C pieces. These results were obtained with mol. 1/10 alcohol. Stronger solutions bring about injury to the pieces in some way and the number of deaths in experiments with such solutions was too great to permit any conclusions regarding their effect on head frequency. In weaker solutions no definite changes resulted. Table 5A gives the high and low extremes obtained with mol. 1/10 alcohol.

Oxygen consumption. The results obtained in measuring the oxygen consumption of pieces in mol. 1/10 alcohol are interesting because of their failure to show any inhibition of the stimulation of section such as one might expect in the light of the data on the effects of the other anesthetics employed. Four experiments were performed, and in every case the rate of the pieces in mol. 1/10 alcohol was slightly higher than that of the control. In eight experiments with stronger solutions the results were indefinite and irregular. In five experiments employing mol. 1/4 and mol. 1/5 alcohol the rate of oxygen consumption of the pieces in alcohol was much above the rate of the controls for the first two hours after section, but fell to a much lower rate the second two hours. In three experiments with the same concentrations the rate of the pieces in alcohol was somewhat below that of the control for the four hours during which the test was carried on. Table 5B gives one result with mol. 1/10, one with mol. 1/5, and one with mol. 1/4 alcohol. All of these solutions produce a marked degree of anesthesia in the pieces.

These results on the measurement of oxygen consumption of pieces subjected to relatively strong solutions of ethyl alcohol are distinctly different in nature from those obtained with the other anesthetics. In all the others employed in concentrations producing a considerable degree of anesthesia the increase of

TABLE 5A

Head frequency—ethyl alcohol. In per cent

SERIES	TIME	PART	NORM	T OPHTH	T MOR	ANOPH	HDL	DEAD
6/28, '17· 18- to 20-mm. worms. Fed three days before cutting								
	hours							
Control..........		A	94	6				0
		B	28	70	2			0
		C	64	26	4	6		0
1. M. 1/10.......	24	A	88	12				·0
		B	78	22				0
		C	78	14				0
3/8, '21· 17- to 19-mm. worms. Fed eight days before cutting								
Control..........		A	100					0
		B	38	46	2	14		0
		C	16	62		22		0
2. M. 1/10.......	10	A	100					0
		B	76	24				0
		C	42	52		6		0

TABLE 5B

Oxygen consumption—ethyl alcohol.

SERIES	OXYGEN CONSUMPTION PER MILLIGRAM PER TWO HOURS	
	First two hours	Second two hours
3/30, '20· 18- to 20-mm. worms. Fed 24 hrs. before cutting. 22 deg.		
	cc.	*cc.*
Control..................................	0.00085	0.00079
M. 1/4..................................	0.00084	0.00075
3/26, '20· 18- to 20-mm. worms. Fed two days before cutting. 20 deg.		
Control..................................	0.00075	0.00078
M. 1/5................................, ...	0.00091	0.00054

8/23, '20· 16- to 18-mm. worms. Fed 24 hrs. before cutting. 20 deg.

	Average oxygen consumption per two hours First ten hours
	cc
Control..................................	0 00047
M. 1/10	0 00058

oxidations attendant upon section was appreciably inhibited. This inhibition of oxidative activity has been assigned a causal rôle in the changes in head frequency produced in such experiments. But in mol. 1/10 alcohol, which produces similar changes in head frequency, the rate of oxygen consumption is higher than that of the control pieces. This fact appears to conflict with the conclusions drawn from the inhibitory effects of the other anesthetics on the increase in rate of oxidations following section unless a strong probability can be shown that the increase in oxygen consumption in pieces subjected to alcohol solutions is different in character from the stimulation of the oxidative processes by section, and that there are sound reasons for believing that this form of metabolic activity in the Y region of the pieces does not exert an influence over the X region and vice versa.

In searching for some explanation of the relatively high rate of oxygen consumption of pieces subjected immediately after section to narcotizing solutions of ethyl alcohol as compared to the rate of the controls, work with intact animals produced some data that are extremely suggestive. Forty experiments were performed. The data are too voluminous to be given here, but may be briefly summarized as follows: If whole animals are subjected to highly narcotizing solutions of alcohol, mol. 1/4 or mol. 1/5, for four hours, their rate of oxygen consumption rises during the first two hours and falls during the second two hours, the fall probably being due to the injurious effect of such solutions, for if the treatment is continued death soon follows. If the animals are transferred to water after a four-hour exposure to such solutions, their rate of oxygen consumption rises very rapidly and the high rate continues for nearly seventy-two hours. This rise after return to water is probably due to the oxidation of a quantity of alcohol retained in the tissues. Cushny ('10) states that in the human body at least 95 per cent of the alcohol taken in is oxidized. If whole worms are starved in the presence of mol. 1/10 alcohol, a concentration producing marked anesthesia for the first few days, their rate of oxygen consumption rises cumulatively for weeks. At the end of six weeks the rate of small worms is 500 per cent that of their control in water.

With large animals ten weeks of starvation in the presence of the alcohol are required to bring about such a great increase in rate. During this exposure to alcohol the animals do not lose weight any more rapidly than their controls in water and, although their rate of oxygen utilization is many times that of the control, they move about less than the control animals. These data indicate that the alcohol is oxidized more and more rapidly as the progress of starvation increases the general metabolism of the animals and that this high rate of oxidative reactions apparently due to oxidation of the alcohol is accomplished without appreciable loss of protoplasm. The observations also indicate that the high rate of the oxidative reactions in the animals subjected to alcohol solutions does not result from nervous stimulation.

That the pieces of Planaria are very considerably anesthetized in solutions of ethyl alcohol up to mol. 1/10 is certain. It is also certain that nerve or other conducting paths narcotized by alcohol do not transmit stimuli. Therefore, any tissue activity concerned in oxidizing the alcohol in solutions strong enough to produce anesthesia is not induced by the stimulation of section. In fact, Winterstein ('14) has shown that alcohol narcosis may be produced in nerve tissue itself with an accompanying increase in oxygen consumption. Furthermore, because of the narcosis of the conducting paths by the alcohol, the tissue activity of the Y region cannot exert any influence on the X region and vice versa. If this is true, the relations of the X and Y regions in pieces in alcohol are thus essentially the same as in the other anesthetics employed.

A strong probability has therefore been established that the relatively high rate of oxygen consumption of pieces subjected to solutions of ethyl alcohol results from the oxidation of the alcohol, and not from the stimulation of section, and that the measurement of the rate of oxygen consumption in such pieces is complicated by the oxidation of the alcohol in the tissues of the pieces. A further complication is introduced by the injury to the tissues in the stronger solutions. There are also sound reasons for believing that the tissue activity induced in the Y

region by exposure to strong solutions of alcohol can exert no
differentiative influence on the X cells and vice versa. One is
justified in stating that head frequency can be controlled to a
limited extent by exposure of the pieces to certain solutions of
ethyl alcohol, but that the probable inhibition of stimulation
resulting from section is masked in the oxygen-consumption
measurements of pieces under such conditions by the amount
of oxygen utilized in oxidizing the alcohol.

EFFECT OF WEAK SOLUTIONS ACTING FOR ENTIRE PERIOD OF REGENERATION

In the group of experiments just described we have seen that
the head-frequency changes in pieces subjected to relatively
strong solutions of anesthetics for short periods after section
result from two sorts of effects of the agents on the differentiative and developmental processes of the X cells: 1) the direct
effect on these processes tending to decrease head frequency
and, 2) what may be called an indirect effect tending to increase
head frequency through elimination by the anesthetics of the
stimulation of the Y region which ordinarily acts as an inhibitor
of head formation. If we employ concentrations so dilute as
not to inhibit to any appreciable degree the stimulation of the
Y region by section, we may expect the direct effect of the agents
on the X cells to be the chief cause of any changes in head frequency that may result. It was with this object in view, i.e.,
a study of the direct effect of the anesthetics on the X cells in
pieces from different levels, that a series of experiments was
undertaken in which A, B, and C pieces were subjected during
the entire period of two weeks to concentrations of the anesthetics so dilute as not to inhibit the stimulation of the Y region
by section.

In all the experiments of this nature so far performed measurements of the oxygen consumed by the pieces in the weak solutions of the anesthetics and that consumed by their controls in
water show that the stimulation of section takes place to the
same extent in these weak solutions as in water. The indirect
effect of the anesthetics tending to increase head frequency by

inhibition of the stimulation of the Y region is therefore more or less completely excluded in these experiments and any changes in head frequency that result must be due chiefly to direct effects on the X cells.

The possible direct effects of dilute solutions of anesthetics on the dedifferentiative and redifferentiative processes of the X cells include, 1) possible stimulation of these processes; 2) direct inhibition of these processes; 3) different degrees of acclimation of the X cells of the A, B, and C pieces to the inhibitory effect of the anesthetics (see foot-note, p. 3).

In event of direct inhibition of the X cells by the anesthetics followed by incomplete acclimation, we should expect decreases in head frequency. The decreases should be less in A pieces than in C pieces because, 1) the X cells of A pieces have a higher rate of metabolism than that of the X cells of C pieces and acclimate more completely to the inhibitory effect of the anesthetic; 2) the stimulation of the Y region in A pieces is slight and has little effect in inhibiting head formation, for the A pieces of control series show head frequencies nearly 100 per cent normal. The decreases in head frequency in the C pieces should be very much greater than in A pieces because, 1) the X cells in C pieces have a lower rate of metabolism than that of the X cells of A pieces and they acclimate less completely to the inhibitory effect of the anesthetic; 2) the stimulation of the Y region by section is greater in C pieces than in A pieces and its inhibitory effect on head formation is added to the direct inhibition of the X cells by the anesthetic. The decreases in B pieces should be intermediate between those in A and C pieces.

These expectations are realized in the results obtained in solutions of mol. 1/3000 to mol. 1/4000 chloroform, mol. 1/800 chloral hydrate, and mol. 1/300 ether. In such solutions there are distinct decreases in head frequency, least in A pieces and greatest in C pieces, and the results in B pieces are intermediate between A and C pieces.

In extremely dilute solutions of chloretone, mol. 1/35000, and chloral hydrate, mol. 1/8000, and in mol. 1/20 ethyl alcohol, increases in head frequency occur in the B and C pieces.

These results suggest the possibility that the developmental processes of the X cells are stimulated by extremely dilute anesthetics.   Numerous cases of stimulation of differentiation and growth by weak solutions of anesthetics have been recorded (Czapek, Biochemie der Pflanzen, Jena, '13, p. 159).

The number of experiments thus far carried out is too small to permit a more complete analysis of the results at this time, and a more detailed consideration of the factors concerned in the control of head frequency by this method and publication of the data are postponed until further work has been completed. The results so far obtained serve to show that head frequency can be controlled by this method and that the effects of weak solutions of anesthetics on head frequency are plainly non-specific.

Interesting results were obtained in measuring the rate of oxygen consumption of pieces in weak solutions of anesthetics throughout the two-week period of regeneration.   In mol. 1/10 ethyl alcohol the rate of oxygen consumption of the pieces is slightly higher than that of the control during the first day after section.   It continues to rise and at the end of two weeks is more than 500 per cent the rate of the control in water.   The pieces in the alcohol solution lose slightly more weight than the controls during the first two days after section, but thereafter the pieces in the alcohol solution decrease in weight at the same rate as the controls (six experiments).   The rate of oxygen consumption of pieces in mol. 1/800 chloral hydrate is approximately the same as the rate of the controls during the first two days, but on the third day it rises above that of the controls.   It continues to rise, and at the end of two weeks is about 50 per cent higher than that of the controls.   The loss of weight takes place at approximately the same rate in both controls and the pieces in the chloral-hydrate solution (six experiments).   The rate of oxygen consumption of pieces in mol. 1/3000 chloretone rises above that of the controls after four or five days, and at the end of two weeks is about 20 per cent higher than that of the controls.   The loss of weight of the pieces in the chloretone solution is slightly greater during the first three days than that of the controls, but thereafter the controls and the pieces in the chloretone solution lose weight at the same rate (six experiments). In mol. 1/3000 chloroform the rate of oxygen consumption of the pieces rises slightly above that of the controls sometime after the fourth day and remains slightly above during the two weeks' record. The loss of weight of the pieces in chloroform is slightly greater during the first three days, but during the later period of regeneration takes place at the same rate as in the controls (six experiments). In the one concentration of ether employed, mol. 1/300, the rate of oxygen consumption and loss of weight of the pieces is approximately the same in both controls and in the pieces subjected to the ether solution throughout the entire period of two weeks (six experiments).

These results are somewhat similar to those recorded by others. Winterstein ('14) found that the oxygen consumption of the spinal cord of the frog was increased during alcohol narcosis. Vernon ('12) found that weak solutions of narcotics increase the oxygen consumption of isolated tissues. Tashiro and Adams ('14) found that weak solutions of urethane and chloral hydrate increase the carbon-dioxide production of nerve fiber. Winterstein ('19) records that Elfving found that the carbon-dioxide production of Salix leaves was increased in certain environments of dilute ether and chloroform. Winterstein also records that von Lauren, Markovine, Kosinski, Gerber, and Zaleski, and others have found that weak solutions of narcotics increase the rate of respiration of plant cells.

DISCUSSION

If head frequency in A pieces is decreased and in C pieces increased by reason of exposure of the pieces for the same period of time to the same concentrations of the same anesthetic, and the head frequency of the B pieces is not altered, or is increased or decreased depending on the length of the period of exposure and the concentration employed, it follows that there must be some difference between the effects of the anesthetic on the three sets of pieces. In analyzing the nature of this difference, we shall consider the possibilities in the A and C pieces, since the effects on head frequency are almost invariably different in pieces from these regions, and regard the B pieces as representing intermediate conditions.

The conception of specific formative substances as determining factors in regeneration advanced by Sachs and offered by Morgan, Loeb, and others (see, for example, Morgan, '05, and Loeb, '16, p. 155) may be dismissed with brief consideration. If a head-forming substance is a chemical entity, it must follow the laws of chemical and physical reactions when acted upon by some external agent. According to the law of mass action, one must assume that it should be acted upon by the agent in the same manner in both A and C pieces. In event of either complete or incomplete reaction between the external agent and the substance, destruction or inhibition of the substance should take place in both A and C pieces. In other words, if head formation depends on the amount of head-forming substance present, head frequency should be decreased in both A and C pieces if the pieces are exposed to an agent under conditions of time and concentration that inhibit head formation. But the evidence presented here that anesthetics decrease head frequency in A pieces and *increase* it in C pieces under identical conditions of time of exposure and concentration is unmistakable, and KNC has been shown to have a similar effect. These results make it impossible to explain head determination on the basis of any specific head-forming substance.

Attempting to explain the experimental results on the basis of the amount of nutritive materials present in the two regions, leads us into a similar conflict with the law of mass action. It is inconceivable that the anesthetics destroy nutritive materials in the A pieces and elaborate them in the C pieces; or that they prevent the utilization of such materials in the A pieces and facilitate these processes in the C pieces.

The idea of a possible different specific action of the anesthetics on the development of the tissues of the A and C pieces is next to be considered. In general, the same body tissues, body wall, muscles, gut, nerves, etc., are present in both regions, although probably in different amounts. The processes involved in the dedifferentiation and redifferentiation of any of these tissues must be much the same in both A and C pieces. The direct effect of an external agent on these processes should therefore be much the same, at least in the same direction, in both. There is, therefore, no reason to suppose that the anesthetics specifically prevent the dedifferentiation and redifferentiation of any one tissue in one region and facilitate these processes in another, either by chemical or physical action. And that any anesthetic could alter the relative amounts of the different tissues present in the two regions is impossible except by differential destruction of the tissues, and there is no evidence of such action.

Can the differential effect of the anesthetics in decreasing the head frequency in A pieces and increasing it in C pieces be due to their specific action on the different cell constituents present in different amounts in the cells of the two regions? It is true that the effects of certain of the anesthetics employed may differ specifically in the two regions because of possible differences in the amount and quality and distribution of certain cell substances, lipoids, proteins, water content, etc. But in view of the width of the range of conditions and agents which bring about similar changes in head frequency, such intracellular effects of certain of these agents cannot be considered the effect *directly* concerned in bringing about the changes in head frequency, but as their characteristic physical or chemical action which brings about certain quantitative metabolic relations. These relations are similar for all the agents and conditions that influence head frequency in the same directions.

A brief examination of the physical and chemical properties of the anesthetics used in this work emphasizes the fact that their effect in altering head frequency is non-specific. Ether and chloroform are highly fat soluble. Chloroform dissolves but very slightly in water; ether dissolves much more readily. Chloretone is soluble in both fats and water. Chloral hydrate is freely soluble in water, but much less so in fats. Chloroform and ether are relatively inert chemically; chloretone and chloral hydrate are rather easily broken up. Thus their widely different physical and chemical properties, when considered in the light of the data, preclude the idea of specific effects and force one to conclude that their common effect on head frequency is due to their common property of anesthesia. The data show very clearly that this common property prevents to at least a marked extent the stimulation of the pieces by section.

But this common effect of the anesthetics in preventing the stimulation of section produces different effects on the head frequency in the A and C pieces. One must therefore look for some sort of difference in the two regions to explain these results. Since morphological and specific differences have been excluded, the difference must be physiological and quantitative. A difference in the rate of metabolism between the A and C regions of intact animals is known to exist. They represent, respectively, regions or levels near the apex and base of an anteroposterior gradient in rate of metabolism. Much evidence has accumulated to show that the head frequency of isolated pieces depends in large part on their original position in this gradient. As outlined in the introductory pages of this paper, the X region of pieces is stimulated directly by the injury of cutting and is also isolated from all regions anterior to it. The Y region is also stimulated at the same time, in part at least, by impulses arising from the severing of the nerve tracts. In pieces of a given length, the more posterior the piece, the greater the stimulation of the Y region; i.e., B pieces are more stimulated than A pieces and C pieces are more stimulated than B pieces. The stimulation of the Y region in all pieces is temporary. The high rate of the X region brought about by the wound is continued by the processes of dedifferentiation and redifferentiation initiated shortly after section. Before a piece is cut from the body of the animal, the cells near the level of the cut and to be included in the piece, i.e., the X cells, function as integrated units of a complete organism. After section the only remaining integrative control of these cells must emanate from the Y region. The more active and intense these controlling factors, the greater their effect in maintaining the specialization of the cells at X as functional units of the old organism. But if for some reason the activity of the Y region is low, or its stimulation by section is slight, or is inhibited by some agent, the stimulated X cells are freed to a greater or less extent from any integrative control or the controlling factors emanating from Y are not adequate to control them completely.

It is a well-known fact that an isolated group of cells from the body of a plant or many of the lower animals undergoes dedifferentiation and development into a new individual if it is not made up of highly specialized tissues and is able to maintain life. The well-known facts of agamic reproduction constitute evidence beyond question that isolation is followed by the changes attendant upon the development of a new individual from an isolated part. The very fact that a piece taken from the body of a planarian will reconstitute a new individual in many cases like those in nature is itself a demonstration of the accuracy of the statement. These processes of development are initiated at the point where the metabolic activity is highest. The point of highest rate in a piece of Planaria is the X region. (For a full discussion and references on the subjects of isolation and the establishing of the new individual, see Child, Individuality in Organisms, Chicago, 1915.)

There are, therefore, two factors that are plainly antagonistic exerting influences on the X cells, one, the factor originating in Y which tends to prevent the dedifferentiation of X, the other tending to bring about dedifferentiation, cell division, growth, and the initiation of development of a new individual. If either of these two factors is inhibited or is inhibited more than the other, the one inhibited less becomes the factor which more or less dominates the fate of the X cells. Consequently, if a single piece is isolated from an animal, whether or not a head forms will depend on the relative activities of these two factors. And in a mass experiment when the pieces are subjected to experimental conditions, the head frequency is increased or decreased in relation to the differential effect of the agent or condition on these two factors. The two antagonistic factors are undoubtedly functions of the rates of metabolism of the X and Y regions, respectively, and may be conveniently described as Rate X and Rate Y, as pointed out above (p. 7).

In the light of this dynamic conception of head determination, the data on the effects of chloretone on head frequency and on oxygen consumption of the pieces when subjected to various concentrations for short periods after section may be explained in detail and serve as a basis for comparison with the effects of the other anesthetics employed. The chloretone results permit the following general statements: First, that, within certain limits, short period exposures are as effective as long in increasing the head frequency in C pieces. The causes of this increase must therefore be exerted very soon after section. The events which are known to occur in C pieces placed in water immediately after section are the wound stimulation of the X region and a relatively great stimulation of the Y region. But the rate of metabolism of the X region is known to remain high, while the stimulation of the Y region disappears after a number of hours. Consequently, we must look to the effect of the chloretone solutions on the Y region in these short-period exposures for the factors causing the increase in head frequency. In the measurements of oxygen consumption the data demonstrate that in the chloretone solutions employed the stimulation of the Y region does not occur. The relation between the increase in head frequency in C pieces after exposure to the chloretone solutions and the prevention of the stimulation of section during the period of exposure is regarded as causal for the reasons already given (Introduction and p. 37). There is no intent here to imply that the wound stimulation of the X region is not also inhibited in C pieces by the anesthetic. Child, in his paper on the effects of KNC on head frequency (Child, '16), gives data showing that the general head frequency of C pieces may be increased by subjection of the pieces to certain solutions of KNC, but that the number of normal heads in such series may be decreased. The conclusions drawn from the data are that the effect of the prevention of stimulation of the Y region by the KNC overbalances the effect on the X region, but that the effect on Rate X shows up nevertheless in the less complete regeneration of the anterior ends of the higher types. In the present work the agents used were not nearly so powerful protoplasmic poisons as KNC, and this effect on Rate X in C pieces is indistinguishable except in a few cases.

Second, decreases in head frequency in A pieces are general, and the longer the time of exposure or the higher the concentration, the greater the decrease. In A pieces the stimulation of the Y region by section is slight and it evidently has little effect in inhibiting head formation, since head frequency is nearly 100 per cent in the A pieces of large animals. Disregarding this slight stimulation, largely or completely prevented by the anesthetic, there remains to be considered the condition of the cells of the X region. If we grant that chloretone has an inhibitory effect on developmental processes, it follows that the decreases in head frequency in A pieces result from the direct inhibition of the processes of dedifferentiation and development of the X region by the chloretone solution.

Third, decreases in head frequency occur in B pieces when the period of exposure is relatively long or the concentration very high, especially in the former case. In other experiments the head frequency of the B pieces is not altered or is increased. These effects, intermediate between the effects on A and C pieces, are to be expected, since B pieces are obviously intermediate anatomically and physiologically between A and C pieces.

The results obtained with chloroform, ether, and chloral hydrate support the conclusions drawn from the chloretone data. Chloroform produces more striking increases in head frequency in the C pieces and less striking decreases in the A pieces than chloretone. This is true to an even greater degree in the experiments with ether. The reasons suggested for these differences in the effects of anesthetics will appear farther on in this paper. I have been able to control head frequency to a limited extent with ethyl alcohol, but have been unable to obtain definite and uniform data showing that solutions of ethyl alcohol inhibit the stimulation of section. These results, when considered in the light of the very definite relations established between the inhibition of the increased oxidations attendant upon section and the changes in head frequency with the other anesthetics used, suggest that the factors, whatever they may be, that complicate the effects of the alcohol on the oxygen consumption of the pieces immediately after section are not greatly concerned in its inhibitory effect on the nervous stimulation of the Y region by section. Winterstein, ('14) has shown that nervous tissue itself may be narcotized without reduction in rate of oxygen consumption. A number of experiments on head frequency were performed with solutions of methyl and iso-butyl alcohol. The results are in agreement with those obtained with chloretone, chloroform, ether, and chloral hydrate, but no measurements of their effects on the oxygen consumption of the pieces were attempted. Further work of this nature with the alcohols is contemplated.

One is justified in stating positively that the results presented are in perfect accord with, and afford strong support to Child's conception of the factors concerned in head determination in Planaria. When one considers that the data on the experimental control of head frequency now include a wide range of conditions and agents, age, nutritive condition, mechanical stimulation, temperature, KNC, and a number of anesthetics of different types, all of which are known to affect quantitatively the physiological gradient of the intact animal or the stimulating effects of cutting and isolation in the pieces and that in most cases the effects of the conditions and agents on the gradient or the metabolic conditions in the pieces after section have been established by measurement, it becomes evident that the relation between head-frequency changes and the effects of the agents or conditions on the metabolic conditions in the pieces is something more than an interesting parallelism. None of the evidence has been refuted and the facts upon which this dynamic conception of head determination is based have become formidable in number and import. Both the problem and its answer have come out of the material, and the control of head frequency has been accomplished by the use of agents or conditions whose quantitative effect on the metabolic conditions has been determined by known chemical methods.

This paper is not primarily concerned with the problem of anesthesia, but with the effects of the anesthetics on head frequency and on the stimulation following section. The experiments on oxygen consumption were not planned to investigate whether or not the state of anesthesia is dependent on reduced oxidations or accompanied by reduction in rate of oxygen consumption, but to investigate the effects of the agents on the stimulation of section with reference to the degree of anesthesia produced. For the sake of completeness, it may be stated that in all the strong solutions employed there was distinct anesthesia of the pieces. It has been sufficient for analysis of the factors concerned in head determination to show that in the concentrations employed these anesthetics, chloretone, chloroform, ether, and chloral hydrate do prevent to an appreciable degree the increase of oxidations attendant upon section. The exception in the case of ethyl alcohol has already been dealt with.

An extended discussion of the physical or chemical action whereby this inhibition of stimulation is brought about would lead us into the prevailing confusion of the numerous theories of anesthetic action and avail nothing. For the question of the nature of the physical or chemical action of the anesthetics in the inhibition of stimulation is of no more importance in determining the effects of the anesthetics on head frequency than that of the action whereby KNC reduces oxygen consumption in Planaria, or the biological processes whereby young animals maintain a higher rate of metabolism than old, or the manner of action of the other agents and conditions that have been shown to control head frequency. It is the quantitative effects on the relative rates of metabolism in the X and Y regions that are of primary importance.

The data indicate that there is some sort of difference between the nature of the stimulation of the X region and that of the Y region, although the two are anatomically and physiologically continuous. It is probable that at X the stimulation is general for all tissues cut by the knife, body wall, gut, nerves, etc., but not necessarily equal in each. The condition at X must be the sum of the excitations of these tissues plus the effects of isolation from all more anterior regions. The stimulation of tissues other than nervous probably undergoes considerable decrement in its transmission to regions more posterior, while the stimuli set up by section of the nerve tracts may be expected to be transmitted with less decrement. (For a discussion of decrement in transmission of stimuli, see Verworn, '13, chap. VI; also Child, '20 c, chap. IV.) One cannot say whether or not all the tissues other than nervous transmit stimuli equally. In any event, the stimulation of the Y region must be the result of some sort of transmission of stimuli arising in the X region, but that the stimuli are not transmitted solely by nerves, or else nervous stimuli in these animals undergo marked decrement is suggested by the fact that short pieces are more stimulated than long pieces from the same region.

There is evidence of a difference in the nature of the changes in the X and Y regions to be found in Child's results on head frequency with KNC and my own results with chloretone, chloroform, and ether. With both KNC and chloretone the decreases in head frequency in the A pieces are marked and may easily be extended to B or even C pieces by extending the period of exposure or increasing the concentration. KNC is a general protoplasmic poison and inhibitor, but not a particularly good anesthetic, while chloretone might be called a general protoplasmic anesthetic, i.e., its effectiveness is in a large measure independent of the degree of specialization of the nervous system. Chloroform and ether, on the other hand, are powerful anesthetics in the strict sense, i.e., they act more intensively on the highly specialized nervous system than on protoplasm in general.

In the pieces of Planaria the changes in the X region following section are general protoplasmic changes which lead to dedifferentiation, cell division, and growth; while the changes in the Y region represent a temporary excitation which is certainly in a large measure nervous in character. We may expect, then, that agents which are general protoplasmic inhibitors will affect the X region more than those which are chiefly nervous inhibitors. The first group, the general protoplasmic inhibitors, should be more effective in producing decreases in head frequency in A pieces and less effective, except perhaps in low concentrations (KNC), in producing increases in the C pieces. The nervous inhibitors, on the other hand, should be less effective in producing decreases in head frequency in the A pieces and more effective in producing increases in the C pieces.   In the changes in head frequency brought about by KNC and chloretone as compared with those produced by chloroform and ether, just these differences appear.   KNC and chloretone are very effective in producing decreases in head frequency in the A pieces and will produce decreases in pieces from the B and C regions with sufficient time of exposure or concentration, while chloroform and ether produce marked increases in the C pieces, but little decrease in the A pieces even when the time of exposure or concentration is very nearly lethal.   In other words, KNC and chloretone inhibit to a considerable extent the intracellular changes in the X region of the pieces as well as the nervous stimulation of the Y region, while chloroform and ether in the concentrations used inhibit the nervous activity to a relatively greater degree than the general protoplasmic activity.   KNC and chloretone are, then, more effective as direct inhibitors of head formation, and chloroform and ether are more effective as inhibitors of the apparently nervous stimulation of the Y region which tends to inhibit head formation.   Further data on the nature of the stimulation of the X and Y regions and the differences in effect between several anesthetics will be presented in another paper.

The criticism may be advanced that when we find certain decreases in head frequency in the A pieces and increases in the C pieces on examination after two weeks of regeneration, we are not dealing with factors determining whether or not a head will form at X, but that the results merely indicate retardation of the heads in A pieces and acceleration of development in the C pieces.   This would be only another way of stating a claim for specific action on pieces from the two regions, and that point of view has been shown to be untenable.   Furthermore, experiments have been performed which show that in such series of regenerated pieces there are no further changes in head frequency after two weeks.

Criticism of the general nature of the data may also be advanced: that controls cut from worms of the same length and nutritive condition at different times do not show the same head frequency; that in many experiments the C pieces in the controls produce more normal heads than the B pieces. There are elements, both subjective and objective, which play a part in the results over which I have no control, although their nature may be recognized. I can cut in immediate succession two series of controls from the same selection of worms and be assured of a very slight difference in their head frequencies. But I cannot cut two series on successive days and expect such a close similarity in results. Diversion of attention in the interim alters one's judgment of the planes of cutting on the second day. That the physiological condition of the animals varies from day to day has already been pointed out. Therefore, one may not compare in detail the head frequency of similar controls cut on different days nor compare a series subjected to an anesthetic with a control cut at another time. But we are justified in comparing the head frequency of a series treated with an anesthetic with its own control and in comparing in a general way the effects of the same anesthetic in other experiments.

As regards the number of normal heads produced in the C pieces in the controls, it may be pointed out that the plane of fission between the anterior and posterior zooids is not a fixed anatomical structure, but a physiological condition which shifts anteriorly or posteriorly with the physiological condition of the animal. If one cuts a C piece so as to include a portion of the posterior zooid, the conditions established in the piece by the act of section will be very like the conditions in an A piece, for the anterior end of the posterior zooid is the apical region of a second metabolic gradient. Consequently, such pieces will show a high head frequency. Since one depends solely on judgment and experience in locating the plane of fission, it is not surprising that some of the pieces are cut to include a portion of the anterior region of the posterior zooid and hence develop normal heads.

In conclusion it may be stated that there are three lines of evidence presented here that demonstrate the non-specific nature of the factors concerned in head determination in pieces of Planaria. First, the forms of anterior ends developed in series treated with the various solutions of anesthetics are identical with the same types developed in the controls, and only those types which are developed in the controls are found in the series treated with the anesthetics; no forms are found which are characteristic of the effects of anesthetics.   Second, all the anesthetics used when employed under comparable conditions of concentration and period of exposure produce similar changes in head frequency. Third, with a single anesthetic increases or decreases in head frequency can be produced almost at will by employing different concentrations and periods of exposure.

SUMMARY

Evidence is presented which shows very plainly that the factors controlling head formation in pieces of Planaria are non specific and strongly supports Child's conclusions regarding the nature of these factors, viz., that head formation is determined chiefly by the relative activities of two antagonistic factors: 1) the tendency of the cells near the anterior cut surface of the piece to dedifferentiate and develop into the head of a new individual; 2) the tendency of the whole of the piece, exclusive of the cells directly concerned in the development of the new head, to maintain the differentiation of the old individual.   This region exerts a certain degree of control over the cells near the anterior cut surface and consequently tends to prevent the development of a new head.

The new evidence presented includes mass experiments in which it is shown, 1) that head frequency in pieces of Planaria can be controlled in either direction by subjecting the pieces for short periods after section to appropriate concentrations of the following anesthetics: chloretone, chloroform, chloral hydrate, ether, and ethyl alcohol; increases or decreases can be produced with a single anesthetic by employing different concentrations and periods of exposure; 2) that in such concentrations of anesthetics the increase of oxygen consumption following section does not occur.   An exception occurs in the case of ethyl alcohol, and evidence is presented which indicates that the increase in oxygen utilization is alcohol solutions is due to oxidation of the alcohol and does not result from the stimulation of section.

The data indicate that the stimulation of the cells near the anterior cut surface of the piece and concerned in the formation of the new head is a general protoplasmic excitation and that the stimulation of the whole of the piece exclusive of these cells is largely of nervous origin.

Anesthetics alter head frequency, 1) by direct inhibition of the processes of development of the cells near the anterior cut surface; this inhibition decreases head frequency; 2) by inhibition of the increase of metabolic activity of the whole of the piece after section; in pieces from certain regions this effect overbalances the direct effect on the developmental processes of the cells near the anterior cut surface, and in such pieces head frequency is increased.

A preliminary account is given of experiments which show that head frequency can be controlled by subjecting the pieces throughout the entire period of regeneration to concentrations of anesthetics so dilute as not to inhibit the stimulation of section.

The oxygen consumption of pieces subjected to weak solutions of anesthetics rises continuously. At the end of two weeks the rate of pieces in ethyl alcohol is several times that of the control. The rate of pieces in chloretone, chloral hydrate, and chloroform solutions is also higher than that of the controls at the end of two weeks.

BIBLIOGRAPHY

BELLAMY, A. W. 1919 Differential susceptibility as a basis for modification and control of early development in the frog. Biol. Bull., vol. 37.

BIRGE, E. A., AND JUDAY, C. 1911 The inland lakes of Wisconsin. Wisconsin Survey, Bulletin no. 22.

CHILD, C. M. 1911 a Experimental control of morphogenesis in Planaria. Biol. Bull., vol. 20.
1911 b Studies on the dynamics of morphogenesis and inheritance in experimental reproduction. I. The axial gradient in Planaria dorotocephala as a limiting factor in regulation. Jour. Exper. Zoöl., vol. 10, no. 3.
1911 c Die physiologische Isolation von Teilen des Organismus. Vortr. und Aufs. uber Ent.mech., Heft 11.
1913 a Studies, etc. V. The relation between resistance to depressing agents and rate of metabolism in Planaria dorotocephala and its value as a method of investigation. Jour. Exper. Zool., vol. 14, no. 3.

CHILD, C. M. 1913 b Studies, etc. VI. The nature of axial gradients in Planaria and their relation to antero-posterior dominance, polarity, and symmetry. Arch. f. Ent. mech., Bd. 37.
1914 a Studies, etc. VII. The stimulation of pieces by section in Planaria dorotocephala. Jour. Exper. Zool., vol. 16, no. 3.
1914 b Studies, etc. VIII. Dynamic factors in head determination in Planaria. Jour. Exper. Zool., vol. 17, no. 1.
1915 a Senescense and rejuvenescence. Chicago.
1915 b Individuality in organisms. Chicago.
1916 Studies, etc. IX. The control of head form and head frequency by means of potassium cyanide. Jour. Exper. Zool., vol. 21, no. 1.
1919 a The effect of cyanides on carbon dioxide production and on susceptibility to lack of oxygen in Planaria dorotocephala. Amer. Jour. Physiol., vol. 48, no. 3.
1919 b Susceptibility to lack of oxygen during starvation in Planaria. Amer. Jour. Physiol., vol. 49, no. 3.
1920 a Studies, etc. X. Head frequency in Planaria dorotocephala in relation to age, nutrition, and motor activity. Jour. Exper. Zool., vol. 30, no. 3.
1920 b Some considerations regarding the nature and origin of physiological gradients. Biol. Bull., vol. 39, no. 3.
1920 c The origin and development of the nervous system. Chicago.
CHILD, C. M., AND McKIE, M. 1911 The central nervous system in teratophthalmic and teratomorphic forms of Planaria dorotocephala. Biol. Bull., vol. 22.
CUSHNY 1910 Pharmacology and therapeutics, or the action of drugs. London, p. 145.
CZÁPEK, F. 1913 Biochemie der Pflanzen. Jena, p. 159.
HYMAN, L. H. 1919 a On the action of certain substances on oxygen consumption. II. Action of potassium cyanide on Planaria. Amer. Jour. Physiol., vol. 48, no. 3.
1919 b Physiological studies on Planaria. II. Oxygen consumption in relation to regeneration. Amer. Jour. Physiol., vol. 50.
1921 a Metabolic gradients of vertebrate embryos. I. Teleost embryos. Biol. Bull., vol. 40, no. 1.
1921 b Physiological Studies, etc. V. In press. Jour. Exper. Zool.
LOEB, J. 1916 The organism as a whole. New York, p. 155.
MORGAN, T. H. 1898 Experimental studies of the regeneration of Planaria maculata. Arch. f. Ent. mech., Bd. 7.
1905 Polarity considered as a problem of gradation of materials. Jour. Exper. Zool., vol. 2.
RANDOLPH, H. 1895 Observations and experiments on regeneration in planarians. Arch. f. Ent. mech., Bd. 5, no. 2.

Resumen por el autor, J. M. D. Olmsted.

El papel del sistema nervioso en la regeneración de los turbelarios policládidos.

Planocera californica, Phylloplana littoricola y Leptoplana saxicola siguen la regla en la regeneración de los policládidos. Pueden restaurar partes perdidas siempre que los ganglios cefálicos estén intactes. Si estos ganglios se destruyen parcialmente o hieren no se produce nuevo tejido nervioso para restaurar el cerebro a su tamaño original, y se se extirpan por completo la regeneración no puede tener lugar anteriormente, aun cuando puede llevarse a cabo posteriormente. Si permanece en el animal una porción del cerebro, tiene lugar una cierta cantidad de regeneración anterior y los ojos se regeneran, pero no se añade bastante material para restaurar la forma original y los nuevos ojos nunca alcanzan el tamaño de los presentes antes de la operación.

Translation by José F Nonidez
Cornell Medical College, New York

THE RÔLE OF THE NERVOUS SYSTEM IN THE REGENERATION OF POLYCLAD TURBELLARIA

J. M. D. OLMSTED

Hopkins Marine Station of Stanford University and Department of Physiology, University of Toronto

NINE FIGURES

The triclad turbellaria have been supposed to differ fundamentally from the polyclads in their powers of regeneration, the former being able to regenerate complete individuals from pieces taken from any portion of the body, the latter being unable to restore anterior parts when the cephalic ganglia are absent. Regeneration in the polyclads is therefore controlled to a considerable extent by the central nervous system. This is in strong contrast to the regenerative powers of the triclad, Planaria maculata, which, according to Morgan ('04), may regenerate an entire individual from a piece entirely devoid of nervous system. Miss Lloyd ('14), however, found that such a distinction between these two classes of turbellaria did not hold, for she discovered a marine triclad, Gunda ulvae, which was incapable of regenerating an anterior end unless the cephalic ganglia were present. This triclad therefore possesses the limited regenerative powers of the polyclads.

Opportunity was afforded at the Hopkins Marine Station of Stanford University on Monterey Bay, California, to study the regeneration of three species of polyclads in order to discover whether any one of them might possess exceptional powers of regeneration, or whether all would show the same influence of the nervous system on the regeneration of anterior parts as other polyclads already reported upon (e.g., Leptoplana tremellaria, Child, '04). The three species were found in sufficient numbers under stones in the water of tide pools or imbedded in sand near the low-water level at low tide, to carry out a series of experi-

49

ments on each. The species used were Planocera californica, Phylloplana littoricola, and Leptoplana saxicola. The species were determined by means of the key and descriptions given by Heath and McGregor ('12).

The worms could be kept for an indefinite time in fingerbowls in the laboratory, provided the water had been taken directly from the bay, and there was present a small amount of a green alga, such as ulva. No food was provided, although judging from the color of the digestive tract some of the green alga was ingested, for the intestines of specimens in dishes without the alga never showed a greenish tinge. The temperature was about 15°C.

In making the more careful operations, the worms were rendered motionless by adding crystals of chloretone to the seawater, but since all except P. californica very frequently disintegrated in the chloretone solution (cf. Heath and McGregor, '12), the majority of operations were performed without anaesthetizing. This disintegration while the worms are still living is a most striking phenomenon. All the cells, epithelial, muscle, digestive, separate from each other and the cytoplasm liquefies, leaving free nuclei, pigment granules, etc. The result is a mass of slime which clings to the dish when one attempts to remove the animal. This disintegration takes place first in the posterior half of the body in the region of the digestive tract. Occasionally this part will entirely drop out, leaving the margins intact except at the tip of the tail. The second region to disintegrate is that immediately anterior to the brain. Figure 1 indicates the position of these two regions.

It was found that all three species regenerated in the same manner, and, with the exception of P. californica, in approximately the same time. P. californica is much larger than any of the other species, its tissue is much firmer, and regeneration proceeds much more slowly. The lengths of time given below for the regeneration of various parts refer to P. littoricola and L. saxicola.

After cutting these polyclads in two by a transverse section anywhere posterior to the cephalic ganglia, the anterior piece restored all the missing parts. New material was evident along the margin of the cut by the fourth day. The digestive tract appeared in the new part by the thirteenth day, and at the end of four weeks the original form was fairly well restored, provided that the cut had been made at a level more than quarter of the animal's length from its anterior end. If the cut were more anterior than this, the worm did not attain its normal shape even at the end of six weeks, but remained as in figure 2. The process of morphallaxis occurs only to a limited extent (cf. Child, '04).

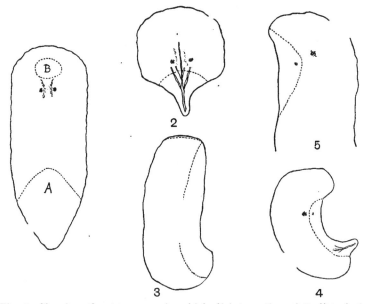

Fig. 1 Showing the two areas in which disintegration of P. littoricola in chloretone takes place, A and B.

Fig. 2 Regeneration of L. saxicola at the end of six weeks when cut was made just posterior to brain. New material below broken line.

Fig. 3 Regeneration of headless lateral piece of P. littoricola after eleven days, showing practically no regeneration along the anterior margin and an accumulation of material at the site of the new tail to the right.

Fig. 4 Regeneration after five weeks of small piece from left side of head of P. littoricola containing the left half of the brain. Accumulation of material to form tail is marked.

Fig. 5 Regeneration of remainder of the worm shown in figure 4, possessing the right half of the brain. Insufficient material has been added to restore completely the original shape.

Probably if the worms had been fed they would have resumed their usual form in a shorter time. At the time of collecting, several worms were found in nature to possess regenerated posterior portions, some having the parts perfectly restored, others with the new part much too small for the rest of the body.

Similarly, all missing lateral parts were restored after a longitudinal cut which did not involve the cephalic ganglia. Diverticula of the digestive tract could be seen in the new material after ten days, and the original form was fairly well restored after a month. It was noticeable that in all the cases of lateral cuts in which a portion of the tail was removed there was a tendency to form a tail regardless of the amount of material available (cf. head-forming tendency in Planaria maculata, Olmsted, '18). Longitudinal pieces always curved toward the cut side, and there was always a greater accumulation of material at the posterior end (figs. 3 and 4), often with the digestive tract clearly differentiated within it.

Restoration of material anterior to the cephalic ganglia was complete in about ten days, though if some of the eye spots were removed they were over two weeks in reappearing.

The regenerative powers are, however, quite different when the cephalic ganglia are injured, and each of the species followed the polyclad rule for regeneration. Posterior pieces after a transverse cut at any level of the body behind the brain were unable to restore the missing anterior parts. One might imagine that since anterior pieces can restore all the missing parts, posterior pieces might be able to restore some tissue at least at the anterior end, even if unable to regenerate a brain. But no matter where the cut had been made, immediately behind the brain or at the very tail, the raw edge was covered with epithelium and only a very narrow white margin of new material extended beyond the old tissue, even after a period of six weeks. Yet if the posterior end of one of these headless pieces was removed, another posterior end began to regenerate promptly and was completed in only a slightly longer time than ordinary. Similarly, the anterior tip of the head in front of the brain was unable to regenerate at all.

In one series of experiments the anterior ends of worms were split as nearly as possible along the midline. If the cut passed slightly to one side of the brain, the edges of the wound closed completely within three or four days, depending upon the length of the cut. But if the brain was injured, in every case except two the cut edges healed together up to about the level of the brain, and from this point remained open indefinitely, the flaps thus formed folding over each other. Examination afterwards showed that the nervous tissue had been so injured that the contours of the cephalic ganglia could not be recognized, and they were not restored at the end of a month. In the two cases referred to, portions of the cephalic ganglia remained. Eye spots and a certain amount of anterior tissue were restored, though not in sufficient amount to restore the normal form. There was no addition to the amount of nervous tissue left in the pieces after the operation.

In one of the two cases the knife passed fairly exactly through the center of the brain with so little damage to the nervous tissue that each piece possessed practically half the brain. The left side of the head with its two sets of eyes and tentacle on the brain was separated from the rest of the body, and the regeneration of both the smaller left piece and the larger right one were followed for over a month. After six days regeneration became evident in both pieces. On the tenth day the smaller one showed the characteristic tendency to accumulate material at the point where the tail should appear, and about the fourteenth day the digestive tract began to be evident in this projection. After twenty-five days eye spots showed the beginnings of a series of right eyes, and the form of the worm was much like that of worms sectioned transversely immediately behind the cephalic ganglia. The final result a little over a month after the operation is shown in figure 4. The larger piece began to show diverticula from the digestive tract in the new material on the left side after twelve days. Pigment spots also appeared at this time where the left eyes should be. The form was never fully restored, since there was not enough material to allow the worm to become straight (fig. 5). When these two pieces were examined to determine the extent of regeneration of the nervous system, it was found that each half of the brain remained in each piece exactly as it was at first (figs. 6 and 7). In no case was a portion of the brain able to restore the missing nervous tissue.

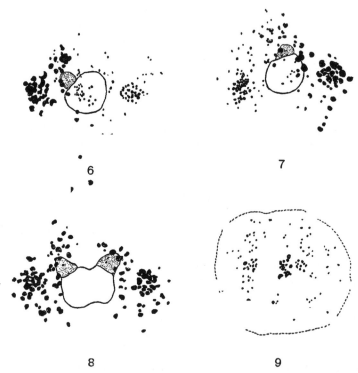

Fig. 6   Eyes and brain of worm shown in figure 4.
Fig. 7   Eyes and brain of worm shown in figure 5.
Fig. 8   Normal eyes and brain of P. littoricola for comparison.
Fig. 9   Eyes of P. littoricola four weeks after removing the brain. Broken line indicates the extent of new tissue. Tendency toward normal grouping of eyes is marked even in absence of brain.

To make more certain that these polyclads were unable to regenerate a brain, a small disc just large enough to include the brain was cut from the worm by means of a thin-walled glass tube. In many cases the anterior margin of the hole left in the worm broke through and left two finger-like projections. In these specimens there was never any regeneration beyond the healing over of the raw edges and the projecting fingers remained indefinitely. When the rim of the hole remained intact, new tissue filled in the space within two days. After a week one could see the parts of the digestive tract joining up in the new tissue so as to form continuous tubes, and at the end of three weeks pigment spots showed that eyes had been regenerated. These eye spots showed a tendency to be grouped into cephalic and tentacle eyes as in the normal individual even in the absence of the brain (fig. 9). When the posterior end of such a brainless specimen was removed, a new tail was regenerated in the usual manner.

The results of these experiments bear out Miss Lloyd's idea that "the mechanism for the restoration of the tail belongs to the body as a whole, while that for restoring the head is an entirely independent one, which may or may not be localized in some part of the body, notably the anterior end." In these three species of polyclad worms this localization of the mechanism for the regeneration of anterior parts is undoubtedly in the cephalic ganglia. The exact nature of this relation of the central nervous system to the powers of regeneration still remains unsolved, and the merits of the various suggestions that it may be a matter of enzymes, flow of organ-forming substances, hormones, or differences in axial gradients have yet to be proved.

SUMMARY

Planocera californica, Phylloplana littoricola, and Leptoplana saxicola follow the rule for polyclad regeneration. They are able to restore missing parts, provided the cephalic ganglia are intact. If these are injured, new nervous tissue is not added to restore the brain to its original size, and if they are entirely removed, regeneration cannot take place anteriorly, though it may do so posteriorly. If a portion of the brain remains, a certain amount of anterior regeneration takes place and eyes are regenerated, but not enough material is added to restore the original form, and the new eyes never reach the size of the old ones.

BIBLIOGRAPHY

CHILD, C. M. 1904 Studies on regulation. IV. Some experimental modifications of form-regulation in Leptoplana. Jour. Exp. Zool., vol. 1,
pp. 95–133.

HEATH, H., AND McGREGOR, E. A. 1912 New polyclads from Monterey Bay,
California. Proc. Acad. Nat. Sci. Phila., Sept., 1912, pp. 455–488.

LLOYD, D. J. 1914 The influence of the position of the cut upon regeneration
in Gunda ulvae. Proc. Roy. Soc. Lond., Series B, vol. 87, pp. 355–366.

MORGAN, T. H. 1904 The control of heteromorphosis in Planaria maculata.
Arch. f. Entw. Mech., Bd. 17 S. 683–694.

OLMSTED, J. M. D. 1918 The regeneration of triangular pieces of Planaria
maculata. A study in polarity. Jour. Exp. Zoòl., vol. 25, pp. 157–176.

Resumen por el autor, J. M. D. Olmsted.

El papel del sistema nervioso en la locomoción de ciertos
policládidos marinos.

Los policládidos Planocera californica, Phylloplana littoricola
y Leptoplana saxicola exhiben cuantro tipos posibles de locomoción. Para los movimientos de natación es necesario que los
ganglios cefálicos estén intactos. La destrucción parcial o la
pérdida del cerebro impide el uso de este método de locomoción.
La acción ciliar no está regulada por el sistema nervioso y prácticamente no juega papel alguno en la locomoción. La locomoción atáxica es un fenómeno puramente local, pero está
regulado por el sistema nervioso, puesto que puede abolirse por
la cloretona. Euryleptotes cavicola se mueve por este tipo de
locomoción solamente. La locomoción retrógrada detáxica está
regulada por los ganglios cefálicos, cada ganglio regulando la
progresión de la olas musculares de su mismo lado. Los cordones
nerviosos sirven como conductores de los impulsos para la formación de las ondas de su mismo lado. Cuando se corta un
nervio las ondas desaparecen en el mismo lado de la operación
al nivel de la sección. La locomoción de los policládidos es
comparable en este respecto a la de los moluscos.

Translation by José F. Nonidez
Cornell Medical College, New York

THE RÔLE OF THE NERVOUS SYSTEM IN THE LOCOMOTION OF CERTAIN MARINE POLYCLADS

J. M. D. OLMSTED

Hopkins Marine Station of Stanford University and University of Toronto

During experiments on the regeneration of polyclad turbellaria of Monterey Bay, California, it was noticed that there were pronounced changes in locomotion following certain operations which involved the central nervous system.

Three species were studied with reference to the control of locomotion by the nervous system, Planocera californica, Phylloplana littoricola, and Leptoplana saxicola, with a few observations on Euryleptotes cavicola. The methods of locomotion are the same in the first three species, the chief differences lying in the rate of progression—L. saxicola being the fastest moving and P. californica the slowest—and also in the relative frequency of each method in the different species.

A general account of the locomotion of polyclads is given by Lang ('84) in his monograph on the polyclads from the Bay of Naples. A more detailed account of the locomotion of Leptoplana tremellaria, also from the Bay of Naples, is given by Child ('04), and of Leptoplana lactoalba var. tincta from Bermuda by Crozier ('18). Child states that there are two chief methods of movement in L. tremellaria, 'swimming and creeping.' The former is an 'undulating movement' of the margins of the body, the latter involves 'both muscular and ciliary activity.' "When the animal is moving quietly—the cilia afford the chief motive power, although the slight muscular movements of the margin of the body are almost constant, portions being lifted from the substratum, brought forward, and again attached." But when strongly stimulated, "movements occur in rapid alternation on the two sides of the body and the similarity between this mode of progression and the use of legs cannot escape the observer." This is a type of locomotion common to gastropods and is called retrograde detaxic (Olmsted, '17). This type of locomotion was observed by Crozier to occur in his Bermudian species. The methods of locomotion described by these authors are seen in the species of polyclads from Monterey Bay, but they are in reality four distinct methods of progression.

In these species the swimming movement involves the whole body, not merely the lateral margins as in Leptoplana tremellaria (Child, '04). P. californica resorts to this method more frequently than the other species. E. cavicola was never seen to release itself from the substrate, and therefore never swam freely. While creeping about a dish, P. californica will suddenly release its anterior end, raise it above the substrate, and initiate a series of waves which, from their resemblance to gastropod locomotion, may be termed retrograde monotaxic (Olmsted, '17). The resulting movement is like that of a rug being rapidly shaken. The waves appear at the rate of two a second, and pass posteriorly at such a rate that two or two and a half waves are present at a given moment. An individual seldom swam more than 2 cm. in this fashion before resuming its creeping. The other two species seldom employed this method of locomotion except when falling through the water after creeping along the under side of the surface film, or when somewhat roughly dislodged from the substrate during active creeping. Swimming was never initiated from rest, but only occurred if the animal were already in motion. In this type of locomotion either the dorsal or ventral surface might be uppermost, the dorsal more often in P. littoricola.

When transverse cuts were made at any level posterior to the cephalic ganglia of these polyclads, this swimming movement could no longer be elicited from brainless posterior pieces, but the anterior pieces containing the brain, no matter how short, exhibited a few swimming movements when falling from the surface. Likewise after splitting the head longitudinally so that the two halves of the brain were separated, an animal could not be made to swim, even if one cephalic ganglion remained in each piece. The brain must therefore be intact for this movement to take place. The wave of muscular contraction in this method of locomotion involves the coordinated movements of both sides of the body, for the wave extends across the entire width of the animal. One might imagine that each cerebral ganglion could control the movements on its own side, and in those individuals with the brain split in halves each side might act independently. But this proved not to be the case. It was necessary that the entire brain be undisturbed. Again, it would seem reasonable that the impulse to start off this method of locomotion might originate from some stimulus at the anterior end of the worm, since preliminary movements are made by this portion before releasing the rest of the body from the substrate in preparation for swimming. But if the anterior ends were removed by a transverse cut immediately in front of the cephalic ganglia, it was found that the anterior brainless pieces were unable to swim, while the posterior pieces retained this power perfectly. The impulse to initiate the wave seems therefore not to occur as a stimulus from the anterior part of the body, but from within the brain itself.

When the cephalic ganglia were removed by cutting out a small disc of tissue in the vicinity of the eyes, and the hole had filled in by regeneration, it was found that the brain could not be regenerated. Such brainless worms were unable to swim although no part was lacking except the brain.

Ciliary movement does not play a prominent part in the locomotion of these four species of polyclads. When they are apparently at rest and making absolutely no progress at all, the cilia can be seen still beating. When the worms are placed in chloretone, the cilia continue to beat, and if a worm becomes detached from the substrate to which it often adheres even when anesthetized, it will be carried along at a very slow rate by the cilia. The speed is the same whether the worm is on its dorsal or ventral surface, and is slower than any method of locomotion by muscular action. I am convinced that the cilia never function as the sole organs of locomotion under normal conditions. They may aid muscular locomotion, but do not act alone. Miss Stringer ('17) has made this same criticism of the accepted view of the means of locomotion in the triclad planarians and Crozier ('18) claims that in the Leptoplana 'tincta' "when muscular waves are absent, no creeping progression can be detected." Pieces of any size and from any portion of the body of the Monterey polyclads show this same ciliary movement in chloretone. This ciliary action is therefore not dependent on the central nervous system.

The slow creeping movement is accomplished by means of the constant slight muscular contractions to which Child ('04) refers. There are no definite waves, but the entire ventral surface appears to be thrown into irregular ripples. This can be especially well observed when the polyclad is creeping under the surface film of the water. E. cavicola is especially favorable for observation, since the animal is a broad oval, some 3 cm. in length by 2 in width. The animal lays down a track of mucus as it proceeds. The midventral region is depressed and fairly quiet, as if this portion were holding on by suction, while the margins are especially active. The rippling motion is due to momentary local release of a portion of the ventral surface from its point of attachment and a shifting of this area by muscular contraction. These worms are able to go both forwards and backwards and to turn to one side. The movement is slow and often irregular. Pieces formed by longitudinal cuts always moved in circles toward the injured side (cf. Child). With P. californica and E. cavicola the rate of progress averaged some 0.5 mm. per second. The other species moved slightly faster, but so seldom in a straight line that accurate measurements were not obtained. Worms with the cephalic ganglia removed and pieces from any region anterior or posterior to the brain exhibited this ataxic (Olmsted, '17) type of locomotion. The rippling muscular movement persisted for some fifteen minutes in a saturated solution of chloretone in sea water, but finally ceased if the animal were undisturbed. The slow creeping movement is therefore under control of that part of the nervous system in the immediate locality of each muscular contraction and is a local phenomenon.

Of the fourth type of movement Child makes the following remark, "The animals appear almost as if walking forward." This ditaxic retrograde locomotion is brought about by the releasing of a small portion of the lateral margin at the anterior end of the worm, the pulling of this bit forward 2 or 3 mm. by contraction of the longitudinal muscles, and its reattachment at a point anterior to its former position. The contraction once started proceeds in a wave down the entire margin to the posterior end. These waves alternate on each side of the body, so that the worm appears to stride along like a biped. This agrees in every respect with the process in gastropods. From records of observations on one large specimen of P. californica, the rate of slow ditaxic locomotion is 0.21 cm. per second, and the worm takes 0.33 step per second. Other records on the same individual gave 0.36 cm. and 0.4 step per second. When disturbed it could proceed at the rate of 0.39 cm. per second in 0.66 step. The rate of locomotion may therefore be varied both by an increase in the number of steps and the length of each step as well. For P. littoricola the ordinary movement was more rapid, i.e., 0.4 cm. per second in 0.11 step. When disturbed, there was very little increase—0.43 cm. in 0.13 step. This type of locomotion was not observed in E. cavicola.

To determine whether the wave was initiated in the margin of the head, this portion was cut away. It was then found that waves started just posterior to the cut, but their rate was practically normal. In another experiment a semicircular cut in front of the brain of L. saxicola severed the connection of the nerves of the anterior margin with the brain. This worm exhibited the walking movement perfectly, the waves starting on a level with the ends of the cut. The cut healed over in three days, and by the fourth day locomotion was perfectly normal, the waves starting at the anterolateral margin of the head. Headless specimens and those with the cephalic ganglia removed were unable to employ this type of locomotion, but anterior pieces from worms transected just behind the brain nearly always use this as their sole means of locomotion.

Splitting the head through the brain usually resulted in loss of this ditaxic type of movement. In two cases, however, parts of the cerebral ganglia remained intact. In one of these the cut removed a portion of the right half of the brain. For a week after the operation this worm performed the walking movement with its left side perfectly and moved in circles to the right. Later, an occasional wave on the right alternated with those on the left. This imperfect coordination continued even after a month, so that the animal was never able to move in a straight line. In the other case the brain was evenly divided. Each piece regenerated all its parts except the missing half of the brain. The smaller piece which contained the left half of the brain moved in circles to the right by means of a single wave which took its origin at the usual point and passed down the left margin. There was never any indication of a wave in the new tissue on the right side during the month it was under observation. In the larger piece in which the left half of the brain was missing, waves passed down the right side quite normally with only an occasional alternate wave down the left side, so that this piece moved in circles to the left. By the end of a week, however, the waves on the left side made their appearance somewhat more frequently and alternated some two or three times with waves on the right side. These waves never appeared in the new tissue, they began posterior to the cut. The worm remained in this condition for more than a month, still circling to the left and lacking perfect coordination between the right and left halves of its body.

These experiments indicate that each cerebral ganglion controls the 'stepping movement' on its own side, each half being independent of the other. But for perfectly coordinated movements involving both sides of the body, the connection between the halves of the brain must be intact. In view of the fact that a few coordinated movements were possible in the two cases just described, where a portion or all of the brain had been removed from one side of the body, it is evident that a wave may appear on a side lacking the brain. But it must be remembered that there was at least one-half the brain present in the body, that the wave appeared only in the old tissue, never in the new, and that it appeared infrequently and only after several waves had passed down the uninjured side. I should explain the appearance of the wave on the side lacking its half of the brain on the ground that through the physiological principles of 'facilitation' and 'summation,' the movement of a wave down the uninjured side was able to serve as a stimulus, which, upon being transmitted through the connecting commissure immediately behind the brain to the other nerve cord, was sufficient to initiate an occasioual wave. (Cf. the diagram of the ventral nervous system of L. saxicola given by Heath and McGregor, '12, fig. 21.)

If the stepping movement in ditaxic locomotion is controlled by the cephalic ganglia, the question arises as to the function of the longitudinal nerve cords. Various operations were tried in order to answer this question. It was found necessary to make observations fairly soon after cutting the nerve cords, for within three or four days a wound extending almost the entire width of the body would be joined together and perfect coordination restored. For example, in a specimen of L. saxicola a horizontal slit was made posterior to the brain, severing both nerve cords. The part anterior to the wound at once moved rapidly by the ditaxic method of locomotion and also attempted to swim freely in the water. The posterior part moved slowly by the ataxic method. The consequence was that the pulling of the anterior end caused the wound to gape in a wide circle. When the worm came to rest, the edges of the wound closed. Two days later the wound was healed and on the next day perfectly coordinated alternate waves passed down the entire length of the worm in a normal manner.

The nerve cords serve as conductors of impulses which cause progression of the waves in ditaxic locomotion. This is nicely shown by cutting a portion from the side of a worm without injuring the nerve cord. A wave starting at the head end of the body stops when it reaches the upper edge of the wound, for the muscles which would carry the wave on have been cut away. But the wave again makes its appearance at the proper time at the lower margin of the wound, and then continues on to the posterior end of the worm. The impulse therefore was carried on through the nerve cord, though there was no visible effect accompanying its passage. But if the wound were near enough to the midline to have cut into the nerve cord, the wave on that side ceased entirely at the level of the wound. According to the diagram of the nervous system of L. saxicola of Heath and McGregor, there is only one cross connective between the two nerve cords, and that is situated immediately posterior to the brain. If the wound is made anterior to this connective an occasional wave will pass down the injured side as explained above, but if the wound is posterior to this connective there can be no conduction of the impulse to form a muscular wave. Local muscular contraction can still take place, however. In one case a long strip was cut from one side of the body of a worm, leaving a narrow bridge of tissue at the anterior end to attach it to the body. The ditaxic waves stopped when they reached the bridge of tissue, and reappeared at the posterior edge of the wound. No wave appeared in the strip itself. One could cause the strip to contract by pricking any portion of it, so that the local muscular response remained. Within two days the strip was attached to the body except at the extreme posterior end, but no waves appeared in it when the worm was 'walking.' On the third day the wound was entirely healed and waves passed two-thirds of the way down the strip. On the following day the animal moved normally. The union of cut ends of the nerve cord or of severed peripheral branches with the brain or nerve cords occurred in a remarkably short time, since perfect coordination was established in a very few days. But if a piece of the nerve cord were actually removed, there was lack of coordination for several weeks, because of the relatively slow regeneration of lateral parts including the nerve cords. Nerve cords do regenerate, while the brain will not.

The locomotion of these polyclad worms is comparable in every respect with that of gastropod mollusks. The types of movement are the same. Retrograde monotaxic, retrograde alternate ditaxic, and ataxic methods are clearly distinguishable. The part played by the nervous system is similar in each group (cf. section of "Physiologie des Nervensystems," by S. Baglioni in Winterstein's "Handbuch der vergleichenden Physiologie"). Slugs are able to move by means of waves on the foot when the connection with the brain is severed. Waves can also appear on isolated pieces of the foot. This is correlated with the presence of an extensive nerve net in the foot with many cross connectives. Ataxic locomotion in the polyclads is likewise carried on in the absence of the brain and in pieces from any part of the body. Their nerve net must be responsible for this movement, since the movement is under the control of the nervous system in the immediate vicinity of the contracting muscles. In snails, such as Helix pomatia, the impulse which causes the normal peristaltic wave arises in the pedal ganglion, and is transmitted by the nerve cords of the foot, each of which serves a definite area. In a similar way both monotaxic and ditaxic locomotion of the polyclads are controlled by the cephalic ganglia, and the nerve cords transmit the impulses to a definite area. This is another instance of evolution along the same lines in two quite different groups.

SUMMARY

The polyclads, Planocera californica, Phylloplana littoricola, and Leptoplana saxicola, exhibit four possible types of locomotion.

For the swimming movement it is necessary that the cephalic ganglia be intact. Injury to or loss of the brain prevents the use of this method.

Ciliary action is not under the control of the nervous system and plays practically no part in locomotion.

Ataxic locomotion is a purely local phenomenon, but controlled by the nervous system since it is abolished by chloretone. Euryleptotes cavicola moves by this type of locomotion alone.

Ditaxic retrograde locomotion is under the control of the cephalic ganglia, each ganglion controlling the progression of muscular waves on its own side.

The nerve cords serve as conductors for the impulses to wave formation each on its own side. Cutting a nerve causes the waves to disappear on that side at the level of the cut.

The locomotion of polyclads is comparable in these respects with that of mollusks.

BIBLIOGRAPHY

CHILD, C. M.   1904   Studies on regulation. IV. Some experimental modifications of form-regulation in Leptoplana. Jour. Exp. Zoöl., vol. 1, pp. 95–133.

CROZIER, W. J.   1918   On the method of progression in polyclads. Proc. Nat. Acad. Sci., vol. 4, pp. 379–381.

HEATH, H., AND McGREGOR, E. A.   1912   New polyclads from Monterey Bay, California. Proc. Acad. Nat. Sci. Phila., Sept., 1912, pp. 455–488.

LANG, A.   1884   Die Polycladen (Seeplanarien) des Golfes von Neapel. Leipzig.

OLMSTED, J. M. D.   1917   Notes on the locomotion of certain Bermudian mollusks. Jour. Exp. Zool., vol. 24, pp. 223–236.

STRINGER, C. E.   1917   The means of locomotion in planarians. Proc. Nat. Acad. Sci., Dec., 1917, pp. 691–692.

WINTERSTEIN, H.   1913   Handbuch der vergleichenden Physiologie, Bd. 4. Jena.

Resumen por el autor, Leonell C. Strong.

Un análisis genético de los factores responsables de la susceptibilidad á los tumores transplantables.

La raza es el factor primario que determina si un individuo
dado ha de presentar crecimiento progresivo o no ha de pre-
sentarle, cuando alberga un tumor transplantable. La susceptibilidad y la ausencia de esta son manifestaciones de la consti-
tución genética del huésped. Varios factores fisiológicos
secundarios, el más importante de ellos la edad, funcionan en la
determinación del resultado de una reacción determinada. Estos
factores pueden denominarse contribuyentes o accesorios. El
factor edad es una expresión del grado del proceso de la adquisi-
ción de la especificidad de los tejidos, regulado hasta cierto punto
por las gónadas. La curva de susceptibilidad de la edad hacia
los tumores transplantables para los individuos normales de una
raza no susceptible presenta una notable semejanza con la
curva de la actividad de las gónadas. El factor sexo (que se
encuentra en los ratones jóvenes) depende por lo menos de dos
causas primarias: 1) del factor edad y 2) de la diferencia en
actividad fisiológica de los sexos en diferentes periodos de la
vida. La extirpación de las gónadas no cambia el tanto por
ciento en masa de las reacciones para los individuos de una raza
no susceptible. La gonadectomía produce en la variedad em-
pleada un aumento significativo del tanto por ciento de las
reacciones hacia ambos tumores, conforme sucede con ciertas
características morfológicas. Mediante extirpación de las
gónadas, la individualidad de los tejidos y el funcionamiento
normal del factor edad pueden influirse. Los dos adenocarcinomas de la glándula mamaria empleados en los experimentos
han retenido una potencia de reacción constante durante los
experimentos.

Translation by José F. Nonidez
Cornell Medical College, New York

A GENETIC ANALYSIS OF THE FACTORS UNDERLYING SUSCEPTIBILITY TO TRANSPLANTABLE TUMORS

LEONELL C. STRONG

Department of Zoölogy, Columbia University

THIRTY-THREE FIGURES

CONTENTS

I. INTRODUCTION

1. Contributions to the genetics of cancer

The first part of the following section reviews those genetic investigations that contribute to an explanation of the conclusions reached in this paper. Most of the work, unfortunately, has been published in periodicals not generally seen by medical investigators. Some of the work has been of such a nature that it has led only to skepticism.

Three contributions to the genetics of cancer appear to be worthy of special significance. These relate to experiments on transplantable tumors, and involve—

a. The recognition of the factor of race.

b. The demonstration of mendelian segregation and recombination of factors that underlie susceptibility to transplantable tumors.

c. The possibility that the causation of spontaneous neoplasms may be due to a process of mutation.

a. Race. Up to within the last few years, most investigators in this field considered the phenomenon of race as a factor underlying susceptibility to transplantable tumors to be of secondary importance. It has become evident, however, that market mice are not reliable for experimental work in cancer. Where the same tumor tissue is inoculated into mice that have been proved to be homogeneous in their genetic constitution, no rhythms of growth activity or fluctuations are observed. (See the work on the transplantable neoplasms of the Japanese waltzing mice discussed below.)

The first work that showed the importance of race was that of Jensen. He discovered the interesting fact that if tumor tissue derived from a Berlin stock was inoculated into Berlin mice and into Hamburg mice there was a decided difference in the results. The percentage of susceptible individuals among the original stock was distinctly greater. He did not attempt any analysis of this difference. This work, however, attracted the attention of several students of genetics

Genetic research is primarily concerned in the recognition of races and variations within a species. The modern geneticist analyzes his variations by the method of hybridization. He must, however, be sure that the races used are homogeneous for all the factors that determine the differences encountered in the character under investigation. This fixation of characteristics is accomplished by inbreeding.

b. Mendelian segregation. The next step was taken by Doctors Tyzzer and Little, working on two transplantable tumors that arose in a closely inbred strain of Japanese waltzing mice. The first (J. W. A.) was a carcinoma of the mammary gland that grew in 100 per cent of all mice (of that particular Japanese waltzing strain) inoculated. The first hybrid generation (F_1, produced by the hybridization of the two parent stocks) produced sixty-two individuals, sixty-one of which grew the transplanted tissue, even faster than animals of the original Japanese waltzing strain. No explanation of this apparent increased activity on the part of the tumor cell was offered. In the general discussion of this paper we shall consider this phenomenon and attempt a genetic explanation. In the second filial generation (F_2, produced by mating the F_1 individuals inter se) only three out of the 183 mice obtained were found to be susceptible to the inoculated tissue. Tyzzer and Little applied the multiple-factor hypothesis to explain these results. This hypothesis postulates that in the production of certain characteristics several independent (or linked) mendelian factors may be necessary.

Little and Tyzzer concluded that the simultaneous presence of from twelve to fourteen mendelian factors which were characteristic of the Japanese waltzing race was necessary for the progressive growth of the transplanted tissue under consideration. They thus placed the inheritance of susceptibility to transplanted tumor tissue on mendelian grounds.

Working with a sarcoma (J. W. B.), Doctors Tyzzer and Little were enabled to verify their first conclusion based on the carcinoma, J. W. A. In this second case the results obtained were simpler and even more convincing. The parent stocks and their F_1 hybrids behaved as before. The F_2 generation, however, gave twenty-three susceptible to sixty-six non-susceptible animals. By the use of back-crossing (produced by crossing the F_1 individuals to the non-susceptible race), Doctor Little ('20) was able to analyze the mendelian factors more fully. The conclusions arrived at by this second analysis were:

(1) That from 3 to 5 factors—probably four—are involved in determining susceptibility to the mouse sarcoma J.W.B.

(2) That for susceptibility the simultaneous presence of these factors is necessary.

(3) That none of these factors is carried in the sex (X) chromosome.

(4) That these factors Mendelize independently of one another.

c. *Mutation hypothesis.* Theorizing from the Japanese waltzing mouse experiments, Doctor Tyzzer, in a general paper on "Tumor Immunity," suggested that the cause of spontaneous neoplasms may be due to some sort of mutational process. The process of mutation is accepted by most modern geneticists as the cause underlying the production of diverse variations within a species. (In 1908, Williams made a statement to the effect that the cause underlying the tumor-cell formation may be analogous to the phenomenon known as mutation. He must receive the credit, therefore, of being the first to suggest such a possibility.) ·

2. Factors underlying susceptibility to transplantable tumor tissue

a. *Race* as a factor has already received sufficient preliminary consideration.

b. *Sex* as a factor underlying susceptibility to inoculable tumor tissue is disputed. Certain investigators have been able to determine a significant difference between the sexes in their receptivity toward transplanted tumors. Others have been unable to determine any significant difference between the sexes.

c. *Age* is a recognized factor underlying susceptibility to transplantable tumors as well as having some causative relation to the origin of spontaneous neoplasms.   1) A very young individual from a susceptible race will sometimes fail to grow the transplanted tissue, although the same individual will do so if inoculated when it is one-half to three-fourths grown. At this age it is more susceptible than at any other period in its life-cycle. 2) Very young animals from a non-susceptible race will sometimes grow a transplanted tumor when inoculated, although no matured animal in that particular race grows the same tissue.

d. *Pregnancy.* Several investigators have concluded that pregnancy has some influence on the rate of growth of the neoplastic tissue.

Leo Loeb has studied several reactions with transplantable tissues in relation to pregnancy. He determined that if carcinoma of the mammary gland be inoculated into pregnant females, the tissue would fail to grow, although no such behavior was encountered in the controls. Later he discovered the interesting fact that after autotransplantation, an adenofibroma of the mammary gland survived, but showed progressive growth only when the host became pregnant. Here we see that the factor of pregnancy can apparently differentiate between malignant and benign growths.

According to Loeb, transplanted normal mammary-gland tissue behaves in the same way as adenofibroma tissue during pregnancy of the host. Further, he recognizes that the normal embryonal tissue reacts like carcinoma in the mouse, but evi-

dently not in the rat.   The result obtained with the transplanted
normal tissue was to be expected—it is a matter of common
knowledge that secretions from a corpus luteum can incite normal mammary-tissue development.

Loeb has concluded that—

(a) there is a specific affinity of the transplanted tissue for a certain
growth substance given off by the ovaries.   This affinity is greatest in
the case of normal mammary gland tissue and of adenofibroma of the
mammary gland; less marked in the carcinoma of the mammary gland
and lacking in the ordinary embryonal tissue.   (b) A factor injurious to
tissue growth operating in pregnancy.   This may be either directly
injurious substance or a shortage of ordinary foodstuffs due to the
growth of the embryo.   There are certain facts which suggest the first
alternative rather than the latter.   (c) Homoiotoxins seem to strengthen
the second injurious factor, while their absence seems to favor the first
aiding factor.   (d) There seems to be variations in the strength of one
or several of these variable factors in various species.

3. Prevalent conceptions concerning peculiarities and characteristics of the tumor cell (derived from investigations on trans- plantable tumors)

a. Rhythms of growth.   Of all the interesting peculiarities
that the cancer cell is supposed to be endowed with, that of
alternating periods of depression and growth is the most interesting.   By plotting curves based upon the percentage indications[1]
as ordinates and the interval elapsed before the tumor reached
the inoculating point as abscissae, Bashford, Murray, and Bowen
concluded that the transplanted tumor cell underwent cyclical
changes of growth activity.   It must be remembered, however,
that these data have been collected from market mice.   The
objection has already been offered (Calkins) that if these were
real rhythms of activity in the tumor cell, it could only be
determined with accuracy by studying the rate of growth in
one host only.

The English investigators referred to maintain that the fluctuating results obtained are due to the varying ability of the cell
to adapt itself to the foreign-host tissue.   This matter will

[1] By percentage indications was meant the relative number of the mice inoculated that grew the tumor mass progressively.

be taken up next under the consideration of 'virulence or adaptation.'

Tyzzer and Little, on their work with the Japanese waltzing-mouse tumors (J. W. A. and J. W. B.), determined that every mouse of that strain inoculated grew the two tumors progressively and at a constant rate of growth (there was no significant difference between the rates of growth of any two transplanted tumors in those series). Evidently, the results observed by Bashford, Murray and Bowen and Calkins may have been largely due to fluctuations in the genetic constitution of the 'market mice.' Where a definitely proved constant homogeneous race of mice is employed, no rhythms of growth activity are encountered throughout a period of over ten years.

b. Virulence or adaptation. It has long been recognized that by continually inoculating tumor tissue into animals of a foreign strain, one can gradually increase the percentage of takes until the maximum is reached. The English observers were the first to determine that this percentage of positive tumor takes fluctuates between the minimum and maximum for any given race. At present two explanations are possible: 1) the tumor cell has an inherent capacity of adapting itself to a foreign host or, 2) the cell possesses fluctuating 'virulence.'

The first explanation has the support of Bashford, Murray, Haaland, Bowen, Cramer, Woglom, and others; the second that of Ehrlich and Apolant.

The arguments for and against these explanations are too well known to be repeated here. We may emphasize, however, that neither theory is entirely acceptable. Each assumption rests on the old idea of the inheritance of acquired characteristics, although several attempts have been made to mask this implication.

The conception of virulence has been analyzed into two elements by Apolant, 1) transplantability, determined by the percentage indication of positive takes, and, 2) proliferative energy, calculated from the rate of growth of the transplanted mass. In the light of modern genetical investigation, both of these terms lose much of their significance. The percentage indications òf reactions of positive growths is determined not by any characteristics of the tumor cell alone, but by its reaction with the genetic constitution of the strain of mice under investigation. Again, the rate of growth of the transplanted mass is constant provided one is dealing with a constant homogeneous race of individuals.

A genetic interpretation of the observed results (that by progressive inoculations into a foreign strain one can increase the percentage of positive indications) would be somewhat as follows: Since all races of 'market mice' have had a common origin, they have therefore some genetic factors that relate to cancer in common. Within a single family individuals are more closely related to one another than to individuals from another family. When an investigator found that an individual of the foreign strain grew the transplanted tumor, he would naturally pick out individuals within the family of susceptible mice and eliminate those from other apparently non-susceptible families. There is thus an unconscious tendency to select, from the stock, individuals more susceptible to the transplantable tissue—more susceptible because they approach more nearly the genetic constitution of the mouse that grew the original tumor spontaneously. An investigator not realizing the full significance of the race factor would believe that he was dealing with only one variable—that of the behavior of the tumor cell itself.[2] This explanation is offered not as conclusive proof of what necessarily was involved, but as what may possibly have occurred in the experiments that gave the conflicting results previously obtained by various workers.

c. *Transitional conditions.* Certain investigators believe that a carcinoma may be transformed directly into a sarcoma and vice versa.

Several investigators have already foreseen the difficulty encountered in such a transformation from epithelial to connective tissue. Among these may be mentioned Ehrlich, Apolant, Ribbert, and, more recently, Woglom. It has been suggested

---

[2] Of course, healthy and vigorous tumor tissue alone must be used, or variations due to infection will enter in.

that the tumor mass under consideration was at the outset a 'mixed' tumor, each element being derived from its embryonic anlage (carcinoma from epithelial cells and sarcoma from the connective tissue). Assuming that connective-tissue products can only be produced by connective-tissue elements, we may conclude that if a tumor mass contained these specialized products it must have also contained functioning connective-tissue elements. Some evidence of this nature has been discovered by Haaland, Slye, Holmes, Wells, and Woglom. Haaland found that in a mixed carcinosarcoma, there were inter- and intracellular fibrils present in the sarcomatous parts.

By the discovery that myxomatous changes may occur in the connective-tissue part of a carcinosarcoma, Slye, Holmes, and Wells indicated that the sarcomatous part must have arisen in the stroma. Woglom in a recent communication concludes that, since cartilage is found in a carcinosarcoma of the mouse, the sarcomatous element of the mass must have been derived from preëxisting connective tissue.

II. EXPERIMENTAL

1. Materials

In order to test how far the conflicting results that have been obtained in investigations with transplantable tumors on the lower animals have been due to the use of various market stocks, careful attention has been given to the strain of mice employed.

Rigorous inbreeding may or may not produce harmful results. It does produce genetically homogeneous races. Relative homozygosity (95 to 99 per cent) is only approached after from eight to ten generations of the most intense method of inbreeding. The third inbred generation by any method can never give an index of homozygosity of more than 87.5 per cent. Counting a generation every three months, this process of inbreeding would consume over two years. The approach toward homozygosity is very slow. The time element and expense are therefore ob viously too great for the patience and resources of most investigators.

a. Races used. The common wild house mouse fulfills the requirement of a homogeneous race[3] to a marked degree. If collected in the same locality (an isolated group of buildings, a small island, etc.), one may be fairly sure that such stock is homogeneous for several reasons. Strange mice very seldom invade any particular location already occupied by a well-established colony of mice. This is evidenced by the fact that slight variations in color and form tend to be restricted to the same corner of a building, etc.

Wild adult female mice will not breed readily in captivity, nor will wild pregnant females (caught wild) usually care for their young when born. It was necessary to find the breeding places (nests) of the wild mice. The young found were reared by foster-mothers from an albino stock. Care must be employed in placing the new young in the nest. The foster-mother must first be removed from the box. After the young have been in the nest long enough for them to acquire the odor of their new surroundings, the foster-mother may be replaced. If this precaution is not taken the females may eat the wild young.

The mice were kept in a closed room in which the temperature did not vary more than a few degrees throughout the day and night. The wild mice are numbered consecutively, their serial number being preceded in each case by the letter W, so that no confusion would arise as to their origin.

The dilute brown (silver fawn) stock. This is a special strain of mice produced by inbreeding during the last eleven years. With the exception of the Japanese waltzing-mouse strain of Mr. Lambert (Boston, Massachusetts), this strain represents, no doubt, the nearest approach to a homogeneous strain of mice employed in cancer research. It has a distinct advantage over the Japanese waltzers in the matter of breeding and of rearing the young. This strain consists of animals containing the three recessive characters for coat color: 1) dilution of pigmentation,

---

[3] Theoretically, there is no question but that homozygosity would be produced by continued inbreeding. In the absence of controlled experimental evidence, it seems better to employ the term 'genetic homogeneity' or merely 'homogeneity' to express this result.

2) black (non-agouti) and, 3) cinnamon (brown agouti). The inbreeding was started by Doctor Little while at the Bussey Institution (1909) and is still being continued at the Carnegie Institution Station, Cold Spring Harbor. This is the parent strain of mice, in that it gave rise to the two adenocarcinomata employed in this investigation. We will refer to this strain as the susceptible race. By a susceptible race we mean one in which there is 100 per cent indications of progressive tumor 'takes' upon inoculation with a bit of the transplantable tissue.

The albino stock. This is one of the non-susceptible races employed. By a non-susceptible race we mean one in which every mouse inoculated with the tissue fails to grow the transplanted tissue to an appreciable size. This stock has been used as a control on the experiments with the wild mice. The albino strain was obtained from Dr. H. J. Bagg, of Memorial Hospital. He has inbred this strain, brother-to-sister matings, since 1912. Mice of this strain were given serial numbers preceded by the capital letter A.

b. The tumors employed. We have used two adenocarcinomata of the mammary gland that arose spontaneously and independently of each other in the pure dilute brown strain. The first arose some three weeks before the other. The first was designated as dBrA; the second, dBrB. Microscopically, the two tumors are indistinguishable, as shown by the accompanying microphotograph (fig. 1, p. 78). The slight difference is not real as might be supposed, but due to a slight defect in the staining of the dBrB tumor   We are in agreement, then, with the conclusion of Dr. James Ewing that these two tumors are histo logically identical.

The preliminary experiments dealing with the inoculation of these two tissues into the pure dilute brown stock were performed by Doctor Little during the spring of 1920. He determined that either tissue grows progressively in 100 per cent of the mice inoculated, regardless of whether the tissue is placed in separate individuals or on opposite sides of the same mouse. There is no appreciable difference in the tumors (rate of growth, etc.). Within limits, there is apparently no effect of one tumor upon the other when growing in the same host. They each show identical growth activities, percentage of indications, etc. So far, one is warranted in concluding that the two are identical. There are, however, as outlined in the following pages, other tests that can be employed.

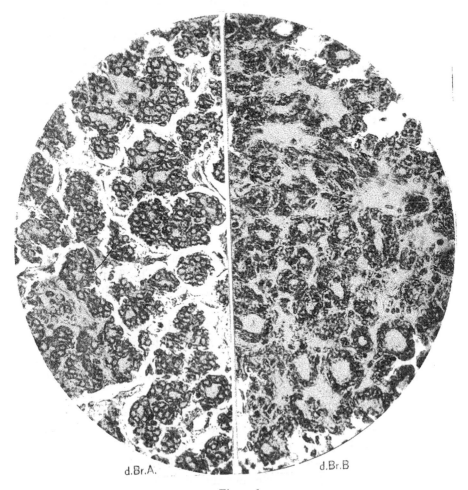

d.Br.A.   d.Br.B

Figure 1

2. Methods

a. Inoculation and observation. The methods used in transferring the tissue from one host to the other are those commonly employed. Ordinary precautions of asepsis were used. It has already been shown (Crocker Laboratory) that tumor susceptibility is not influenced by the manipulation of the tissue during the transfers. The mouse possessing the tumor was first carefully shaved around the site of the proposed incision. The approximate volume of the mass was then determined by palpation and recorded. The instruments employed were sterilized by being placed for a few moments in boiling water. The mouse was etherized as lightly as was possible consistent with relaxation. A straight incision of about three-fourths of an inch was then made by means of curved scissors. By manipulating two pairs of forceps around the tumor, the connective-tissue strands that anchor the tumor to the skin and body wall of the mouse can be severed. A small amount of hemorrhage usually occurs. The blood may be removed by a moistened piece of absorbent cotton. The operation was performed with as much speed as possible—the time element being one of the important factors of success, especially when the mouse has two tumors to be removed. Complete extirpation of the mass was frequently obtained, this being a distinct advantage, since the mouse can then be used for breeding after its susceptibility has been tested. The tumor mass was then placed in a weighed sterilized Syracuse watch-glass to determine the actual weight of 'type' masses.

b. Measurement of size of tumor. Several methods have recently been employed in determining the rate of growth of the tumor mass. The older method (used by Bashford, Murray, etc.) was to compute the interval between inoculation and the time at which the tumor reached the inoculating point. The procedure gives a rough approximation concerning the time required to reach a given size when a series of tumors are to be compared. It does not, however, give any evidence of the successive growth points before this end point is reached. By this method it is impossible to differentiate between the behavior of a tumor that developed fast for the first few weeks and then became practically stationary and one that showed a slow initial impulse and a rapid progressive advance during the latter part of the experiment. A modification of this method has been used by Tyzzer and others. By weighing the tumor mass at definite intervals of time, it is possible to compute the rate of growth, provided that the rate of size increase was progressive. It is, however, a matter of common observation that transplantable (or even spontaneous) tumors do not progress uniformly. This method does not give relatively accurate growth rates.

By palpation one can estimate the relative increase (or decrease) week by week. The method is an advance over the two older methods, but has a few disadvantages. Tumor masses are usually irregular in outline. The time element therefore involved in computing the volume of such a mass is considerable. The method, moreover, does not take into consideration that there may be necrotic or hemorrhagic areas present in the tumor mass.

By weighing of 'type' masses from 0.01 gram to 10 grams we have endeavored to eliminate the disadvantage of the previous methods. The entire history of the mouse under consideration was kept on one chart of coördinate paper. By comparing each individual mass (determined by palpation) with the series of type masses, we were able to estimate the approximate weight of the tumor under investigation from week to week. Calculations of rates of growth were only made on tumors which were firm. The method is not as accurate with recurrent masses for the reason that ulceration and hemorrhagic areas are apparently more prevalent than in the original inoculated growth. The data thus obtained were used in the determination of rates of growth (both progressive and regressive), studies on virulence, etc., as outlined below. The method gives only an approximation to the true state of affairs. By the use of large numbers of mice, however, closer approach toward accuracy can be made than by any other practicable method.

The trochar method of implantation was employed throughout the experiment. A small piece of the tumor is packed snugly into the base of the neck of the trochar with the blunt plunger. The assistant holds the mouse firmly with the left hand by the ears and with the right by the tail, by taking hold of the skin of the mouse near the iliac region with the right hand. The instrument is then pushed forward, placing the tissue in the axillar region. The trochar is sterilized in hot water between every inoculation. No ether is necessary for this process. By this method fifty or sixty mice can be inoculated in about forty-five minutes.

c. *Gonadectomy*. The term 'gonadectomy' signifies the removal of the sex glands of either male or female. Wherever 'castration' is used in this paper we mean the removal of the male gonads only. 'Spaying' will be employed for removal of the female gonads.

Gonadectomy should not consume more than three minutes of actual operating time for the male and not more than four minutes for the female. Ordinary aseptic precautions suffce.

Castration. The mouse is etherized until almost all voluntary action ceases, then placed on the table ventral side up. The assistant holds the mouse by pressing two fingers of the right hand down on the hind legs; with the left hand it is possible to manipulate a small etherizing bottle periodically and to keep the mouse stretched out by pressing lightly on one of the front legs. The hair on the posteroventral portion of the mouse can be removed by means of straight scissors. With a pair of forceps, the testes are then pushed back out of the scrotum into the body cavity. All instruments used from this point on should be sterilized in boiling water for several minutes. A single longitudinal incision about midway between the umbilicus and penis is then made by means of a pair of small curved scissors. This opening need not be more than three-eighths of an inch in length. By pressing the forceps between the outer skin and the body wall, these two layers are separated in the region surrounding the incision. A cut (about three-eighths of an inch in length) is then made through the body

wall.   This second incision should be made a little to the right
of the midventral line.   The edge of the cut is held up slightly
with a pair of small forceps in the left hand.   Another pair of
forceps is inserted into the opening and pushed caudad, care
being taken to avoid touching the urinary bladder.   The testis
is easily withdrawn through the incision.   By holding the
testis with one forceps and manipulating the other pair of
forceps freely, the testis can be severed from its normal connections.   Bleeding can be prevented by pinching, before
severing with forceps, the few cut blood vessels.   The adipose
tissue together with the sperm duct can then be pushed back
into normal position.   Care should be employed not to touch
the alimentary canal (or any other structure in that region
except the testis), as adhesions may result.   There should be
practically no bleeding.   The inner wall of the body cavity is
sutured by means of silk, one stitch is usually sufficient.   The
outer wall is sutured by means of two stitches.   The area can
then be moistened by a piece of absorbent cotton that has been
dipped in warm water.   The mouse is then kept warm until
it has completely recovered from the effects of the ether.

Spaying.   The operation for removing the ovaries is somewhat
different.   The hair is first clipped from an area of about one-
half inch square on the dorsolumbar region.   The assistant
then places the mouse dorsal side up, the right hand holding
the hind legs stretched out at an angle of about 45°.   With
the left hand the small etherizing bottle can be manipulated.
The mouse is kept in position by holding the ears with the
left hand.   The mouse is very easily smothered by the slightest
pressure in the region of the neck.   A single longitudinal incision (about three-eighths inch long) is then made in the middorsal line about one-third the distance between the base of
the tail and the neck.   A pair of forceps is worked around under
the skin near the incision, thus making the skin freely movable
from the body wall for about a square inch.   The skin is pulled
over to one side until the incision coincides with a line drawn as
a continuation of the outer surface of the hind leg.   The point
thus determined should be directly over the position of the

ovary. A small incision about one-eighth of an inch is made by means of curved scissors through the body wall. The edge of the incision is held by one pair of forceps while a second pair is inserted into the opening. With a little practice the ovary together with the oviduct can be pulled out through the incision on the first trial. The ovary is now separated from the coiled oviduct. The tubes and adipose tissue are replaced and the inner incision sutured. The skin can then be pulled over to the other side until the outer incision is in a direct line with the outer surface of the other hind leg. The operation is then repeated on the left side. A little difficulty is first encountered by the proximity of the spleen. This difficulty can be eliminated by making the inner incision slightly more caudad than the corresponding one on the right side. The outer incision. is pulled back into its original position before being sutured. The single outer incision is of advantage because about one-half minute of operating time is saved and because there is not a direct opening from the outside to the inside, which reduces the chances of infection.

These methods of gonadectomy are applicable to very young mice as well as to adult ones. When young mice are used they should be placed more completely under the influence of ether. Unless this is done, the viscera may be extruded by muscular contraction. If this happens, there is no use trying to save the mouse. Young females are best placed for operating on a warm convex surface. Care should be employed not to press too tightly on the legs. Paralysis of the hind legs is frequently the result. Every trace of blood must be carefully removed. Diluted collodion is probably the best suturing material that can be used. Just enough of the flexible collodion is placed on the skin to hold the two sides of the incision in place. The skin is too soft to permit the use of suturing silk. The young mouse should be held in the palm of the hand several minutes until completely out of the ether. It can then be replaced into the nest. The mother should be removed from the box for an hour to permit the young to reacquire the odor of the nest. If these precautions are employed the mortality can be kept at a minimum.

3. Results

a. *Difference between dBrA and dBrB.* The first experiment consisted in comparing the reactions of the two adenocarcinomas in the same medium. Over two hundred wild mice were used for this purpose. If there is a variable factor present it must be in the tumor cell itself. Not one mouse out of all those inoculated ever grew the dBrB tumor progressively and only one ever grew the dBrA (this will be discussed in experiment 2). We are therefore dealing with a race definitely proved to be non-susceptible. There were, however, several positive indications of the host-tumor reaction,[4] as plotted in figure 2.

The ordinates represent the percentage of positive apparent reactions for any given observation period. The abscissae represent the successive weekly periods of observation beginning with the second week after the inoculation. It will be noted that the dBrB tumor gave the greater percentage of reaction indications throughout the experiment. Each tumor produced a characteristic number of initial reactions (dBrB 14.79 per cent $\pm$ 1.69; dBrA 4.39 per cent $\pm$ 1.01). The number of visible reactions decreased with every observation until none (after ten observations in a few cases) remained.

[4] In this paper the term 'reaction' is used to denote a palpable mass occurring for a certain length of time at the site of the inoculation. It is, in all probability, the implanted tissue surrounded by a reaction zone set up by the host. 'Indication' cannot be employed, since this implies a definite growth on the part of the tumor cell. In a few cases we examined a small nodule microscopically, which proved to be actual tumor tissue. All nodules could not be so examined, as this procedure would interfere with the continuation of the experiment. We have some indirect evidence to show that the palpable mass was (in most cases at least), actual tumor tissue. When all normal individuals are lumped, we obtained a percentage of visible masses of 5.97 $\pm$ 0.40%; for gonadectomized individuals, 3.75 $\pm$ 0 24% The percentage for gonadectomized individuals is too low, due to the depressing effect of the operation (as discussed later in this paper), so that the two classes gave practically the same number of 'reactions' Normal individuals were able to eliminate every reaction mass in a few weeks, whereas some gonadectomized individuals that gave initial reactions continued to grow the tumor mass progressively. This result tends to indicate that the reaction mass in normal individuals is due to an actual initial growth or indication on the part of the tumor cell, which is then eliminated.

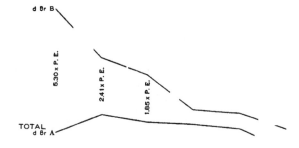

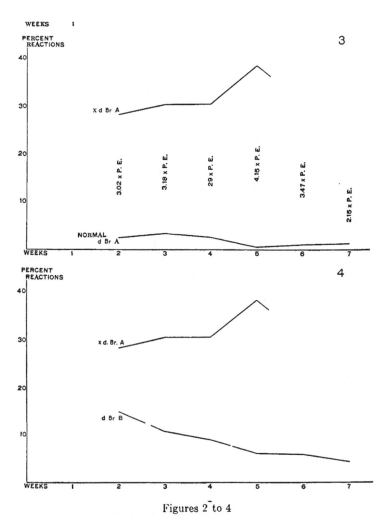

Figures 2 to 4

The method employed for testing the data was by comparing the probable errors of the massed observations of each tumor According to this method, any' result that is three times its probable error is usually considered significant.[5]   In the present experiment the results obtained were as follows:

    1. dBrB 786 negative : 80 reactions ±5.74 or 9.23% ± 0.67
    2. dBrA 760 negative : 38 reactions ±4.05 or 4.76% ± 0.52
                Difference   4.47% ± 0.85
    The difference is thus 5.25 times its probable error.

In the case of these two tumors, therefore, physiological reactions are independent of histological characteristics. It will be remembered (p. 77) that the two tumors appeared histologically identical.

b. Constant reaction potentiality. The objection may be raised that perhaps the reaction difference is due to the fluctuating activity of only one type of tumor cell—the dBrB type being in one phase, the dBrA in another phase of a large rhythmical process of change. To test out the disputed cyclic changes of tumor activity, the whole data for the dBrB tumor were grouped into three periods, making the number of observations in each class as nearly equal as possible without splitting up the data from any one chart.   The three groups represent successive periods in the course of the experiment.

The data bearing on supposed fluctuations of the tumor cell are as follows:

    1st period 198 negative : 26 reactions ±3.23 or 11.61% ± 1.44
    2nd period 309 negative : 32 reactions ±3.63 or 9.38% ± 1.06
    3rd period 282 negative : 22 reactions ±3.05 or 7.23% ± 1.01

Comparing the differences between the three periods we obtain:

    Difference between:
    1st and 2nd 2.23% ± 1.79 or 1.24 times probable error
    2nd and 3rd 2.15% ± 1.46 or 1.47 times probable error
    1st and 3rd 4.78% ± 1.76 or 2.48 times probable error

---

[5] The formula used in the present experiment in computing the probable error of a given ratio is the one proposed by Pearl in his paper on human sex ratios ('08).

There is no significant difference between any of the three periods. It will be noted, however, that there is a gradual decline in the percentage of reactions from the first period through the third. This decline is not significant for the entire period of the experiment which consumed about a year. Whether this decline will continue in the future is problematical and will be referred to later. It may be well, however, to emphasize the fact that this decline is primarily caused by the use of numbers of individuals from the different age groups, more very young individuals being employed at the beginning of the experiment than at any other time. All we can conclude from the results is that the adenocarcinoma dBrB has probably retained a constant reaction potentiality throughout the present experiment.

c. Exceptional dBrA. The same phenomenon was encountered with the dBrA tumor, with one exception. Experiment N (started August 16, 1920) gave such a large number of positive reactions that the question arose whether we were dealing with the same tumor reaction that we had observed up to that time. In fact, this one experiment gave more indications of growth than did all the other experiments involving the dBrA tumor combined. This experiment also included the only wild mouse, adult male W238, that ever grew the dBrA tumor progressively. Comparing the percentage of reactions of the exceptional dBrA tumor (xA) with all the other dBrA (nA), we obtain the graph on page 85 (fig. 3).

With the exception of the last point, it will be seen that every point in the exceptional dBrA reaction curve is significantly greater than the corresponding point in the normal dBrA curve. The data relative to this matter may be combined as follows:

1. Exceptional dBrA 55 negative : 23 reactions ± 2.72 or 29.48% ± 3.49
2. Normal   dBrA 705 negative : 15 reactions ± 2.58 or 2.08% ± 0.36

Difference   27.40% ± 3.51

The difference is thus 7.81 times its probable error

The difference between the two dBrA tumors is therefore mathematically significant.

Was there a mistake made in the exceptional experiment N by inoculating the dBrB tumor into the right axilla and then using these data as being derived from the dBrA tumor reaction? We can check with the normal dBrB reaction (fig. 4, p. 85).

The exceptional dBrA reaction is even significantly greater than the normal dBrB:

1. Exceptional dBrA 55 negative : 23 reactions ±2.72 or 29.48% ± 3.49
2. Normal   dBrB 786 negative : 80 reactions ±5.74 or 9.23% ± 0.67
                                    Difference   20.25% ± 3.55
          The difference is thus 5.70 times its probable error

We have not, therefore, made a mistake in tabulating the data. Evidently the exceptional dBrA reaction was neither produced by the normal dBrA tumor nor by the dBrB type. The dBrA tumor must have undergone a significant change. It is no longer the same as the normal dBrA tumor that has retained a constant reaction potentiality during a year's observation.

By microscopical examination the exceptional dBrA tumor was found to be histologically identical with the original dBrA (or even the dBrB for that matter). What produced this significant increase in the reaction capacity of the tumor cell? If the same phenomenon of a sudden appearance of a change in the somatic or physiological characteristic of a normal cell was encountered by the geneticist, he would maintain that the variation was produced by the process known as 'mutation.' There seems to be no objection to using a similar mutational process to explain the origin of the observed significant different reaction capacity.

The exceptional dBrA is not the original normal dBrA. We have no right, therefore, to include it in the same category. Comparing the corrected dBrA curve (after subtracting the exceptional dBrA) with the dBrB curve, we obtain the curve shown on page 89 (fig. 5).

Every point of the dBrB curve is significantly greater than the corresponding point of the NdBrA (normal dBrA) curve with the exception of the last one. Since both curves are approaching zero, there would naturally be convergence. For

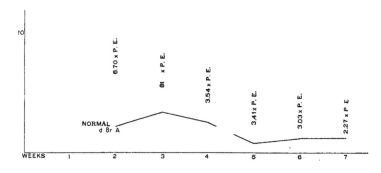

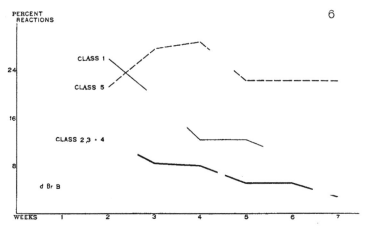

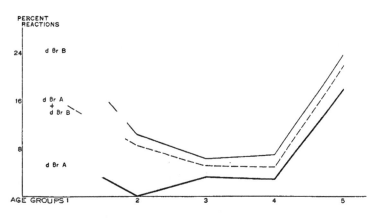

Figures 5 to 7

that reason the difference between corresponding points at the latter end of the curve would not be as great as between points at the other end of the curve. The slight rise in the dBrA curve is not real, but due to the dying off of several negative individuals during an epidemic in the summer of 1920.

1. Total dBrB 876 negative : 80 reactions ±5.74 or 9.23% ± 0.67
2. Normal dBrA 705 negative : 15 reactions ±2.58 or 2.08% ± 0.36
Difference   7.15% ± 0.76
The difference is thus 9.40 times its probable error

d. Age at inoculation (dBrB). All mice were separated into five groups according to age. The five classes arbitrarily selected were as follows: class 1, 0 to 3 weeks old; class 2, 4 to 7 weeks old; class 3, 7 weeks to three months; class 4, adults, and class 5, old mice. The mice were so classified so as to include groups to represent the approximate life-cycle of a mammal (youth, puberty, adolescence, maturity, and senescence). The dBrB tumor may be considered first and the data for the three middle classes (2, 3, and 4) compared.

The data may be summarized:

Class 2 87 negative : 10 reactions ±2.02 or 10.30% ± 2.09
Class 3 146 negative : 10 reactions ±2.06 or 6.41% ± 1.32
Class 4 523 negative : 43 reactions ±4.25 or 7.59% ± 0.76
The difference between the percentages are:
    2–3 3.89% ± 2.47 or 1.57 times probable error
    2–4 2.71% ± 2.22 or 1.22 times probable error
    3–4 1.18% ± 1.52 or 0.77 times probable error

The three classes are remarkably similar, there being no significant difference between any two classes. We may therefore group these three classes into one general class, so as to simplify the comparison with class 1 and with class 5.

Individuals from class 1 (0 to 3 weeks old) gave a very high initial percentage of palpable reactions. Unfortunately, the positive individuals did not survive the whole length of the experiment, so that the class 1 curve drops to zero with the seventh week. We have made the curve end with the sixth week because we feel that that is the only way to represent the true state of affairs. Very young wild mice, if they showed positive reactions, usually died, possibly because the experiment was performed during the epidemic previously mentioned. The young mice either died of the infection or because they were deprived of their foster-mother, or because the reaction (host-tumor) was too virulent for the young mice. But even taking this complicating factor into account, there is still a significant increase in the percentage reactions of the very young (class 1) individuals (fig. 6, p. 89).

Class 1	134 negative : 30 reactions ±3.34 or 18.29% ± 2.03
Class 2, 3 and 4	714 negative : 54 reactions 4.73 or 7.03% ± 0.61
	Difference 11.26% ± 2.12

The difference is thus 5.32 times its probable error

The numbers for the very young wild mice are small, but even in spite of this, the difference between class 1 and 2, 3, and 4 is significant. The fact that very young individuals of a non-susceptible race are more susceptible to transplantable tissue has been recognized by several investigators, among whom may be mentioned Tyzzer, Little, and Loeb. We may next consider the reaction of old mice to the tumor.

As far as known to us, the following experiment is the first reported observation on susceptibility to transplantable tissue in very old mice of a non-susceptible race (class 5). Contrary to what we expected, susceptibility increased from maturity toward old age.

Figure 6 shows the three middle-class curve (heavy solid line) compared with that for the old (class 5) mice (dotted line) as well as that for very young (class 1) individuals (light solid line).

Tabulated data

Class 5	57 negative : 18 reactions ±2.49 or 24.00% ± 3.30
Class 2, 3, and 4	714 negative : 54 reactions ±4.73 or 7.03% ± 0.61
	Difference 16.97% ± 3.35

The difference is thus 5.07 times its probable error

Two points of especial interest may be emphasized at this time: 1) The susceptibility toward transplantable neoplastic tissue is as great in very old mice as it is in very young mice

(24.00% ± 3.30 compared with 18.29% ± 2.03) and 2). The susceptibility curve for old mice does not approach zero during the usual time limit of seven weeks as do all the other suscepti- bility curves so far given. Old mice are extremely slow in elimi- nating the last traces of the tumor reaction mass. As a matter of fact, some old mice took from eleven to twelve weeks to over- come finally this reaction (host-tumor). It may be recalled that some reaction masses become encysted and remain apparently quiescent for a considerable period. The time element for the actual reaction can only be determined in those cases in which there is some change in the relative size of the mass from week to week.

e. *Age Susceptibilty (dBrB)*. We are now able to plot the relative percentage reactions for the five age groups. The curve shows at a glance the susceptibility of animals of a non-susceptible race at different periods in their life (fig. 7, light solid line, dBrB).

When individuals (of the non-susceptible race) are very young they are highly susceptible to the transplanted tissue—about one-fifth of all the observations for the six weeks of the experi- ment are positive visible reactions. As the individual grows older there is at first a rapid decline in susceptibility until the second class is reached, then a slower decline until the mouse is about three months old. The lessened susceptibility then remains fairly uniform throughout the adult stage. With senes- cence and old age, the susceptibility rapidly increases until the original maximum of about 25 per cent is reached. The two points in this curve call for emphasis. 1) Individuals that are one-half to three-fourths matured are less susceptible to trans- plantable tissue than are individuals from any other age groups. This is the exact converse of what has been recognized by some investigators as applying to a susceptible race. 2) The age susceptibility toward a transplantable tissue bears a remarkable parallelism to the activity of the gonads.

Before the gonads mature, susceptibility toward tumor tissue is very high (although we are dealing with a non-susceptible race). During the period of the development of the sex organs (classes 2 and 3), susceptibility at first rapidly, and then gradually, decreases to a minimum.   During maturity, the gonads are relatively constant in their activity; at this time susceptibility is also relatively constant.   With the decreased physiological activity of the gonads and accompanying the onset of senescence, susceptibility toward transplantable tissue increases in the reverse order—slowly at first and then rapidly.

One more experiment dealing with the age groups of the mice for the dBrB tumor may be given here.   Eliminating the recognized significant differences between class 1 and the three middle classes on one hand and between the three middle classes and class 5 on the other, we have attempted to analyze further the so-called rhythmic activity of the tumor cell.   In the following chart only individuals belonging to the three middle classes were employed.   There is no significant difference in the susceptibility of these animals, so the objection cannot be raised that any rhythmic reaction encountered may be due to individuals from different age groups being used from time to time.   Figure 8 represents the three time groups, calculated from the same data given on page 86 with the one correction of eliminating the very young and very old mice.

Tabulation

Time period (1) 167 negative : 15 reactions ±2.49 or 8.24% ± 1.36
Time period (2) 274 negative : 20 reactions ±2.90 or 6.80% ± 0.99
Time period (3) 273 negative : 19 reactions ±2.83 or 6.51% ± 0.96
Difference between:
Time period (1) and (3) 1.73% ± 1.66 or 1.04 times probable error
Time period (2) and (3) 0.29% ± 1.38 or 0.21 times probable error
Time period (1) and (2) 1.44% ± 1.68 or 0.86 times probable error

This experiment furnishes still clearer results than those given on page 86.   We are therefore justified in our previous conclusion that the dBrB tumor has retained a constant reaction potentiality throughout the experiment.

f. Age at inoculation (dBrA).   As with the dBrB tumor, the three middle-age groups (classes 2, 3, and 4) are classified together to simplify the calculations.   Comparing the middle-age group with class 1 individuals, we obtain the following graph (fig. 9, p. 95.)

Tabulation

Class 1	76 negative : 5 reactions ±1.46 or 6.17% ± 1.81	
Classes 2, 3, 4	631 negative : 9 reactions ±2.01 or 1.40% ± 0.32	
	Difference	4.77% ± 1 84

The difference is thus 2.59 times its probable error

This difference approaches mathematical significance, although a difference of three times the probable error is not obtained.[6] It is safe to maintain that the difference would be significant when larger numbers are obtained. The important point to notice is the fact that the very young mice produce a susceptibility curve that is on the same side of the middle-group curve as was found to be the case with the dBrB tumor.

Although the numbers are rather small for class 5, there is a significant increase of positive reactions. In the following chart bearing on this point the exceptional dBrA tumor was included to increase the number of observations, but the results are the same even if this exceptional tumor is eliminated (fig. 10).

Class 5	55 negative : 12 reactions ±2.12 or 17.91% ± 3.16	
Classes 2, 3, and 4	673 negative : 26 reactions ±3.36 or 3.71% ± 0.49	
	Difference	14.20% ± 3.20

The difference is thus 4.43 times its probable error

The first point of the class 5 curve is possibly too low.

g. Age susceptibility (dBrA). Susceptibility to the dBrA tumor for the different age groups is given in figure 7 (heavy solid line, dBrA). The curve is of the same type as that for the dBrB tumor (fig. 7), although it is not so striking. (This was to be expected, since the dBrA tumor produced fewer reaction masses than did the dBrB tumor.)

h. Composite age susceptibility. The dotted line on figure 7 represents the total observations for both the dBrA and dBrB tumors.

We are justified, therefore, in concluding that the susceptibility at different ages to transplanted tissue is independent of the tumor tissue employed.

[6] The odds against chance alone being a factor when a difference of three times the probable error is obtained is only one to 22.26, while for 2.6 times the probable error it is one to 11.58.

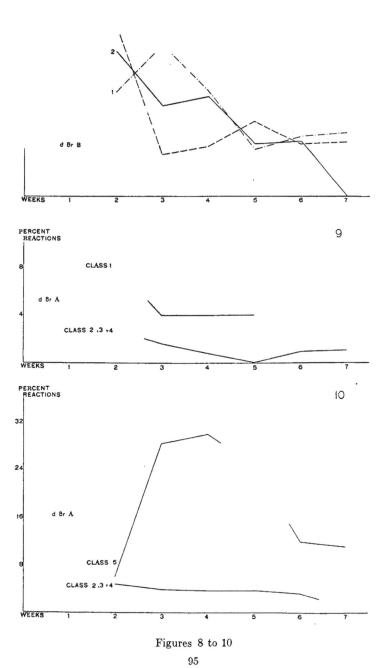

Figures 8 to 10

95

The decrease of susceptibility from youth to adult age has been recognized by several investigators. In order to check still further the increase of susceptibility with old age, one more chart was drawn. The percentage reactions for class 5 with the dBrB tumor were added to those with the dBrA tumor, and likewise classes 2, 3, and 4 for the dBrA tumor reaction. (This means merely the addition of the corresponding curves on figures 6 and 10—a curve of the first chart being added to its analogous curve of the second chart.) Figure 11 shows the results thus obtained.

Class 5   112 negative : 30 reactions ±3.27 or 21.12% ± 2.30
Classes 2, 3, 4 1387 negative : 80 reactions ±5.86 or 5.45% ± 0.40
                         Difference   15.67% ± 2.35
        The difference is thus 6.67 times its probable error

It will be noted that the two curves begin about the same point—that is, that the initial percentage reactions for middle-aged individuals (classes 2, 3, and 4, 8.17% ± 1.0) are about the same as these for older mice (class 5, 13.5% ± 3.7). Individuals in the full vigor of maturity are readily capable of counteracting the tumor tissue. The number of positive reactions gradually decrease to zero after the eighth observation (last two points are not included in this chart, since only the mice showing positive reactions were kept after the sixth observation). Old mice, however, apparently cannot counteract the foreign tissue as readily as adult mice. The percentage reactions gradually increase for the first few weeks, then gradually decrease, until at the end of the sixth observation the old mice show a greater number of positive reactions than they did at the beginning of the experiment. As before stated, it is only after several additional observations that the last palpable mass from old mice disappeared. The increase in susceptibility is real and not the result of a peculiarity of the tumor tissue employed. It is evident that the old mice are approaching a condition under which they are able to harbor the foreign tissue better than they did when they were mature and reproductively active.

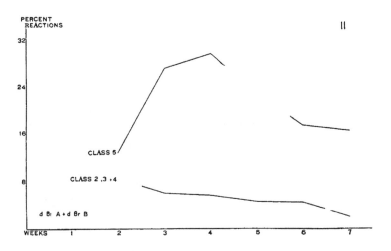

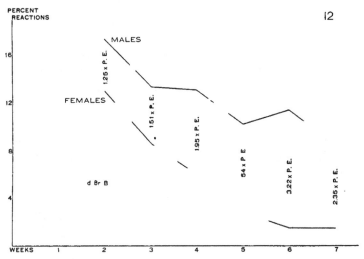

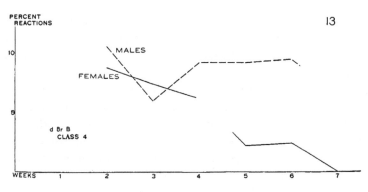

Figures 11 to 13

i. Sex (dBrB). Whether sex is a determining factor underlying susceptibility to transplantable tumor is a disputed question.

By dividing all the data into the classes according to the sex of the mice, we obtained the following result (fig. 12, p. 97):

Tabulation

Males   313 negative : 46 reactions ±4.27 or 12.81% ± 1.19
Females 476 negative : 34 reactions ±3.80 or 6.66% ± 0.75
Difference   6.15% ± 1.40
The difference (6.15% ± 1.40) is 4.39 times its probable error

There is, therefore, a significant difference between the sexes, the males giving a preponderance of reactions throughout the experiment.

Because certain investigators, with carefully controlled experiments, were not able to determine any difference between the sexes as a factor underlying susceptibility to transplantable tumors, it seemed advisable to analyze our findings relative to this point more completely.

Analyzing the data from class 4 (adults) individuals only, it appears that there is no significant difference between the sexes in that particular group (fig. 13, p. 97).

Tabulation

Males   177 negative : 17 reactions ±2.66 or 8.76% ± 1.37
Females 304 negative : 17 reactions ±2.70 or 5.29% ± 0.85
Difference   3.47% ± 1.61
The difference is thus 2.15 times its probable error

We are, therefore, confronted with the phenomenon that there is no significant difference between the sexes for adult individuals, whereas, if all five age classes are considered together, there is a significant difference.

Two statements relative to this point will suffice at this time:

1. According to a genetic interpretation, sex may or may not be a contributing factor underlying susceptibility. Indeed, sex may be an important factor for certain types of tumors or be entirely negligible for others. There may or may not be elements in the tumor cell that react differently in the presence of sex hormones, etc. We can expect, therefore, that certain trans-

plantable tumors would show a different reaction for the two sexes. But this does not explain the results outlined above that there was a significant difference between the sexes when all age classes are taken into consideration, whereas no such difference is encountered with adult mice alone.

2. Female mice mature earlier than male mice from the same litter. If this is taken into account, the sex difference behavior of the dBrB tumor may be partly explainèd as a result of the age factor alone. Female mice maturing faster than males would be more able to counteract the tumor tissue (susceptibility decreases with age up to maturity). In other words, the males would show a larger percentage of positive reactions, thus giving a significant difference between the sexes, although the result is more probably due to the factor of age. The age factor is also able to explain the divergence of the two curves in figure 13. Male mice grow old faster than females. The adult females are able to resist the tumor tissue until by the end of the sixth observation no positive palpable reaction can be observed, whereas the males growing old faster (first stages of senescences) are unable to withstand the tumor reaction as readily. These two facts are of primary genetic importance.

k. Sex (dBrA). Contrary to the findings for the dBrB tumor, there is no significant difference between the sexes for the normal dBrA tumor reactions when all age groups are massed.

Tabulation

Females 388 negative : 10 reactions ±2.10 or 2.51% ± 0.53
Males   282 negative : 5 reactions ±1.49 or 1.74% ± 0.52
                    Difference   0.77% ± 0.74
The difference is thus 1.04 times the probable error

If individuals from age group no. 4 only are compared we get the following results:

Males   154 negative : 2 reactions ±0.94 or 1.28% ± 0.70
Females 281 negative : 1 reaction ±0 67 or 0.35% ± 0.24
                    Difference   0.93% ± 0.74
The difference is thus 1.26 times the probable error

There is, therefore, no difference between the sexes in the reaction toward the normal dBrA tumor, either when animals of all age groups are considered or when adult mice only are compared.

A very remarkable result appears when we compare the sexes in their reaction to the exceptional dBrA and to the normal dBrA tumor (fig. 14).

Tabulation
Males   307 negative : 28 reactions ±3.42 or 8.36% ± 1.02
Females 418 negative : 10 reactions ±2.10 or 2.33% ± 0.49
                        Difference   6.03% ± 1.13
The difference is thus 5.34 times the probable error

Again, only considering individuals from the fourth age group (adults) we obtain (fig. 15):

Males   178 negative : 19 reactions ±2.71 or 9.64% ± 1.42
Females 299 negative : 1 reaction ±0.67 or 0.33% ± 0.23
                        Difference   9.31% ± 1.05
The difference is thus 8.86 times its probable error

The exceptional dBrA tumor shows a significantly different reaction between the sexes, whereas the normal dBrA tumor did not. This change of reaction in the exceptional dBrA may possibly have been the result of the mutational process that produced it. In other words, the mutation has produced a tumor, the exceptional dBrA, of greater reactive capacity. The reaction between the sexes is so marked that even the changing factor of age has little or no influence on it.

l. Sex composite dBrB and dBrA. Grouping the males and females for figure 12 (dBrB for all classes) and similar data for dBrA for all classes together, figure 16 (dotted lines) was obtained.

Tabulation
Males   595 negative : 51 reactions ±4.62 or 7.89% ± 0.71
Females 864 negative : 44 reactions ±4.37 or 4.84% ± 0.49
                        Difference   3.05% ± 0.86
The difference is thus 3.55 times its probable error

Adding figures 12 (dBrB for all classes) and 14 (both dBrA for all classes) together, figure 16 (solid lines) was made.

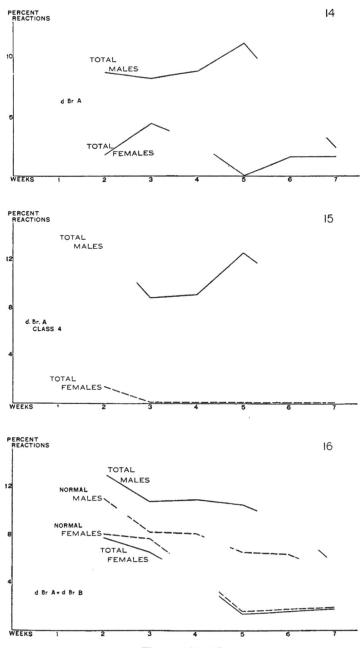

Figures 14 to 16

Tabulation

Males   621 negative : 72 reactions ±5.41 or 10.38% ± 0.79
Females 894 negative : 44 reactions ±4.37 or 4.69% ± 0.47
Difference   5.69% ± 0.92
The difference is thus 6.18 times the probable error

From the foregoing data we may conclude that the rôle of sex as a factor underlying susceptibility to transplantable tumor is influenced by two entirely distinct agents: 1) The characteristic of the tumor cell (inherent within itself) and, 2) the changing age factor which determines the degree of activity of the gonads and possibly of physiological changes in general.

m. Gonadectomy. Since sex is a secondary factor underlying susceptibility to the transplantable tissue, as shown when the data are massed, it becomes important to find out what effect gonadectomy will have. Figure 17 gives the data for the first inoculation series after the gonads had been removed.

Males   308 negative : 16 reactions ±2.63 or 4.94% ± 0.81
Females 319 negative : 12 reactions ±2.29 or 3.62% ± 0.70
Difference   1.32% ± 1.07
The difference is thus 1.23 times the probable error

There is therefore no significant difference between gonadectomized individuals of either sex for percentage of reactions. The two curves, however, are quite different in type. The female curve gradually drops to the base line, whereas the male curve continues to rise. In fact, those males (for this experiment) that gave positive reactions at the end of the six-week observation period continued to grow the tumor mass progressively. We never encountered progressive growth in a normal wild individual. Gonadectomy must therefore be considered to have had two effects: 1) on the number of percentage reactions and, 2) on the progressive growth reaction of the tumor cell.

The same gonadectomized individuals were reinoculated with the same two tumors some time after the end of the first six-week observation. Figure 18 gives the data for this second inoculation.

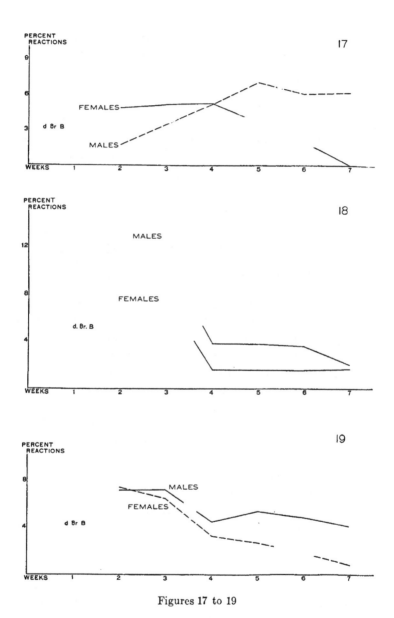

Figures 17 to 19

Tabulation

Males   299 negative : 20 reactions ±2.92 or 6.27% ± 0.91
Females 353 negative : 15 reactions ±2.55 or 4.07% ± 0.70
                    Difference   2.20% ± 1.15
        The difference is thus 1.91 times the probable error

There is still no significant difference between the sexes, although the males are slightly more susceptible. The curves for the inoculation are quite different from the analogous curves for the second inoculation, although not significantly so, as will be discussed later. Figure 19 (p. 103) is a composite for figures 17 and 18.

Tabulation

Total castrated males 607 negatives : 36 reactions ±3.93 or 5.60% ± 0.61
Total spayed females 672 negatives : 27 reactions ±3.43 or 3.86% ± 0.48
                       Difference   1.74% ± 0.77
        The difference is thus 2.26 times the probable error

The male curve is still slightly above that for the females, but not significantly higher. The initial susceptibility of the sexes is approximately equal. The females are, however, able to resist the action of the tumor cell better than the males, and almost entirely eliminate it within the six-week period (one spayed female, however, continued to grow the tumor progressively). The castrated males, on the other hand, have a harder struggle. They fail to eliminate some but not all of the transplantable tumors, since a few individuals continue to grow the tissue mass progressively.

Comparing the castrated males (first inoculation only) with the normal males, we obtain an entirely different type of curve. The normal male curve is probably significantly higher than the castrated one. (The last three points are, however, not different; figure 20, heavy solid lines.)

Tabulation

Normal males   313 negatives : 46 reactions ±4.27 or 12.81% ± 1.19
Castrated males 308 negatives : 16 reactions ±2.64 or 4.93% ± 0.82
                     Difference   7.88% ± 1.44
        The difference is thus 5.48 times the probable error

The normal male curve approaches zero, whereas the castrated curve after rising remains at a fixed level, a few males continuing to grow the tumor tissue progressively.

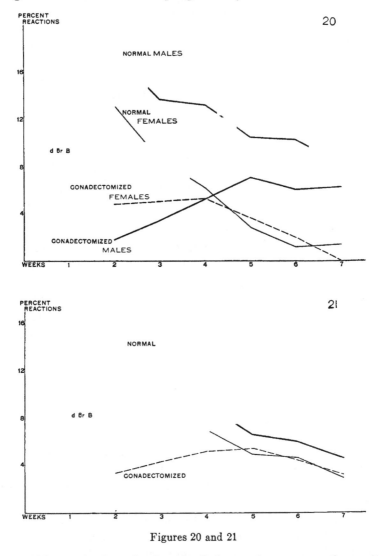

Figures 20 and 21

In this particular, the females behave the same as the males, although they are not so striking (fig. 20, light, solid, and dotted lines).

Tabulation

Normal females 476 negatives : 34 reactions ±3.80 or 6.66% ± 0.75
Spayed females 319 negatives : 12 reactions ±2.30 or 3 62% ± 0.70
Difference   3.04% ± 1 02
The difference is thus 2.98 times its probable error

Since, as has already been shown, there is no significant difference between the sexes of gonadectomized individuals, we are justified in combining the data together. Figure 21 gives the total observations for both sexes for both normal and gonadectomized mice for the first inoculation series.

Tabulation

Normals   789 negatives : 80 reactions ±5.75 or 9.20% ± 0.67
Gonadectomized 627 negatives : 28 reactions ±3.49 or 4.27% ± 0.54
Difference   4.93% ± 0.86
The difference is thus 5.74 times its probable error

The significant difference between the two classes lies only in the first two points of the curve (two and three weeks). No mice from class 5 (old age) group were castrated. In order to get a better comparison, therefore, between the controls and the gonadectomized classes, class 5 individuals should be subtracted from the control group. Figure 21 also shows the results ob tained after this correction has been made (light solid line).

Tabulation

Normal (−class 5) 732 negatives : 62 reactions ±5.10 or 7.82% ± 0.63
Gonadectomized 627 negatives : 28 reactions ±3.49 or 4.27% ± 0.54
Difference   3.55% ± 0.83
The difference is thus 4.28 times its probable error

The normals still show a significantly greater percentage of reactions. The significance (5.3 × P. E.), however, is confined to the second-week observation period only (first point in the curves).

The observations on the reinoculated gonadectomized mice can now be given. (The reader will remember that these reinoculated mice are approximately ten to twelve weeks older than when first inoculated.) (Fig. 22.)

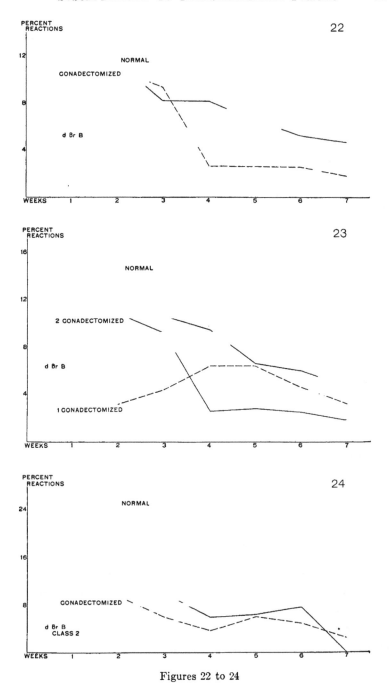

Figures 22 to 24

Tabulation

Normal (−class 1) 741 negatives : 62 reactions ±5.10 or 7.72% ± 0.64
Gonadectomized 652 negatives : 35 reactions ±3.89 or 5.09% ± 0.57

Difference   2.63% ± 0.86
The difference is thus 3.06 times its probable error

Class 1 age group was subtracted from the normal controls, since no reinoculated individual of the operated series could fall into this class. No two homologous points in the curves differ significantly, although when all six points are massed the normal group shows a clearly greater number of reactions (3.06 times its probable error).

Reinoculated gonadectomized individuals therefore approach more closely the susceptibility of the control mice than do mice when first inoculated (4.38 times its probable error). How is this result to be explained? Are reinoculated mice more susceptible to subsequent inoculations of the same tissue or is the increase due to the wearing off of the ill effects of the operation for gonadectomy? Figure 23 shows the comparison of the first and second inoculations series for gonadectomized mice and the normal controls.

Tabulation

Second inoculation 652 negatives : 35 reactions ±3.88 or 5.09% ± 0.59
First inoculation   627 negatives : 28 reactions ±3.49 or 4.27% ± 0.54

Difference   0.82% ± 0.80
The difference is thus 1.03 times its probable error

There is no significant difference between first and second inoculations. The most striking difference, however, lies in the first observation period. The first point for the second reinoculation series is probably significantly greater than the corresponding one for the first inoculation series. The other five points of the two curves are very similar. This tends to show that the first point for the first inoculation series is too low. The only conflicting element that may be present to lower the first point in the first and not in the second inoculation is the after effects of the operation. The complicating factor introduced by the shock, etc., of the operation will be considered again later.

n. Age susceptibility in normal and gonadectomized mice. The next experiment was to determine whether there is any differential effect of gonadectomizing individuals from the different age groups. Unfortunately, the number of individuals from class 1 and class 5 that were operated on were small, so that no valid comparison can be made for these groups. Figure 24 (p. 107) shows that the percentage reactions for group 2 individuals is somewhat lower than the controls, but not significantly so.

Tabulation

Class 2 normals   87 negatives : 10 reactions ±2.02 or 10.31% ± 2.08
Class 2 Gonadect. 468 negatives : 28 reactions ±3.47 or 5.65% ± 0.69
                                  Difference   4.66% ± 2.19
The difference is thus 2.13 times its probable error

If the depressing effect of the operation is taken into consideration, there is, in this age group, no difference between nor mals and gonadectomized individuals in their susceptibility to transplantable tumor tissue. Class 3 individuals gave quite unexpected results. The gonadectomized individuals showed a significant increase over the controls. In this one age group (class 3) there was apparently no depressing effect of the operation encountered (fig. 26, p. 111).

Tabulation

Gonadectomized 197 negatives : 28 reactions ±3.34 or 12.44% ± 1.49
Normals   146 negatives : 10 reactions ±2.06 or 6.41% ± 1.32
                               Difference   6.03% ± 1.98
The difference is thus 3.05 times its probable error

It may be recalled that the normal class 3 individuals gave the minimum percentage reactions. Just why we should get the greater increase in percentage indications for individuals in this group of gonadectomized mice is not clear at present, but will be considered again later.

With individuals of group 4 just the reverse is found. Very few gonadectomized adult mice show any indications whatever of a favorable reaction (fig. 25, p. 111).

Tabulation

Normal   481 negatives : 34 reactions ±3.80 or 6.60% ± 0.74
Gonadectomized 453 negatives : 7 reactions ±1.77 or 1.52% ± 0.39
                              Difference   5.08% ± 0.83
                The difference is thus 6.12 times its probable error

The operation on adult individuals therefore decreases the percentage reactions significantly.

We can study the results better if we compare the life-cycle susceptibility for gonadectomized individuals with that for normal mice (fig. 27).

The first point for the gonadectomized curve is determined by 102 observations; that for the fifth point, 42. The data for the other three points have been given in the preceding three charts. It is needless to emphasize the fact that the two curves are directly each other's opposite. Whether this result is due to the absence of the gonads or to the effects of the operation alone remains to be proved. It is hard to understand how the effect of the operation (shock, etc.) could have such a differential effect on mice from the various age groups.

o. Influence of operation. In an attempt to analyze the effect of the operation alone upon the susceptibility toward transplantable tumor tissue, the mice are classified, regardless of age, according to the length of time which had elapsed between operation and inoculation. For convenience the arbitrary classes of 0 to 5 days, 5 to 10 days, and 10 to 15 days after operation were chosen. This experiment includes only individuals inoculated for the first time.

Tabulation

Controls   789 negatives : 80 reactions ±5.75 or 9.21% ± 0.66
Group 1, 0– 5 days 177 negatives : 14 reactions ±2.43 or 7.33% ± 1.27
Group 2, 5–10 days 207 negatives : 4 reactions ±1.32 or 1.90% ± 0.62
Group 3, 10–15 days 194 negatives : 8 reactions ±1.87 or 3.96% ± 0.93
    Differences:
    Controls–Group 1 1.88% ± 1.43 or 1.31 times probable error
    Controls–Group 2 7.31% ± 0.91 or 8.10 times probable error
    Controls–Group 3 5 25% ± 1.14 or 4.60 times probable error
    Group 1–Group 2 5.43% ± 1 41 or 3 85 times probable error
    Group 2–Group 3 2.06% ± 1.11 or 1.86 times probable error
    Group 3–Group 1 3.37% ± 1.58 or 2.13 times probable error

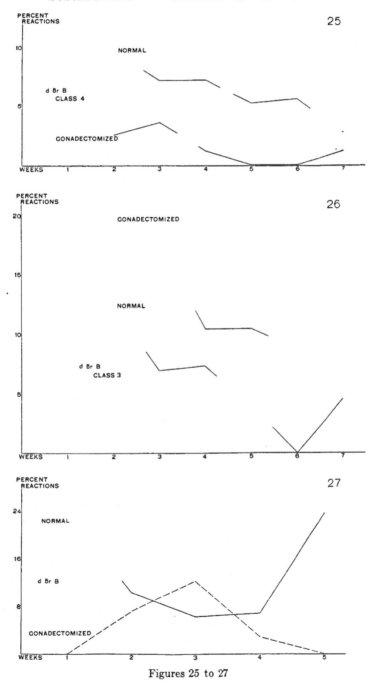

Figures 25 to 27

The most important point to emphasize in this connection is the fact that if individuals are inoculated five to ten days after operation, they give a significantly smaller number of percentage reactions than if inoculated at the other times. There is no significant difference between controls and the first-time group (0 to 5 days).

By including the data on second-inoculation individuals as well as first inoculations (since there is no significant difference between first- and second-inoculation observations when the data are massed), we are able to study further the apparent deleterious effects of the operation (fig. 28).

The data for the first three points of the curve are the same as those given on page 110. The data for the additional points are as follows:

(Group 4) 16 da.– 60 da. 106 negative : 4 reactions ±1.33 or 3.64% ± 1.21
(Group 5) 60 da.– 90 da. 177 negative : 10 reactions ±2.08 or 5.34% ± 1.12
(Group 6) 90 da.–120 da. 250 negative : 17 reactions ±2.68 or 6.36% ± 1.07
(Group 7) 120 da.–150 da. 93 negative : 4 reactions ±1.32 or 4.12% ± 1.36
(Group 8) 150 da.–180 da. 73 negative : 1 reaction ±0.67 or 1.35% ± 0.91
    Differences:
    Controls–Group 4 5.57% ± 1.37 or 4.07 times its probable error
    Controls–Group 5 2.87% ± 1.30 or 2.21 times its probable error
    Controls–Group 6 2.85% ± 1.26 or 2.26 times its probable error
    Controls–Group 7 5.09% ± 1.51 or 3.37 times its probable error
    Controls–Group 8 7.86% ± 1.15 or 6.83 times its probable error

Three items concerning this curve will be emphasized here: 1) Individuals inoculated from five to ten days after operation are significantly more resistant to the tumor tissue, 2) from sixteen days on there is a gradual increase of susceptibility until the maximum is reached with individuals inoculated about 110 days after operation, and 3) the result obtained by keeping individuals for about 150 days and longer between operation and inoculation (irrespective of the age at which they were gonadectomized) is the same as that obtained by inoculating individuals that had been operated on when they fell in age class 5 (old mice)—there is an approach toward zero susceptibility in each case. It may be well to recall that in normal individuals susceptibility to transplantable tissue increased with old age.

The dBrA reaction for gonadectomized mice. The results obtained with the dBrA tumor were practically the same as those with the dBrB. Only a few charts will be given for the dBrA tumor—enough for comparisons between the two tumors.

There is no significant difference between the susceptibility of the castrated and of normal males for the dBrA tumor, as shown in figure 29 (both first and second inoculations included).

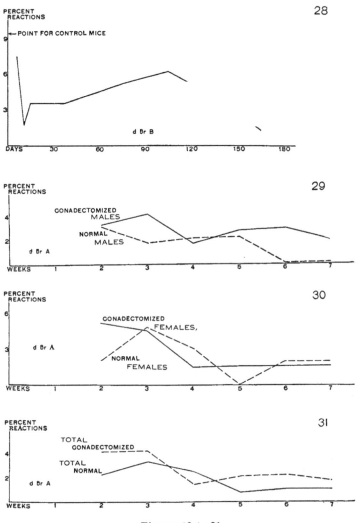

Figures 28 to 31

Castrated males 640 negative : 19 reactions ±2.89 or 2.88% ± 0.44
Normal males　282 negative :　5 reactions ±1.50 or 1.74% ± 0.52
　　　　　　　　　　　　　　　Difference　　　　　1.14% ± 0.68
　　　The difference is thus 1.68 times its probable error

The curve for castrated males does not approach zero, as does the curve for the normal males, because of the fact that a few castrated males continue to grow the tumor mass progressively (this was also found to be the case in the dBrB tumor).

A similar result is obtained when the normal females are compared with the spayed individuals (fig. 30, p. 113).

Normal females 388 negative : 10 reactions ±2.10 or 2.51% ± 0.53
Spayed females 721 negative : 21 reactions ±3 05 or 2.83% ± 0 41
　　　　　　　　　　　　　　　Difference　　　　　0.32% ± 0.67
　　　The difference is thus 0.48 times its probable error

It may be well to repeat again that the rise in the normal female curve near the end is not real, but due to the dying off of several negative individuals.　The spayed female line remains parallel with the base line for the last three points.　This indicates a similar condition to be found in the castrated male line, namely, that a few spayed females continue to grow the tumor tissue progressively.

By adding the observations for both males and females together we are able to demonstrate more completely this slight increase in percentage reactions of the gonadectomized individuals over the controls for the latter end of the curve (fig. 31, p. 113).

Operated 1361 negatives : 40 reactions ±4.20 or 2.86% ± 0.32
Controls　670 negatives : 15 reactions ±2.58 or 2.19% ± 0.38
　　　　　　　　　　　　　　Difference　　　　　0.67% ± 0.49
　　　The difference is thus 1.37 times its probable error

The gonadectomized individuals gave a slight increase of positive reactions, which is not, however, mathematically significant.

The following figure 32 shows the effects of gonadectomy upon mice, regardless of the tumor employed.　It is constructed by adding the percentage reactions of both tumors for both the normal and operated individuals.

Normal   1491 negative :   95 reactions ±6.37 or 5.99% ± 0.40
Gonadectom. 2640 negative : 103 reactions ±6.71 or 3.75% ± 0.25
                        Difference   2.24% ± 0.47
        The difference is thus 4.77 times its probable error

There is no significant difference between any two corresponding points of the curves, although when the data are massed

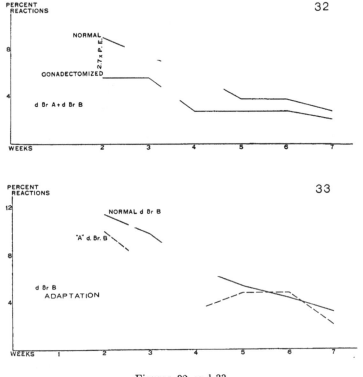

Figures 32 and 33

there appears to be a significant lower percentage reactions on
the part of the gonadectomized individuals. This difference,
however, is mainly confined to the first part of the curve. Indeed, this decrease of positive reactions is confined to the first
inoculation series (just after the operation). It is justifiable
to maintain, therefore, that there is no visible significant difference between the percentage reactions for normal and gonadec-

tomized individuals when all age groups are considered. The primary effect of gonadectomy is not the changing of the percentage reactions, but that some few individuals are then able to grow the transplanted tissue progressively. This question will be taken up later in the General Discussion.

p. '*Adaptation.*' The following experiment has to do with a supposed peculiarity of the tumor cell, namely, that of its ability to adapt itself to a foreign environment (host). The tumor tissue derived from gonadectomized individuals was inoculated into normal and into other gonadectomized individuals of the wild house mouse. Figure 33 includes the susceptibility curves for both the normal dBrB tissue and for the dBrB tissue that had already succeeded in growing for at least eight to ten weeks in wild individuals (gonadectomized).

In order to increase the number of observations in one comparison, both normal and gonadectomized individuals are included in a single curve.

Normal dBrB (ndBrB) 1768 negative : 137 reactions ±7.62 or 7.19% ± 0.40
Adapted dBrB (adBrB) 386 negative : 27 reactions ±3.39 or 6.54% ± 0.82
                                 Difference   1.06% ± 0.92
The difference is thus 1.15 times its probable error

There is no significant difference between the potency of the normal dBrB tumor cell and the 'adapted' dBrB tumor cell. In other words, when a relatively homogeneous race of mice has been employed, the phenomenon of adaptation (or virulence, for that matter) is not seen. The special 'A' tumor does not show any particular adaptation. This will be discussed later. If there is any difference (however slight) between the two reactions, it is in favor of the normal dBrB tumor that had never grown in wild house mice.

The evidence against the theory that a transplantable tumor possesses variable powers of adaptation derived from the use of the dBrA tumor is not so conclusive, but bears out the general result obtained from the dBrB tumor. A total of three dBrA tumors have grown in wild mice. The first one in the normal adult male W238 referred to previously and the other two in gonadectomized individuals (one a male, the other a female).

Forty-five wild mice were inoculated with the three tumors. Not one of these mice ever grew the tumor progressively. Unfortunately, the tumor in this case was inoculated into the iliac region. It was found too late that this region is unsuited for the palpation of small masses, so that the end result alone can be utilized in reaching a conclusion for the dBrA tumor.

Probably, neither tumor, therefore, possesses the power of adaptation when inoculated into a relatively homogeneous non-susceptible race.

III. GENERAL DISCUSSION

1. *Value of proved stocks*

Several results here recorded are quite distinct from those usually obtained in experiments dealing with transplantable tumor tissue. Since we have used as homogeneous a race of mice as is at present available, it seems that the conflicting results of other investigators in this field must have been in a large degree due to the different kinds of mice that they employed. It has already been demonstrated (Little and Tyzzer) that the susceptibility to transplantable tumor tissue depends on a complex of mendelian relations. That a mouse will grow the transplanted tissue progressively is as much the result of the genetic constitution of that mouse as it is of any inherent characteristic of the tumor cell. Both these factors, however, must be taken into consideration for the determination of the final outcome of implanting a bit of tumor into a given mouse.

Investigators, as a rule, have neglected the host factor. Most of them still maintain that the host employed is of only secondary importance.

Doctor Woglom, in a recent publication on "Virulence of Adaptation," evidently witnessed a phenomenon similar to that seen by us in our first experiment (figs. 2 and 5). This is the fact that apparently identical neoplasms (derived from multiple mammary carcinoma primarily) have different reaction potentialities when inoculated into mice from the same dealer. He described this difference to the presence or absence of the power of adaptation on the part of the tumor cell; different tumor cells are supposed to be endowed with different capacities for adaptation. He admits, however, that proliferative energy may be a secondary factor. From the genetic viewpoint, several objections appear to be valid: 1) all the results which he obtained are supposed to be explained by adaptive characteristics of the tumor cell alone; 2) "mice from the same source" introduce a third variable; this, however, he does not take sufficiently into consideration; 3) the supposedly fluctuating power of adaptation on the part of the tumor cell appears up to the present time to have been based upon the old conception of the inheritance of acquired characteristics and is therefore unsupported by experimental evidence.

In our own experiment we found that identical histological adenocarcinomas (derived from two very closely related individuals) possess different physiological reactions when placed into the same mice (not individuals derived from one dealer). Our last experiment (outlined in fig. 33) demonstrated that in these tumors, at least, there was no power of adaptation present. Evidently another explanation is therefore necessary.

"It is of course true that cells which are not morphologically different in any visible way may show themselves by their behavior to be physiologically different, so that the absence of visible differentiation in the cell is not proof that the cell is completely unspecialized" (Child, Senescence and Rejuvenescence, p. 48).

We have some evidence to show that the two apparently histologically identical tumor cells differ from one another in one or more mendelian units. (This point can only be tested by raising an F_2 generation between a susceptible and a non-susceptible race. This experiment is now in progress. The numbers obtained so far are not yet large enough, however, to give conclusive proof for this contention.)[7] The two tumors have different characteristics: 1) They possess different reaction potentialities and, 2) they differ in their genetic constitution. Are we justified in drawing the conclusion that the difference in their reaction capacities within a non-susceptible race

[7] Thus we find that some F_2 animals grow both tumors, some one or the other, and some neither, indicating a genetic difference in the tumors, the underlying elements of which are segregating.

somehow or other correlated with the genetic difference which the tumors themselves show?

2. Fluctuations in 'growth' energy or 'adaptation'

That tumor cells fluctuate remarkably in their rate of proliferation is commonly acknowledged. Such changes, however, are not always uniform (either increasing or decreasing), as might be expected if the tumor cells underwent rhythmical activities. The sudden appearance of an increased activity on the part of the tumor cells cannot be explained by assuming that the cells possessed the capacity of adaptation for any particular host. If this is the case, we have given both tumors ample opportunity to express their inherent potentiality. With one exception, neither tumor ever grew progressively in a normal wild mouse, although hundreds have been inoculated. 'Adaptation' certainly ought to have expressed itself more than once in that time. Again, absence of proliferative energy alone cannot be the cause for the failure of the wild mice to harbor the tumor tissue. Individuals from a susceptible strain of mice were inoculated periodically throughout the experiment and they grew both tumors in 100 per cent of all inoculations. The one exception among the wild mice occurred in an adult male individual, no. W238. For some unknown reason, this mouse alone of all the controls inoculated grew the dBrA tumor progressively.[8] The same tissue that was employed for W238 was also inoculated into several other individuals. This single experiment (fig. 3) gave more percentage reactions for this 'exceptional' dBrA tumor than for all the other experiments dealing with the dBrA tumor combined. The mice are probably of the same general genetic constitution as those used in all other experiments with wild mice (discussed on page 76). The progressively growing tumor (derived from W238) was inoculated on both sides in the inguinal region of thirty other wild mice. Not one of the mice ever showed a perceptible growing mass. The possibility,

[8] Male W238 may have possessed a genetic constitution which would have allowed at least temporary growth of the unaltered dBrA tumor—the exceptional dBrA showing successful temporary growth in a number of mice was in him able to continue its growth.

therefore, that this exceptional case was produced by the 'adaptability' of the tumor cell is not very great. The W238 dBrA tumor should have possessed an increased power of adaptation, since it had already grown in one foreign soil for several weeks. The inference has already been drawn that perhaps the reaction potentiality of the tumor cell is an expression of its genetic make-up. If this is the case, we can explain the remarkable increase of growth capacity in the W238 dBrA tumor by maintaining that there must have been some marked change of the tumor cell. We wish to emphasize the fact that this change could have been produced by some process analogous to that giving rise to 'mutations,' that occur in all normal types of animal and plant tissues.

In our own experiment we have been able to verify the previous findings of Tyzzer relative to the growth rate, and in addition can suggest an explanation. The F_1 hybrids obtained between dilute brown (susceptible) and albino (non-susceptible) mice grew the two transplantable tumors, dBrA, dBrB, progressively. The growth curve for the F_1's produce about three to four times the mass of tissue that the controls do in the same length of time. Two interpretations suggest themselves: 1. That hybrid individuals grow the tumor tissue faster because of the phenomenon of heterosis (hybrid vigor). Heterosis is a common occurrence encountered in the crossing of many diverse strains of individuals within a species. The exact mechanism that brings about this increased vigor is still in dispute (East and Jones, Inbreeding and Outbreeding). The observed fact is, that by crossing two diverse strains, individuals are frequently produced that are more active and vigorous than either parent strain that entered into the cross. Thus, F_1 hybrids are found to be more hardy, healthy, and more resistant to diseases. It is, however, not evident why the increased growth rate on the part of the tumor cell should be correlated with the increased vigor of the F_1 host unless the general physiological condition of the animal affects also the fate of tumor transplants on it.[9]

[9] Health and other factors bearing upon the physiological condition of the host do affect the transplantation of tissues.

2. That the growth rate of the transplanted tumor cell is correlated with the number of genetic factors involved in the reaction. This point has been previously discussed. This hypothesis can be tested only by studying the growth rates for back-crossed individuals[10] where the factor of heterosis has been partly eliminated. The last matter is still under investigation.

The next point we desire to discuss is that outlined on pages 86 and 93 (fig. 8). The two tumors here described have been under observation for over a year. In that time it has been repeatedly inoculated into wild house mice. During the whole course of this experiment there has been no significant variation in the reactive potentiality of either tumor (except Ex. N for dBrA). The numbers employed are large enough to be significant and the results cannot be disregarded on grounds of insufficient data. We find no evidence for a rhythmic activity of the tumor cell. In this connection we may recall the findings of Tyzzer and Little. When the definitely proved homogeneous race of mice are employed (Japanese waltzing mice), there has been no significant fluctuation in the activity of the tumor cell in over ten years. This tumor (carcinoma J. W. A.) remained true to type throughout the whole length of the experiment.

Several investigators, using laboratory stocks, claim to have witnessed rhythmic fluctuations of the tumor cell, among whom may be mentioned Haaland, Bashford, Murray, Bowen, and Calkins. It may be suggested that the cyclic phenomenon witnessed was the result of the second variable—that of the genetic constitution of the varied races employed.

3. Age

One of the most interesting results that we obtained was from the study of the susceptibility of individuals of different age groups. The result is a confirmation of the previous conclusions of Little ('20). When a non-susceptible race is employed, susceptibility toward transplantable tissue decreases with age up to the period of maturity. We have, however,

[10] Produced by crossing the F_1 generation hybrids back to the susceptible dilute brown stock.

expanded the experiment to include the whole life of the individual. The five age groups were selected so as to include individuals from the five important periods of the life (infancy, puberty, adolescence, maturity, and senescence). The interesting point discovered in our experiments is that susceptibility to transplantable tumors increases with old age. The age susceptibility toward transplantable tumors (for a non-susceptible race) has a remarkable relation to the activity of the gonads (fig. 7, p. 89). Susceptibility decreases at first rapidly, then gradually, from birth through the period of puberty and adolescence up to the beginning of gonad activity (maturity). The minimum percentage of growth reactions are obtained with mice one-half to three-fourths grown.

Susceptibility remains fairly uniform throughout the period of maturity. With senescence, however, and the accompanying decreased activity of the gonads, susceptibility rapidly increases. This increase in susceptibility in old mice is mainly confined to males and will be discussed under the heading of sex. The same explanation offered by Little for the first part of the life curve can be applied in our own case. Susceptibility to transplantable tumors depends considerably upon the degree in which the tissues of the host have attained a physiological specificity, when fully developed will characterize the particular individual in question. The most characteristic physiological activity of the individual is attained at the period of maturity. With maturity the animal manifests its most pronounced biological characteristics. The susceptibility to transplantable tumors, being a complex mendelian phenomenon, can only be tested by using mature individuals.

The susceptibility curve suggested the possibility of analyzing the factors which underlie tissue specificity more carefully by removing the gonads from individuals in all the different age groups before inoculating them with the transplantable tumor. This experiment will be discussed after the section on the sex factor.

4. Sex

The disputed factor of sex as underlying susceptibility to transplantable tissue has already been briefly discussed (p. 98). Some recent investigations relative to the sex factor have not entirely verified the previous conclusions of Loeb, Ehrlich, and the members of the Crocker Laboratory. The above-mentioned investigators have not observed any differential effect of sex upon susceptibility to transplantable tumor tissue. In the following paragraphs an attempt will be made to reconcile the two opposing theories by pointing out that the sex and age factors as underlying susceptibility to transplantable tumors may be explained as the result of a common group of causative agents affecting the physiological state of the tissues.

The first work of importance that disagreed with the findings of the various investigators already mentioned was done by Little. This work is a continuation of his and Tyzzer's previous investigations with the Japanese waltzing mouse sarcoma J. W. B. Little predicted that the first back-cross generation individuals (produced by crossing the susceptible F_1 hybrids between the Japanese waltzing mice and the non-susceptible common stock back to the non-susceptible parent strain) would behave similarly to the second filial generation of hybrids. This he later demonstrated, although, as expected, fewer individuals of the back-cross generation than of the F_2 generation were susceptible to the continued growth of the tumor cell. The percentage of positive growths expected can be determined beforehand by mathematical calculation from mendelian principles, and is therefore the most accurate work that has been done with transplantable tumor tissue.

In this partially susceptible strain (B.C) Little determined that the percentage of growth increased with age (4.6 × P. E.) (when very young mice were employed). This difference is mainly confined to the female sex. When a definitely proved non-susceptible race was used, susceptibility toward transplanted tissue decreased with age up to the attainment of sexual maturity. Here again this difference was confined mostly to the female sex.

One conclusion (no. 9) was reached by Little that is of primary importance to our own investigation: ". . . . females of the upper age group in both series are, during the later periods of observation, at an age when sexual maturity is attained. Sexual maturity by the awakening activity of the ovary means further differentiation and further development of individuality of the tissue. Such assumption of individuality leads to elimination of the tumor in non-susceptible animals and to encouragement of its growth in animals inheriting all or parts of the factors which are contributed by, and which characterize susceptible animals of the Japanese waltzing race in which the tumor originated."

In our own experiment we were able to verify the previous conclusion of Little and to continue the study to include all age groups. The interesting fact was determined that there is no significant difference for adult mice between the sexes in susceptibility to transplantable tumors, whereas when all age groups are massed there is a significant difference. In other words, we are confronted with the fact that young males give a greater number of visible reactions than do young females, although the sexes are very uniform in their reactions to a foreign tissue when only adult mice are considered.

Age is, perhaps, the most important factor underlying susceptibility to transplantable tumors. Can we explain this observed sex difference by the age factors? It has long been recognized that female mice mature faster than their litter brothers. When the tissues of the female affected by the secretions of the gonads have commenced to attain their adult specificity due to the approach of sexual maturity, the male of similar age is distinctly more immature. Little recognized the fact that with approaching maturity, susceptibility toward transplantable tumors for a non-susceptible strain rapidly decreases. The elimination of the tumor is perhaps correlated with the differentiation of the tissues depending upon several functioning elements, one of which is the gonads. This suggests that the young females are less susceptible to the tumor reaction because they are maturing (growing older) faster than the males. When sexual maturity is definitely reached, the sexes remain fairly uniform in their reactive capacities because maturity is a period of relatively slight physiological change then, in both sexes. We should therefore expect, under this assumption, that adult individuals would show the same reactive potentiality regardless of sex (which is the observed result).

By the use of a preponderant number of adult individuals one would, in any experiment, be unable to detect the influence of a sex factor; the result would be that large numbers of animals of all ages would mask an actual sex difference for individuals of the special age groups where there is a significant difference between the sexes.

What is to be expected when senescence sets in? Do the gonads begin to decrease in physiological activity at the same period of life in both sexes? The period of maturity for male mice is relatively shorter than the corresponding period for females. It has already been pointed out that in a non-susceptible race susceptibility increases with old age. This increase in our experiments is almost exclusively confined to the male sex, so that if old mice alone are inoculated there would be a significant difference in susceptibility between the sexes. The sex factor can therefore be largely explained as the result of the primary factor of age.

The phenomena encountered in studying susceptibility in a non-susceptible race are mostly the exact opposite of those found to hold true for susceptible races. For example, susceptibility to transplantable tissue decreases with age up to maturity for non-susceptible individuals, whereas it increases for susceptible animals for the same age group. We may therefore expect that in both cases the sex-factor difference for very young and very old individuals will be partly explained by the degree of physiological specificity of the tissue which the individual possesses at the time of inoculation. Female mice, maturing faster than males, are able more quickly to counteract the activity of the tumor.

From a genetic view-point, the tumor cell as well as the host element must be taken into consideration. It is reasonable to suppose that certain tumors should show a differential reaction between the sexes for adult mice also. Certain tumors must be differently affected by sex hormones, and other internal conditions characteristic of one or the other sex. The neoplasms that have given rise to transplantable tumors are decidedly more prevalent in females than in males. For that reason, primarily, they probably, in some cases, vary from those arising from male tissue.

The dBrA tumor used in this experiment appears to be sluggish. When it is grown together with the dBrB in the same individual of a highly susceptible race (either an F_1 or a first backcross animal), it is always handicapped by the greater growth vigor of the dBrB mass. Because of its sluggishness, primarily, we believe that it was unable to show a differential effect of sex (as did the dBrB tumor) when all age groups were massed. One point of importance need be emphasized—when the dBrA tumor underwent the probable mutational process in the experiment N, previously referred to, it not only approached the greater reactive potentiality of the dBrB tumor, but also showed a similar differential effect upon the sexes (fig. 14). In fact, it not only exceeded the dBrB tumor in vigor of growth, but also produced a differential effect upon the sexes when only adult individuals are considered (fig. 15). This result can be explained on the assumption that the increased activity of the dBrA tumor was so great that it could even differentiate between the slight metabolic differences in adult mice.

To sum up the factor of sex underlying susceptibility to transplantable tumors, we may say: 1) That the influence of sex is probably a secondary phenomenon of the age factor. 2) That the rôle of sex in determining susceptibility or non-susceptibility may also have some relation to the attainment of physiological specificity of the tissues. 3) That the inherent capacities of the tumor cell itself may determine to some extent the degree of influence which sex has on the fate of the implant.

5. Gonadectomy

The discussion of this part of the paper should have followed the experiments dealing with age, but since the sex factor is primarily a secondary phenomenon of age, it was thought best to discuss that subject immediately following the age factor.

Several recent experiments have been performed dealing with the effects of gonadectomy upon susceptibility to transplantable tumors. As an example may be mentioned the work of Sweet, Corson-White, and Saxon ('13). They employed adult male individuals only. The conclusion was reached that castration had two effects upon susceptibility to transplantable tumor: 1) the number of individuals that grow the tumor progressively was significantly increased and, 2) the rate of growth of the mass is materially increased. It is well to point out, however, that market mice of unknown ancestry were employed; the results are, therefore, in all probability, not very simple to interpret. It is also very difficult to conclude anything about female behavior to gonadectomy from the study of male individuals alone.

In our own experiments we have been unable to verify the results obtained by these investigators. In the first place, we determined that somehow or other the effects of the operation alone had some unexpected influence upon tumor susceptibility. Individuals that are inoculated five to ten days after operation give a significantly lower percentage of reaction masses than do individuals inoculated at any other time interval after the operation. The percentage reactions of the dBrB tumor for individuals inoculated from zero to five days after operation was found to be 7.33% ± 1.27; that for individuals inoculated from five to ten days, 1.90% ± 0.62, and that for animals from ten to sixteen days, 3.96% ± 0.93. The first time group (0 to 5 days) is not significantly different from that obtained for the control mice, 9.21% ± 0.66 (difference 1.88% ± 1.43 or 1.31 times probable error). Individuals from the five-to-ten-day time interval groups gave remarkably few reactions. They are distinctly more resistant to the transplantable tissue than are the controls (difference in favor of the gonadectomized 7.31% ± 0.91 or 8.10 times its probable error). The third group mentioned above is also slightly below the control group in percentage reactions, but not nearly so resistant as are those from the intermediate group (5 to 10 days). The result is probably due to the operation itself, and not to the removal of the gonads as such. The operation itself may have several effects, one or more of which must explain this increased resistance of animals inoculated five to ten days after operation. 1) Ether is very harmful to the health of the animals. This, however, cannot wholly explain the observed result, since the effect of ether must wear off gradually beginning with the first day after operation. 2) The shock and disturbing elements of the operation may be so great that the normal physiological activities of the animal are greatly disrupted. Recovery from such an effort would of necessity be slow and gradual. 3) The third probable effect of the operation that may have an influence on the result obtained is the phenomenon of leucocytosis. After an operation, an individual commonly reacts by producing a large number of leucocytes to serve as protective agents. This reaction reaches its height in about seven to ten days. So that in the arbitrarily chosen time group 2 (5 to 10 days), we have included those individuals that are at the height of the 'leucocytic' reaction. The leucocytes probably function in protection against all foreign bodies. Is it not reasonable to suppose, therefore, that an increased leucocytosis (as a result of the operation) is the chief causative agent in the temporarily acquired resistance to the tumor cell? The lymphocytic theory as an all important explanation for tumor immunity is fast losing ground. The recent conclusion of Sittenfield,''.   that neither increase nor reduction of the lymphoid elements in the blood had any influence upon either resistance or susceptibility to tumor growth,'' is being strengthened by the work of several investigators, especially those at the Crocker Laboratory.

There is a gradual increase of percentage reactions obtained by keeping individuals from sixteen days to 120 days after operation before inoculation. If animals are kept longer than 120 days before inoculation, there is a gradual decline in the number of visible reactions. The mouse slowly recovers its normal physiological activity, and with this gradual recovery it is more capable of producing an increasing percentage of reactions. By keeping individuals for 120 days or longer between operations and inoculation, the process of senescence becomes mechanically an important factor (some of these individuals were gonadectomized when adult). The decreased metabolic activity of old age is probably an explanation of the decrease in the percentage reactions, as stated above (it will be recalled, however, that for normal individuals susceptibility to transplantable tumors increased with old age—just the opposite from that obtained with gonadectomized old mice. This point will be considered later).

Previous investigators have not taken into consideration the deleterious effect of the operation. It has already been pointed out that results obtained with non-susceptible mice are the exact opposite of those obtained with susceptible strains. Corson-White and other investigators have encountered an increased percentage indication by castration (working with a partially susceptible race). Could not this result also be explained by maintaining that the operation alone had a stimulating effect on susceptibility?

Our next experiment dealt with the actual effect of gonadectomy itself. After correcting for the harmful effect of the operation alone, we have been unable to observe any difference in the number of percentage reactions between the controls and gonadectomized individuals (fig. 21, p. 105, and others). Gonadectomy neither increases nor decreases the number of reactions for the six-week observation period, when all age groups are massed.

This result is to be expected. The number of susceptible individuals in any given race of mice is determined by genetic factors primarily. There are, however, several secondary factors (such as age, etc.) that determine whether the reaction to the tumor is to be prolonged or shortened.

The one noticeable effect of the removal of the gonads is the fact that a certain small number of animals continue to grow the tumor mass progressively. We cannot conclude from this result that gonadectomy increases the rate of tumor growth. The number of operated individuals that continue to grow the tumor progressively is approximately the same as those in the control series that showed transitory masses only. The primary effect of gonadectomy therefore lies in the fact that the source of origin (gonads) of certain physiological checks inhibiting the development of the tumor mass in normal non-susceptible individuals has been removed. An individual of the non-susceptible race is non-susceptible because it possesses physiological characteristics that inhibit progressive growth on the part of the tumor implant.

This result suggests very strongly that the internal secretions from the gonads somehow or other produce some of the physiological checks referred to above. As above stated, tumor susceptibility is correlated with the phenomenon of assumption of tissue specificity. The gonads are probably one of the most important agents in bringing about these processes. A non-susceptible strain is non-susceptible because the highly specific physiological condition of adult individuals of that strain produces a soil unfavorable for tumor implantation. By removing the gonads we can prevent, to some extent, this process of assuming high degree of specificity of the tissues. Very young individuals of a non-susceptible race are sometimes susceptible to implanted tissue, although no adult individual of the same race will grow the tumor. Gonadectomy therefore allows the progressive growth of certain of those tumor masses that would have regressed in a normal (non-susceptible) individual.

A most interesting result was obtained by comparing the age-susceptibility curve for gonadectomized individuals with that for the controls (fig. 27, p. 111). Several very striking results were obtained. It will be noticed that the two curves are entirely different.

We started out with the idea that the younger the mice employed for operation, the more effect the removal of the gonads would probably have, due to the fact that differentiation would never have the opportunity to proceed as far as in normal individuals. This is not the case, however. If very young individuals are gonadectomized, no apparent reaction masses of the implanted tumor can be detected, no matter how many times the mouse is inoculated, nor how long the mouse is kept between operation and inoculation. If very old animals are castrated, the same result is obtained. Operated old individuals give no visible reaction masses whatever. The maximum effect of the removal of the gonads is produced on those individuals that are at the moment of acquiring sexual maturity. It may be recalled that age group 3 for the controls gave the minimum number of reaction masses. The age-susceptibility curve for operated individuals is not the same as the age-susceptibility curve for the controls of a non-susceptible race. Gonadectomized individuals (from a non-susceptible race) give a susceptibility curve resembling that characteristic of individuals from a susceptible strain. The last end of the age-susceptibility curve for a susceptible race has not been definitely determined, but it seems likely that it will be found that in such a race susceptibility to a transplantable tumor decreases with old age. The greatest effect of gonadectomy is produced on individuals that are not only undergoing sexual maturity, but also the correlative process of physiological differentiation.

The failure of old gonadectomized individuals to show any reactions to implants may be explained, partly on the decreased physiological activity cf old animals, and partly as a result of the genetic composition which they in common with all adult animals of their race possess. Certain physiological agents cease to function with the onset of senescence. The shock of the operation is severe enough to reduce considerably the normal physiological activity of the organism. Before the shock of the operation is over, so that the normal physiological activity of old mice can be restored, the final lowering of this activity due to old age has set in. The animal is therefore merely able to keep alive for a short time and cannot nourish the implanted tissue. The host has therefore become again a relatively passive element, and the tumor, no longer calling forth any reaction, necessarily dies.

Another point of minor importance is the fact that the effect of gonadectomy upon the two sexes is not different, thus indicating that the gonads are a common factor in the general processes of assumption of tissue specificity for both sexes.

IV. CONCLUSIONS

The conclusions to be drawn from the work here reported naturally group themselves under two headings: A. The activity of the tumor cell; B. The reaction of the host.

A. The activity of the tumor cell

1. There is, in some cases at least, a uniform reaction, providing the tumor is transplanted into individuals of the same age and sex of a relatively homogeneous series of hosts. In other words, no rhythms of tumor growth are encountered.

2. A transplanted tumor grows progressively (within limits) at a fairly uniform rate of development if placed in definitely proved homogeneous mice (dBr stock and F_1 hybrids). Sudden fluctuations in growth activity may sporadically occur, due possibly to a process analogous to mutation.

B. The reaction of the host

1. Race is the primary factor that determines whether or not a given individual shall or shall not grow the tumor mass progressively. Susceptibility and non-susceptibility are manifestations of the genetic constitution of the individual.

2. Several secondary physiological factors, among which age is the most important, function in determining the outcome of a given reaction. These may be called contributory or accessory factors.

3. The age factor is an expression of the degree of the process of the assumption of tissue specificity controlled to some extent by the activity of the gonads.

4. The age-susceptibility curve towards transplantable tumors for normal individuals of a non-susceptible race bears a remarkable similarity to the curve of activity of the gonads.

5. The sex factor (encountered especially with young mice in development) depends upon at least two primary causes, 1) the age factor and, 2) the difference in metabolic activity between the sexes, at the different age periods of life.

6. Removal of the gonads does not change the massed percentage reactions for individuals of a non-susceptible race.   This bears out the previous conclusion that the number of percentage reactions in a given strain depends upon the genetic constitution of the individuals.

7. Gonadectomy produces, in the stock employed, a significant increase in percentage reactions in mice attaining sexual maturity (age class 3).

8. Gonadectomy causes an approach towards a 'neutral' type (loss of characteristic differences between sexes) in the percentage of reactions towards both tumors used, just as it does in the case of morphological characteristics (Hatai and others).

9. By the removal of the gonads, the individuality of tissues and the normal functioning of the age factor can be interfered with.

10. A severe shock caused by such an operation as gonadectomy produces, in some cases at least, a resistant state to transplantable tumors that is at its maximum from five to ten days after the operation.

Without the assistance of several institutions and individuals, the present experiment would have been impossible.

I am indebted to the Department of Zoology, Columbia University, for a grant of the John D. Jones Scholarship Fund during the summer of 1920.   The Carnegie Institution of Washington, through the Department of Genetics, has very kindly assisted the experiments from a financial standpoint during the winter of 1920–1921.   To Dr. C. C. Little I am deeply indebted for numerous favors, especially for the suggestion of the problem, his kind supervision throughout the experiment, and his painstaking criticism and help in the preparation of the manuscript. Drs. G. N. Calkins, F. C. Wood, and T. H. Morgan have offered valuable suggestions.   Through the kindness of Drs. James Ewing and H. J. Bagg, the technical preparation of the material for histological examination and the making of the microphotograph were done at the Memorial Hospital, New York City.   To the

Misses Jennie Hubbard, Elizabeth Jones, and Dorothy Newman I am indebted for technical assistance in operating on and recording the mice. My wife has kindly attended to the clerical work of preparing the manuscript.

PROMPT PUBLICATION

The Author can greatly assist the Publishers of this Journal in attaining prompt publication of his paper by following these four suggestions:

1. *Abstract.* Send with the manuscript an Abstract containing not more than 250 words, in the precise form of The Bibliographic Service Card, so that the paper when accepted can be scheduled for a definite issue as soon as received by the Publisher from the Editor.

2. *Manuscript.* Send the Manuscript to the Editor prepared as described in the Notice to Contributors, to conform to the style of the Journal (see third page of cover).

3. *Illustrations.* Send the Illustrations in complete and finished form for engraving, drawings and photographs being protected from bending or breaking when shipped by mail or express.

4. *Proofs.* Send the Publisher early notice of any change in your address, to obviate delay. Carefully correct and mail proofs to the Editor as soon as possible after their arrival.

By assuming and meeting these responsibilities, the author avoids loss of time, correspondence that may be required to get the Abstract, Manuscript and Illustrations in proper form, and does all in his power to obtain prompt publication.

THE JOURNAL OF EXPERIMENTAL ZOOLOGY, VOL 36, NO. 2
AUGUST, 1922

Resumen por la autora, Ruth L. Phillips.

El crecimiento de Paramecium en infusiones de contenido
bacterial conocido.

La regulación del contenido bacterial de los medios empleados
para el cultivo de Paramecium puede llevarse a cabo perfecta-
mente sin una técnica demasiado laboriosa. Durante un estudio
del crecimiento de Paramecium en infusiones de contenido bacterial conocido, ha encontrado la autora que, en adición a la
rapidez de la fisión, una consideración de los tantos por ciento
de divisiones abundantes y escasas, y la cantidad de individuos
muertos son valiosas. La aplicación del "factor de signifi-
cación" a los datos de este tipo es muy útil, haciendo posible
el llegar a decisiones con relación a los efectos del alimento y
los medios, que de otro modo no pueden obtenerse. La afirm-
ación de Hargitt y Fray que los cultivos puros de bacterias
son, en general, un alimento poco satisfactorio para Paramecium ha sido comprobada por la autora. En un solo caso
en el cual se usó un cultivo puro como alimento durante un
periodo largo, la cantidad de metabolismo de Paramecium
fué constantemente menor que la que presenta cuando se
emlea una mezcla. Las mezclas de bacterias parecen ser el
alimento más satisfactorio. Las mezclas artificiales son difi-
ciles de hallar, y entre todas las probadas solamente dos han
sido satisfactorias. Cuando se alimenta a Paramecium con
bacterias conocidas, tienen lugar pequeñas fluctuaciones en la
marcha de la división, las cuales son independientes de la endomixis, y a semejanza de esta, no son fácilmente influidas por el
medio ambiente. Las pruebas acumuladas tienden a favorecer
la constancia de la alimentación más bien que su cambio frecuente.
La influencia de los medios ordinarios no produce prácticamente
efecto alguno. Paramecium no puede utilizar las substancias
alimenticias disueltas en tales medios.

Translation by José F. Nonidez
Cornell Medical College, New York

THE GROWTH OF PARAMECIUM IN INFUSIONS OF KNOWN BACTERIAL CONTENT[1]

RUTH L. PHILLIPS

ONE FIGURE

CONTENTS

INTRODUCTION

The work of Hargitt and Fray ('17) on feeding pure lines of Paramecium with pure cultures of bacteria initiated a new phase in the investigation of protozoan metabolism. This work outlined the methods necessary for such a study and summarized the probable effect of pure cultures of bacteria, or mixtures of

[1] Contributions from the Zoological Laboratory, Liberal Arts College, Syracuse University; C. W. Hargitt, Director.

several cultures, when used as food by Paramecium. However, these data were not extensive, and it has seemed worth while to carry the investigation further, in order to test the conclusions announced and to extend the experiments over a wider field.

Since food is so important a factor in the growth of animals and since the food of Paramecium is so variable, it would seem strange that so few investigations of this kind have been made, were it not that it is a difficult matter to treat adequately both the bacteriological and the protozoological sides of the subject. A complete analysis of the hay infusion would be very desirable, but the present knowledge of the saprophytic bacteria is so incomplete and the bacteriological work is so time-consuming, that it is impracticable for one person to carry it on together with the study of Paramecium in cultures of bacteria isolated from such infusions. The work described in this paper has accordingly been limited to a study of the continued growth of Paramecium aurelia in pure cultures and mixtures of the bacteria isolated from hay infusions or other infusions used for growing Paramecium.

I wish to express my thanks to Prof. George T. Hargitt, of the Department of Zoology, for his valuable suggestions and criticisms during this work; to Prof. Henry N. Jones, of the Department of Bacteriology, for suggestions and for the use of apparatus for the bacteriological investigations; to Miss Lucy J. Watt for the classification of the bacteria used in feeding experiments; to Mr. Clifton E. Halstead for the determination of the hydrogen ion concentrations of the media, and to Dr. Vasil Obreshkove for suggesting certain biometric methods.

METHODS

a. Media

Since the purpose with which this work was undertaken was to study the effect of food upon Paramecium, it was desirable that the media used be as uniform as possible. To secure such uniformity, a sufficient amount of each medium was made to last throughout the course of the experiments. These stock solutions were sterilized and set aside to be diluted as needed. Hargitt and Fray ('17) found that the high temperatures of the autoclave so altered the nature of the hay infusion that Paramecium could not live in it.   For this reason, the three-day intermittent method of sterilization in the Arnold steam sterilizer has been used for all media in which it was desired to grow Paramecium.   This method proved to be perfectly reliable.

The different infusions were prepared as follows:

1. The standard hay infusion of Jennings ('10) was made by allowing 10 grams of chopped timothy hay to boil for ten minutes in a liter of tap-water.   This was cooled, brought up to volume with tap-water, filtered, and then sterilized in the Arnold for three successive days.   From this stock infusion a 0.1 per cent solution was prepared from time to time as needed.   The 0.1 per cent solutions were sterilized in a similar manner and used as the standard culture medium for inoculation with bacteria and for growing Paramecium.

2. An infusion was made from uncured swamp hay according to the formula used for the standard hay infusion.   This with its 0.1 per cent solution was sterilized in the same way and put aside for further use.

3. Similarly, a third infusion was prepared from the common moneywort, Lysimachia nummularia L., which had been carefully dried and used at once without further curing.

b. Reactions of the media

The hydrogen ion concentrations of these three infusions were determined at the close of the experiments by Mr. Halstead according to the Gillespie ('20) drop method.   The reactions were found to be as follows: For the 0.1 per cent standard timothy hay infusion pH = 8.2; for the 0.1 per cent uncured swamp hay infusion, pH = 8.2; for the 0.1 per cent moneywort infusion, pH = 8.4.   Since the neutral point has been determined as pH = 7, increases over this point indicating alkalinity and decreases acidity, it will be seen that the reactions of these media were rather markedly alkaline.   No attempt had been made to keep the reactions constant by the addition of buffers, so it is impossible to say what was the reaction of the media at the beginning of the work.

A rough determination of the reaction of a watch-crystal culture in which Paramecium had been growing for a few days with the streptothrix C' as food showed an increase in alkalinity over the sterile medium. Although the amount of fluid available was too small to determine the exact amount of this increase, the reaction obtained indicated that the presence of the organism C' in hay infusion tends to increase the alkalinity of the medium. A similar test was made with a like amount of infusion in which Paramecium was living upon the bacterial mixture, A'B'C'. In this case the reaction tended to be more acid than the control sterile infusion. Whether this increase in acidity was due to the carbon dioxide excreted by a large number of Paramecia or to the reaction of the bacteria with the media or to a combination of these factors, it is impossible to say. It would seem that an investigation of the hydrogen ion concentrations of media used for cultivating Protozoa might be helpful in understanding some phases of the activities of these animals.

The fact that the hydrogen ion concentrations of the three types of media proved to be practically identical does not necessarily indicate a similarity in chemical constitution, and it is, I believe, safe to assume that they were chemically unlike. On this assumption it was, therefore, possible to vary the environment of Paramecium by the use of these different media as well as by changing the food.

c. Isolation of the bacteria used for feeding Paramecium

A little preliminary work demonstrated that a complete bacterial analysis of the hay infusion was impracticable in connection with an investigation of the reactions of Paramecium to food. It was, therefore, thought best to inoculate a sterile hay infusion in such a manner as to secure the types of bacteria ordinarily found in the hay infusions used for growing Paramecium in the laboratory. This was done with all precautions necessary to exclude bacteria from sources other than hay.

Agar plates were poured and inoculated from the infusions every day for the first three days, and again at the end of nine

days when the initial rapidity of bacterial growth had somewhat
subsided. In order to secure bacteria from an old culture supposedly unfavorable to Paramecium, plates were made from a
laboratory culture from which the pure lines of Paramecium used
in this work were taken, but in which, for some reason, the animals were no longer abundant. The usual technique for making
bacterial plates was followed, and only such colonies selected as
were predominant on the plates after they had remained at laboratory temperature for forty-eight hours. It is believed that the
bacteria secured in this way were fairly representative of those
which furnish the bacterial food of Protozoa introduced into such
an infusion. Pure cultures of the bacteria from the predominating colonies were grown upon agar slants in the usual way and
served as the source of the material used in testing the behavior
of Paramecium with reference to its food. That these bacteria
were representative of those occurring in fresh, middle-aged and
old infusions may be safely assumed.

d. Characteristics of the bacteria used for feeding Paramecium

When Miss Watt attempted to classify the bacteria used in
these experiments, she found, as did Hargitt and Fray, that these
saprophytic forms are very difficult to identify positively. The
classification of Chester ('14) proved more satisfactory for this
investigation than the more recent work of the Society of American Bacteriologists (Winslow, '20). Chester's classification is
based upon reactions with media as well as morphological characteristics, and the classification of the American Society of Bacteriologists lacks completeness in this respect. Neither work is
satisfactory in classifying saprophytic bacteria beyond genera;
species, therefore, have not been given except tentatively in one
or two cases. The organisms have been placed in groups according to their reactions with media, and each organism has been
given the letter used in the experimental work.

With the exception of one group of bacteria, the organisms
used in the feeding experiments were isolated from infusions
which had been inoculated from hay or actually used for growing
Paramecium. One set of experiments was undertaken in which

the organisms used for food were B. coli, B. cereus, and B. proteus.   These were chosen because of the common occurrence of B. coli in bodies of water in which Paramecium is found in nature, and because B. cereus and B. proteus are such frequent saprophytes and might, therefore, be expected to be occasional components of the food of Paramecium.   The pure cultures of these organisms were obtained from the laboratory stock of the Department of Bacteriology.   The following descriptions of the bacteria used include such features as morphology, nature of the agar colonies, staining reactions, and reactions with various media.   These features should be an aid to one who wishes to work with such bacteria in determining whether the types he has isolated are the same as those here described.   It is fully realized that with the present lack of detailed knowledge of the sapro phytic bacteria, such a description can only be an aid in identi fication and cannot be considered as a key to these types of bacteria.

Indol formation could not be determined, owing to trouble with the reagents.

Organism A.   Belongs to Bacterium class III, group II, the Rhinosclermatis group of Chester.   Isolated from a one- to three-day hay infusion.

Morphology: short rod, nearly spherical, grows in chains, nonmotile.

Agar colonies: round, moist, glistening, granular, edge sharply defined and broadly lobed as in colon type, non-chromogenic.

Staining reactions: capsulated, pleomorphic, show bipolar staining, Gram negative.

Gelatin: not liquefied, growth uniform, line of puncture beaded.

Litmus milk: neutral to slightly alkaline.

Bouillon: sediment at bottom, cloudy.

Dextrose bouillon: no acid, no gas, heavy membrane at bottom, cloudy.

Lactose bouillon: no acid, no gas, slight sediment at bottom, cloudy.

Saccharose bouillon: no acid, no gas, cloudy.

Mannite bouillon: no acid, no gas, cloudy.

Organism B.   Did not grow well after the first few transfers, and died before the characteristics could be determined.

Organism C.   Belongs to class XV, group II B, the subtilis group of Chester.   Isolated from a nine- to fourteen-day hay infusion.

Morphology: short rod, some short chains noted, endospores central, markedly motile.

Agar colonies: amoeboid, spreading, smooth, glistening, growth rapid, non-chromogenic.

Staining reactions: Gram positive.

Gelatin: rapidly liquefied, sacculate, growth best at top, plumose to villous.

Litmus milk: slightly decolorized at bottom of tube after four days, slowly becoming entirely decolorized, completely peptonized after thirty days.

Bouillon: pellicle formed.

Dextrose bouillon: no acid, no gas, membrane formed.

Lactose bouillon: no acid, no gas, membrane formed.

Saccharose bouillon: no acid, no gas, membrane formed.

Mannite bouillon: no acid, no gas, flaky sediment at bottom.

Organism J. Belongs to class VIII, B. subflavus type of Chester. Isolated from a one- to three-day hay infusion.

Morphology: short rod, almost spherical, markedly motile.

Agar colonies: compact, slight tendency toward radiation, slightly chromogenic, yellow.

Staining reactions: capsulated, Gram positive.

Gelatin: not liquefied, growth spreading over top, growth best at top, filiform, beaded along edge.

Bouillon: clear, membrane at top.

Dextrose bouillon: no acid, no gas, membrane present, flocculent sediment, medium later becomes cloudy.

Lactose bouillon: no acid, no gas, membrane present, flocculent sediment, medium later becomes cloudy.

Saccharose bouillon: no acid, no gas, membrane present, flocculent sediment, medium later becomes cloudy.

Mannite bouillon: no acid, no gas, membrane present, flocculent sediment, medium later becomes cloudy.

Organism K. Class IV of Chester. A micrococcus. Isolated from a one- to three-day hay infusion.

Morphology: small spheres, may occur in chains, cells vacuolated, non-motile.

Agar colonies: grow slowly, moist, edge sharply defined, non-chromogenic.

Staining reactions: Gram positive.

Gelatin: no liquefaction after eleven days, growth uniform and filiform.

Litmus milk: becomes slowly alkaline, slightly decolorized, slightly peptonized, later is completely decolorized and peptonized.

Bouillon: clear.

Dextrose bouillon: no acid, no gas, cloudy.

Lactose bouillon: no acid, no gas, cloudy with sediment.

Saccharose bouillon: no acid, no gas, cloudy, heavy sediment.

Mannite bouillon: no acid, no gas, cloudy with small colonies scattered throughout the medium, cloudiness slowly disappears to be replaced by sediment.

Organism L.   Class XV, group II B, the subtilis group of Chester. Isolated from a nine- to fourteen-day hay infusion.

Morphology: long rods, sometimes single, sometimes in chains of two, endospores central, motile.

Agar colonies: round, moist, glistening, edge sharply defined, slightly indented, granular, non-chromogenic, medium strongly discolored, brown.

Staining reactions: Gram positive.

Gelatin: liquefied rapidly, growth sacculate, filiform.

Litmus milk: initial coagulation followed by decolorization and peptonization.

Bouillon: flocculent cloudiness.

Dextrose bouillon: no acid, no gas, pellicle at top.

Lactose bouillon: no acid, no gas, pellicle at top.

Saccharose bouillon: no acid, no gas, pellicle at top.

Mannite bouillon: no acid, no gas, growth cloudy, especially below membrane.

Organism A'.   Bacterium class VI, group II, of Chester, possibly B. Luteum.   Isolated from a nine- to fourteen-day hay infusion.

Morphology: small rods, non-motile.

Agar colonies: round, convex, moist, opaque, edge slightly irregular, chromogenic, deep yellow.

Staining reactions: dark staining granules at one end, Gram positive.

Gelatin: no liquefaction at first, liquefied after two months, growth best at top, yellow, filiform.

Litmus milk: slightly alkaline at first, slowly decolorizing and peptonizing.

Bouillon: yellow pellicle at top, medium fairly clear with flocculent sediment

Dextrose bouillon: no acid, no gas, slight pellicle.

Lactose bouillon: no acid, no gas.

Saccharose bouillon: no acid, no gas.

Mannite bouillon: no acid, no gas.

Organism B'.   Class VII, group II, of Chester, possibly B. latericium or B. havaniensis.   Isolated from an old culture in which Paramecium were not at the time thriving.

Morphology: short rod, non-motile.

Agar colonies: very slow growing, round, moist, raised, opaque at center, edge indefinite, chromogenic, brick red.

Staining reactions: Gram negative.

Gelatin: not liquefied, growth best at top, spreading, beaded along line of stab.

Litmus milk: neutral at first, followed by decolorization and alkalinity.

Bouillon: red sediment, red growth at top.

Dextrose bouillon: no acid, no gas, flocculent sediment, red growth at top.

Lactose bouillon: no acid, no gas, red growth at top.
Saccharose bouillon: no acid, no gas, red growth at top.
Mannite bouillon: no acid, no gas, red growth at top and red sediment.

Organism C'. A streptothrix belonging to group II, Streptothrix of Chester. Isolated from a nine- to fourteen-day hay infusion.

Morphology: long filaments, fruiting bodies produced by multiple division of a filament, unbranched.

Agar colonies: myceloid, aerial hyphae formed, edge not well-defined, growth rather slow, chromogenic, shell pink, color easily lost.

Staining reactions: Gram negative.

Gelatin: liquefies very slowly, becoming evident after two months, growth slow over top and along line of stab where it is villous with radiatingly extending processes.

Litmus milk: alkaline and slowly peptonizing.

Bouillon: growth flaky.

Dextrose bouillon: no acid, no gas, membrane at top which drops to bottom if shaken, flocculent sediment.

Lactose bouillon: no acid, no gas, growth granular with round colonies gathered at top.

Saccharose bouillon: no acid, no gas, growth granular with round colonies gathered at top.

Mannite bouillon: no acid, no gas at first, but showing tendency to gas formation beneath the membrane which forms at the top.

Organism J'. B. coli. Isolated from a laboratory stock culture.
Organism K'. B. cereus. Isolated from a laboratory stock culture.
Organism L'. B. proteus. Isolated from a laboratory stock culture.

e. Method used for growing Paramecium in pure cultures of bacteria

It is obvious that if we are to cultivate Paramecium in pure cultures of bacteria, the animals must first be rendered bacteria-free, and the cultures must be so handled as to exclude contamination with foreign bacteria throughout the course of the experiments. Hargitt and Fray ('17) devised a method for freeing Paramecium of bacteria by washing the animals individually with sterile water in depression slides placed in Petri dishes. The Petri dishes also served as moist chambers in which the animals could be grown and observed without danger of contamination. Their results show that depression slides are better than ordinary watch-crystals for washing, and that they could rely upon five successive washings for sterilizing a given animal.

Their method is undoubtedly reliable, for their tests were thorough and showed that no bacteria would be carried over by the animal after the last washing, thus excluding the contamination of pure cultures of bacteria from this source. This method was modified by using small watch-crystals, 1 inch in diameter, in place of depression slides. From five to seven of these can be easily baked in a single Petri dish. Five was the number used for carrying the cultures from day to day, whereas seven were employed for washing each animal. These watch-crystals were kept from slipping about by asbestos mats cut so as to fit exactly the bottom of a Petri dish. Each mat was perforated with as many holes as there were watch-cyrstals to be placed in the moist chamber. In this way the individual cultures were secured from slipping, and the entire apparatus could be baked repeatedly. The asbestos, furthermore, served as a sponge to keep the air within the Petri dishes moist.

The use of this apparatus made it necessary to wash each animal seven times instead of five, but this slight disadvantage was so outweighed by the advantages of the method that it was employed throughout the course of the work. A further change in the method of Hargitt and Fray consisted in the use of sterile 0.1 per cent standard hay infusion for washing the Paramecia instead of sterile water. The infusion is easily kept in stock, and its use as a washing fluid lessens the danger of injury during the sterilizing process by avoiding a change of medium at this time.

The pipettes used in handling the Paramecia were made by drawing out soft glass tubing to capillary fineness. The tip of each was annealed so that an animal coming in contact with it would not be injured. The upper end of each pipette was plugged with cotton to prevent contamination with the bulbs which were not sterilized. A large number of these pipettes was kept on hand, and a fresh one used for the transfer of each ani mal from one watch-glass to another. Mason jars were found more convenient holders than the ordinary pipette boxes, which were too long. These jars with their contents were carefully baked before using and, in addition to this, each pipette was carefully flamed before it was introduced into a culture fluid.

The efficiency of this method of washing was tested by plating the contents of each watch-crystal within a washing chamber. It was found that the process of plating could be simplified by pouring the agar directly into the watch-crystals. In this way exceedingly striking demonstrations of the efficiency of the method were secured. It was possible to determine the exact point at which the animals became free from bacteria by absence of colonies on these miniature plates. Thorough testing of this method showed that while colonies would not be present after from three to five washings, it was not safe to rely upon this number, for if the animal were plated with the washing fluid of the fifth plate, an occasional colony would sometimes develop. This did not occur after the sixth or seventh washing, and repeated trials convinced me that seven washings could be relied upon to free Paramecium aurelia from bacteria.

The animals were grown in pure cultures of bacteria or in mixtures of such cultures in Petri-dish moist chambers containing five watch-crystal cultures as just described. The safety of such a method was completely demonstrated by the tests made while these cultures were under observation. The method employed was to take samples of the fluid from each watch-crystal of a group of five, dilute these with sterile tap-water, and plate in the usual way. These plates were allowed to develop at room temperature for from three to four days according to the temperature of the laboratory. In this way it was possible to know whether sufficient contamination was occurring to make rewashing the animals necessary. An occasional foreign colony was not considered evidence of serious contamination, since the great mass of bacteria would still be of the original stock.

Since tests at intervals as great as thirty days showed no evidence of serious contamination, it is probable that it is not necessary to wash the animals more frequently. However, in actual practice it was customary to wash them as often as every three or four weeks, in order to be absolutely sure that contamination was being kept out of the cultures. It is interesting to note that a test made after actual observation of the cultures had been discontinued, and the great care in handling lessened, showed a rather remarkable freedom from contamination after an interval of forty-six days. One of the bacterial mixtures showed complete absence of contamination when tested by plating at this time. These tests demonstrate that it is perfectly possible to avoid undue contamination of pure cultures of bacteria used for feeding Paramecium, and that it is even possible to do this for some time without particular care. In practical work, however, one should always flame each pipette carefully before it is used, and take every possible precaution by sterilizing all media and glassware.

The Paramecia used in all feeding experiments were isolated from a laboratory culture rich in both Paramecium aurelia and caudatum. Paramecium aurelia was chosen because of the greater ease with which it can be handled in small amounts of culture fluid. The usual method of growing pure-line cultures was employed, and stock lines were kept in vials of hay infusion.

Perfectly fresh cultures of bacteria for feeding were secured by inoculating small vials of sterile 0.1 per cent infusions with pure cultures of bacteria. This was done each day, so that a series of bacterial cultures, each twenty-four hours old at the time of using, was obtained. Such cultures insured vigorous, rapidly growing bacteria, and were as free as possible from an accumulation of bacterial decomposition products. Mixtures of these cultures were made as desired in the watch-crystals in which the Paramecia were to be grown. By making the mixtures in this way instead of allowing the bacteria to grow in mixed culture in the vials of infusion, the possibility of unfavorable reactions of the bacteria with one another was reduced to the minimum.

Workers with infusions of unknown bacterial content have supposed that the bacteria carried over in transfer of a Paramecium to fresh medium would provide an ample supply of food. This assumption appears to have been justified. It was desired to use infusions of this sort as controls, but it seemed that this method might allow the gradual introduction of harmful bacteria and that the decline so frequently noted in the vitality of cultures of Paramecium might be due to such a change of food. To avoid this possibility in the chance mixtures, hay infusion was inocu-

lated with a portion of a head of timothy hay every day and
used when twenty-four hours old, thus securing fresh cultures of
bacteria on the hay.   It was believed that in this way a chance
mixture of bacteria would be secured which would be more comparable to that of the fresh infusions in which Paramecia thrive
best when grown in mass culture.   It turned out in actual practice that this assumption was probably incorrect.   Infusions
seeded in this way were excellent for the growth of Paramecium
for four months.   After this the division rate sharply declined
and the death rate increased until the lines in this mixture died,
and it was impossible to replace them.   The possibility is suggested that the actual change in the bacterial content of the
chance mixture was greater than if dependence had been placed
upon inoculation with bacteria carried over with the Paramecia.
This failure of the chance mixture could not have been foretold
during the course of the experiments, since its behavior up to
the actual time of death of the animals was not unlike that of
certain of the cultures of known bacterial content.   Because of
the failure of this chance mixture during the latter part of the
work, the culture C′ was used as the control, the chance mixture
serving as an additional index during the time it was furnishing
adequate food.   This line C′ was chosen as the control because
of its uniform behavior throughout the course of the observations, and because it was a component of all infusions of known
bacterial content used during this time, save in those cases where
it was desired to make a complete change of food.

From five to ten lines of Paramecium were grown in each culture of known bacterial content.   An equal number of animals
were grown in the chance mixture.   Daily transfers were made
to fresh fluid except in a few cases during the winter when the
temperature of the laboratory was so low that bacteria would
not grow rapidly enough to furnish a good medium.   But a single
animal was carried over in each transfer, so that a new series of
cultures was started each day.   Contamination from the air
during these transfers was avoided by raising the cover of the
Petri dish only enough to admit the passage of a carefully flamed
capillary pipette.   Since it was not deemed necessary to wash

the animals more frequently than every few weeks, the only precautions taken against contamination with foreign bacteria in
the daily transfer were those which insured sterile glassware and
sterile media and avoidance of air-borne bacteria during transfer. This procedure enabled me to keep accurate daily records
of at times as many as eighty cultures without infringing upon the
routine of other work.

Careful record of the rate of division was kept, and each death
noted, as it was considered important to have a record of the
death rate as well as the rate of fission in determining the effect
of food upon the animals. Previous work of Woodruff ('11) has
established the fact that one may disregard the effect of light,
ordinary ranges of temperature, and barometric pressure in such
experimental work as this. It was, therefore, not considered
necessary to keep records of these features.

A series of tables was compiled from the daily records by
averaging for three-day intervals. The very minor fluctuations
in division rate which appear in the daily observations, and which
depend so much upon the personal equation of the observer,
disappear when the three-day intervals are used; but the five-day
interval, which was also tried, causes too much removal of small
differences in rate of fission for a thorough study of changes in
metabolism due to the influence of food. The tables show the
average number of divisions for the time covered by a given set
of observations, the percentage of high divisions, the percentage
of low divisions, and the percentage of deaths. These percentages were found to be helpful in the interpretation of results.
The graphs were made from tables in which the averages were
computed for the three-day intervals.

OBSERVATIONS

*a. Growth of Paramecium aurelia in certain pure cultures of
bacteria in 0.1 per cent standard timothy hay infusion*

Nine pure cultures of bacteria and twelve mixtures of these
were fed at different times to pure lines of Paramecium. These
bacteria have already been described as to cultural characteris-

tics, and the letters given in this description will be used in the following account.

During the period from August 18 to August 27, 1920, the bacteria A, B, and C were fed in pure cultures and in the mixtures AB, AC, BC, and ABC. A chance mixture made by inoculating sterile infusion with timothy hay was used as a control. The letter M has been given to this type of mixture throughout the description of the feeding experiments.

The average division rate for M for the entire period was taken as the mean from which the percentages of high divisions, low divisions, and deaths were computed. Percentages of mean and zero divisions do not appear to be of any great help in interpreting results, and have, therefore, been omited from all accounts and tables which follow. Throughout this account the term 'interval' is employed to indicate the short groups of three days which were used in computing the averages of division rates, etc., whereas the term 'period' is applied to the longer durations of time during which the animals were subjected to a given set of conditions. Woodruff ('11) has defined a rhythm as "a minor rise and fall of the fission rate, due to some unknown factor in cell metabolism from which recovery is autonomous." It is in this sense that the term is used in this paper.

The bacteria A, B, C and their mixtures proved to be unsatisfactory food for Paramecium. In no case did the division rate exceed 1.5. The death rate was high in all save A, and it was impossible to get an active metabolism by the use of these bacteria in pure culture or in mixtures. Meanwhile, the control, M, maintained a high metabolic rate.

On August 22, 1920, feeding with the pure cultures, J, K, L, and the mixtures JK, JL, KL, and JKL, was started. This group of cultures was under observation until August 31, 1920, during which time the result was comparable to that gained in attempting to feed A, B, and C. Although the results of feeding mixtures were slightly better than those obtained by the use of pure cultures, the slight advantage thus gained could not be continued, and the animals eventually died.

A third series of cultures was started on September 3, 1920, with the bacteria J', K', L', J'K', J'L', K'L', and J'K'L' as food. The bacteria J', K', and L' were B. coli, B. cereus, and B. proteus respectively. These organisms were unsatisfactory as food. One hundred per cent of deaths occurred during the first interval with L' (B. proteus) as food, and the same result took place when K' (B. cereus) was added to L'. The addition of J' to the mixture seemed to prevent this total mortality.

The results of feeding these three groups of bacteria in pure culture seem to support the contention of Hargitt and Fray ('17) that pure cultures of bacteria, as a rule, do not prove to be satisfactory food for Paramecium. The ability of one type of bacterium to neutralize the harmful effect of another, as illustrated by the combination of J' with L', and the stimulating effect of combining bacteria are interesting problems raised in this work. Reactions of this type will be pointed out as they occur, leaving their analysis for future investigation. The outstanding result of these preliminary experiments, in which nine pure cultures of bacteria were used and twelve combinations of these cultures, is that in no case were they able to maintain the life of Paramecium more than a few days. Moreover, of those cultures which did sustain the life of the animals for a few days, none were able to produce a normal rate of metabolism, and were, therefore, very unsatisfactory food.

b. Behavior of Paramecium aurelia when fed upon pure cultures, and mixtures, of the bacteria A', B', C' under varying experimental conditions

On August 28, 1920, feeding was started with the bacteria A', B', C' and the mixtures A'B', A'C', B'C', A'B'C'. Observations were continued upon certain of these lines until March 7, 1921, with only such interruptions as the experimental conditions demanded. The original pure lines of Paramecia with which this series of experiments was conducted was still in existence in the mixed food A'B'C' and in the pure culture C', on June 2, 1921. Although neither A' nor B' proved to be satisfactory when fed in pure culture, they were exceptionally well adapted to this particular line of Paramecium when fed in combination with C', with the exception of B'C'. During the course of the work with these organisms, the chance mixture M failed to provide suitable conditions for the life of Paramecium. The records for M will be given, but after careful consideration, it has seemed best to take the culture C' as a control for the entire series of observations. C' showed greater uniformity in metabolic rate throughout the entire time than any other culture and is particularly well fitted to serve as an index for comparing the effect of feeding Paramecium with the other bacteria of this group. The chance mixture M serves as an additional index during the time when it was supplying an adequate food.

The tables which follow show the averages for the entire time from August 28, 1920, to March 7, 1921, and for the separate periods included in this time. The observations have included a variety of experimental conditions. These were, 1), the behavior of Paramecium when fed upon the bacteria in pure culture and in mixtures in 0.1 per cent standard timothy hay infusion during the entire length of time from August to March; 2) the effect of change in food without changes in medium; 3) the effect of changes in medium without changes in food; 4) the effect of changes in food and medium; 5) the effect of sterile media uncombined with food. It would be interesting to add to these another condition, namely, the feeding of dead bacteria of known types in sterile infusions. These various experimental conditions will be described in the order given.

1. The behavior of Paramecium when fed upon pure cultures of the bacteria A', B', C' and the mixtures A'B', A'C', B'C', and A'B'C' in 0.1 per cent standard timothy hay infusion from August 28, 1920, to March 7, 1921. The results obtained during this time are summarized in table 1. In addition to the averages of the division rates and percentages of high divisions, low divisions, and deaths, the number of days each line was under observation is given. Differences in duration of the observations of A'C' and A'B'C' are due to the fact that, owing to the time necessary for washing the animals, it was not possible to start all lines at once. In the case of the chance mixture, M, the shorter time is due to the failure of this mixture to support the life of Paramecium during the last two months of the observations. The mean division rate for C′, 1.033, is somewhat lower than is characteristic for Paramecium aurelia when living under favorable conditions. It was therefore thought best to take the mean rate for M, 1.444, as the basis for computing the various percentages.

An examination of table 1 confirms the evidence of the preliminary experiments, namely, that Paramecium does not ordinarily thrive upon pure cultures of bacteria. That it may sometimes do so is evident from the results of feeding with the

TABLE 1

Summary of the results of feeding Paramecium with the bacteria, A′, B′, C′, the mixtures A′B′, A′C′, B′C′, and A′B′ C′, and the chance mixture M, from August 28, 1920, to March 7, 1921

FOOD	AVERAGE NUMBER DAILY DIVISIONS	PER CENT HIGH DIVISIONS	PER CENT LOW DIVISIONS	PER CENT OF DEATH	NUMBER DAYS
A′	0.6	0	37.5	1.2	3[1]
B′	0.333	0	18.2	45.5	4[1]
C′	1.033	18.3	36.3	5.2	130
A′ B′	1.175	11.8	17.6	29.4	3[1]
A′ C′	1.793	40.3	33.7	3.3	131
B′ C′	1.305	37.2	26.4	11.8	27[2]
A′ B′ C′	1.530	29.2	35	7.6	127
M	1.444	22.4	39.7	9.5	92

[1] Died at end of time designated.
[2] Discontinued because unsatisfactory.

streptothrix C′. This organism was used with good effect for a period covering six and one-half months. Nor should it be forgotten that Paramecia from this line were doing well in pure cultures of C′ on June 2, 1921—a total of over nine months.

A transitory stimulating effect on Paramecium was had by combining the bacteria A′ and B′, as is seen by the increase in division rate and percentage of high divisions and the decrease in low divisions and deaths over those for these organisms when fed in pure culture. That this stimulation was very fleeting is evidenced by the fact that the actual duration of life of animals fed upon this mixture was no longer than for those fed with A′ alone.

An entirely different picture is presented by the combination of
A' or B' with C'.   In the combination A'C', a mixture was obtained which proved to be the most satisfactory of all the possible
combinations of these three bacteria.   The average division rate
for the entire time, 131 days, was 1.793.   During that time the
percentage of high divisions was 40.3—a higher mark than was
reached by the other combinations.   The percentage of low
divisions, 33.7, is somewhat higher than some of the others, but
this is more than compensated for by the exceedingly low death
rate of 3.3 per cent.   We have here a very striking example of
the stimulation of the metabolism of Paramecium when fed with
a mixture of a bacterium in itself incapable of maintaining a
high rate of metabolism in Paramecium with one which, when
fed alone, failed utterly to support life.   That this increase in
the metabolic rate is of real significance is shown by computing
the probable error of the constants of the two uncorrelated series,
C' and A'C', according to the formula $\sqrt{(E_1)^2 + (E_2)^2}$.

Here, E_1 and E_2 represent the probable errors in the mean
division rates of C' and A'C', respectively, when computed according to Peter's formula for the probable error (Huntington,
'18), namely:

$$E = \frac{0.08453}{n\sqrt{n-1}} (v_1 + v_2 + v_3 + \qquad\qquad v_n)$$

It is well known in biometry that if the probable error of the
difference in the constants of two uncorrelated series is contained
more than three times in the numerical value representing the
difference in the means under the two experimental conditions,
the results are of statistical significance.   For such a quotient I
have used the expression 'significance factor,' a term suggested by
Doctor Obreshkove.   It is evident that the greater this factor,
the greater the value of the numerical data.   In the case of C' and
A'C', the difference in the mean division rates is 0.760.   The
probable error of the mean rate of division for C' is 0.0462; that
for A'C' is 0.0624.   Applying the formula, $\sqrt{(E_1)^2 + (E_2)^2}$,
the probable error of the difference in the constants of these two
series, C' and A'C', is 0.775.   This is contained in the difference

in the mean division rates (0.0760) 9.832 times. This whole reaction may be expressed as follows:

$$\frac{1.793 - 1.033}{\sqrt{(0.0462)^2 + (0.0624)^2}} = \frac{0.760}{0.0775} = 9.832.$$

Therefore, the increase in metabolism, as indicated by the increased division rate for A'C', is significant, and demonstrates that the addition of A' to C' produced a food for Paramecium of much greater value than C' alone. The data from which the significance factors were derived are given in table 2.

TABLE 2

Mean division rates with their probable errors, differences in mean division rates with probable errors of the differences, and significance factors for the data summarized in table 1

FOOD	MEAN DIVISION RATE	PROBABLE ERROR	DIFFERENCE IN MEAN DIVISION RATE		PROBABLE ERROR OF DIFFERENCE	SIGNIFICANCE FACTOR
C'	1.033	0.0462	C' & A' C'	0.760	0.0775	9.832
A' C'	1.793	0.0624	C' & A' B' C'	0.497	0.0863	5.758
A' B' C'	1.530	0.0732	C' & M	0.411	0.110	3.736
M	1.444	0.0997	A' C' & A' B' C'	0.263	0.0960	2.635
			A' C' & M	0.349	0.121	2.88
			A' B' C' & M	0.086	0.122	0.704

When B' was combined with C', a food was obtained, which, although far more satisfactory than B' in pure culture, nevertheless failed to maintain a high rate of metabolism in Paramecium. This mixture was, therefore, discontinued after twenty-seven days' trial.

When to C' were added both A' and B', a fairly satisfactory food was obtained, the average daily division rate being 1.530 for 127 days. The significance factor for A'B'C' and C' is 5.758, showing the mixture to be quite superior to C' alone. If we compare A'C' and A'B'C', we find a significance factor of 2.635, which indicates no real difference in the relative food values of these two mixtures. Other data than division rate suggest that A'C' may be slightly better than A'B'C'.

The mixture M appears to be less favorable than either A'C' or A'B'C', for the metabolic rate of Paramecium is lower than with either of the artificial mixtures, the daily rate for M averaging 1.444 for ninety-two days. But the differences between these mixtures are not sufficiently great to have any meaning, since we find the significance factors as follows: M and A′C′, 2.880; M and A′B′C′, 0.704. If we consider the control mixture M only for the first three periods, during which time it supported a higher metabolic rate in Paramecium, we find no ground for changing the conclusions already given. For these three periods M was superior to C′, while M, A′C′, and A′B′C′ did not show sufficient difference to give a significance factor equal to 3 in any case.

These figures thus make it clear that the mixtures of bacteria maintain a higher rate of metabolism in Paramecium than do the pure cultures of the bacteria comprising the mixtures. There appear to be differences in the mixtures, for the daily rate of division would indicate that A′C′ was best, then comes A′B′C′, and M last. But the biometrical tests applied showed the differences to be too small to warrant any belief in the marked superiority of one mixture over the others.

The time during which the various cultures of the bacteria A′, B′, and C′ were used for feeding was divided into five periods, because of the necessity for occasional washing of the animals and because, for the purposes of this investigation, it seemed wise to avoid experimental work during endomixis. The phenomena of endomixis are accompanied by a marked slowing of the division rate, and it was thought that the effect of the different kinds of food would not be so evident at these times. The occurrence of fragmentation of the macronucleus as shown in stained specimens was taken as evidence of endomixis, and this was determined as taking place in this strain of Paramecium every thirty to thirty-five days.

It was found to be well-nigh impossible to revive the animals after the first endomixis period, which began September 28, 1920. Vigorous, rapidly dividing animals were not secured until October 10th. I have been at a loss to account for this condition of low vitality, save that it was in some way the result of environment. The animals passed through this endomixis in the control mixture

M and only a single animal survived. All the Paramecia used in the subsequent experiments were derived from this single animal. It is interesting to note that no such period of marked depression occurred later among the animals fed upon the artificial mixtures of bacteria, which seemed to be efficient in maintaining the normal rhythm emphasized by Woodruff and Erdmann ('14). The question immediately arises: Are we dealing in the case of the artificial mixtures with a type of food, capable in itself of preventing the periods of marked depression noted by Calkins ('02 a), and avoided by Woodruff by the use of a 'varied culture medium'? This could be determined only by attempting to feed Paramecium such a mixture many months, and is a problem in itself.

Since, as is indicated in table 1, the bacteria A', B' and the mixture A'B' did not prove to be satisfactory as food, they have not been included in the tables and graphs which follow.

TABLE 3a

Summary of results of feeding Paramecium aurelia in 0.1 per cent standard timothy hay infusion with the bacteria C', A' C', B' C', A' B' C', and M, for period I, from August 28, 1920, to September 28, 1920

FOOD	AVERAGE NUMBER DAILY DIVISIONS	ER CENT HIGH P DIVISIONS	PER CENT LOW DIVISIONS	PER CENT OF DEATH	NUMBER OF DAYS
C'	0.967	20.2	43.5	3.3	31
A' C'	1.712	61.1	13.3	3.3	31
B' C'	1.305	37.2	26.4	11.8	27
A' B' C'	1.710	52.9	28.8	1	28
M	2.047	81.1	6.6	0.4	31

TABLE 3b

Summary of results of feeding Paramecium with bacteria C', A'C', A'B'C', and M, in 0.1 per cent standard hay infusion for period II, from October 20, 1920, to November 24, 1920

FOOD	AVERAGE NUMBER DAILY DIVISIONS	PER CENT HIGH DIVISIONS	PER CENT LOW DIVISIONS	PER CENT OF DEATH	NUMBER OF DAYS
C'	1.006	20.3	32.5	5.6	35
A' C'	1.65	46.7	29.4	2.2	35
A' B' C'	1.433	35	28.5	10.4	34
M	1.214	27.3	28.6	12.3	32

TABLE 3c

*Summary of results for period III, from December 5, 1920, to December 28, 1920.
Food and media as in table 3b*

FOOD	AVERAGE NUMBER DAILY DIVISIONS	PER CENT HIGH DIVISIONS	PER CENT LOW DIVISIONS	PER CENT OF DEATH	NUMBER OF DAYS
C′	1.13	21.6	33 6	4.8	21
A′ C′	2.108	59.8	16.3	7.7	21
A′ B′ C′	1.883	50.8	18.2	11.3	21
M	1.098	23.4	36.3	7.3	20

TABLE 3d

*Summary of results for period IV, from January 12, 1921 to February 5, 1921.
Food and medium as above*

FOOD	AVERAGE NUMBER DAILY DIVISIONS	PER CENT HIGH DIVISIONS	PER CENT LOW DIVISIONS	PER CENT OF DEATH	NUMBER OF DAYS
C′	0.906	15.5	34.8	7	20
A′ C′	1.303	38	29.3	3.8	21
A′ B′ C′	1.168	28.3	28.3	4.9	21
M	1.058	9.4	10.8	37.7	6[1]

TABLE 3e

*Summary of results for period V, from February 13, 1921, to March 7, 1921. Food
and medium as above*

FOOD	AVERAGE NUMBER DAILY DIVISIONS	PER CENT HIGH DIVISIONS	PER CENT LOW DIVISIONS	PER CENT OF DEATH	NUMBER OF DAYS
C′	1 254	32.2	37 4	4 3	23
A′ C′	2.084	69.1	15 2	1	23
A′ B′ C′	1.384	40 2	7.1	10.2	23
M	0 727	23.1	15.3	15.4	3[2]

[1] Died and restarted after a week's interval, then died again.

[2] Started three days before end of period. Showed low vitality.

Tables 3, a to e, demonstrate that the chance mixture M showed a steady decline in all factors considered.   This decline, however, was not evident during the daily course of the work, for, as the graphs show, such fluctuations as occurred would naturally be attributed to normal rhythms, and were not noticeably different from those occurring in the artificial mixtures.   It was only when all evidence was collected and the final averages made that it could be seen that there was in reality a steady decline in the power of the chance mixture M to maintain a normal rate of metabolism in Paramecium, and that the seemingly abrupt failure of this mixture was, after all, only apparent.

The streptothrix C', the only organism of those tested which proved to be a satisfactory food when used in pure culture, is a filamentous organism, the individual filaments of which are too large for Paramecium to ingest. The animal was, therefore, forced to subsist upon the fruiting bodies or upon very young filaments. The questions might be raised: Was not the low metabolic rate of the animàls fed with this organism due to an insufficient amount of food? Was the supply of fruiting bodies adequate? It is probable that the food was quite adequate in amount and the increases in activity noted for these animals were not due to any increase in the reproductive power of the streptothrix. It is far more likely, since they occurred synchronously with similar increases in other cultures, that they were associated with rhythms and had no direct relation to food.

The mean division rate for C' for the entire series of observations was 1.033. An inspection of table 3, shows that the division rate fell below this mean three times; at the beginning of the experiments and during the second and fourth periods. During the other periods it was slightly in excess of the mean rate. On the whole, it showed less variation from the mean than the other two artificial mixtures. The fourth period, during which the rate dropped below the mean, was marked by a fall in metabolic rate in all cultures. It is possible that the greater range in temperature combined with an unsatisfactory condition of M was sufficient to cause the death of the animals in this mixture.

When $A'C'$ is compared with C', it is evident that the addition of A' to C' caused a marked acceleration or stimulation of the metabolic rate of Paramecium. The division rate exceded that of C' during all the periods. The percentages of high divisions, low divisions, and deaths show an increased metabolism. There was more fluctuation in the death rate for $A'C'$, and it was greater than for C' during the third period. It was found in computing the percentages for the three-day intervals that an increased death rate sometimes accompanied an increased metabolism. This suggests that an accelerated rate of metabolism may be paralleled by an excessive mortality, as if the unusual expenditure of energy was destructive to weak individuals. With the exception of these occasional increases, the death rate of A'C' remained less than that for C'. The significance factor for the difference between C' and A'C', 9.832, indicates a marked acceleration of metabolism in animals fed with A'C'.

The analysis of the conditions obtaining in the feeding of the above mixtures is aided by a series of graphs of the division rates averaged for three-day intervals during the five periods of observation. An inspection of these graphs shows the superiority of the A'C' mixture. It is true that for the first ten intervals, the average division rate, 1.712, was less than that for M, which was in excess of 2 (table 3 a). A fission rate greater than 2 is unusual for Paramecium aurelia, yet the rate for A'C' exceeded 2, twelve intervals out of a total of forty-four. The rate for the mixture A'B'C', used for a total of forty-two intervals, exceeded 2 nine times, thus approximating A'C'. The division rate of the control, C', did not exceed 2 in any of the averages for three-day intervals. Although the daily rate rose above this figure or equaled it 174 times between August and March, these high rates did not occur with sufficient frequency to bring any single average for a three-day interval up to 2.

Certain minor fluctuations in the rhythms are to be seen on an inspection of the graphs. These fluctuations are independent of those for the three-day intervals and of the occurrence of endo mixis. They are evidence of an alternating elevation and depression of the metabolic rate which seems to be independent of food. They occur in all cultures and are fairly synchronous irrespective of the type of food used. The temperature of the laboratory was fairly constant; that for the liquid in which the animals were living was necessarily more constant than the air of the room, and this even temperature obtained throughout the course of the work save for a part of the winter when there was some variation. Since these minor fluctuations occurred during the time of even laboratory temperature as well as when it was more variable, it would seem that they cannot be entirely due to this factor. In the case of a satisfactory food like A'C', the degree of fluctuation tends to be greater than with a food like C', which is incapable of sustaining a high metabolic level in Paramecium. Aside from this fact, no definite correlation seems to exist between the occurrence of these fluctuations and the type of food used. They appear to be due to intrinsic factors affecting the metabolism irrespective of food or to slight variations in the environment not under control.

It is of interest to note that the chance mixture M, which throughout the first period appeared to be the most satisfactory kind of food used, showed the least degree of fluctuation of any of the cultures under observation during this period. That this mixture was comparable at this time to any laboratory infusion in which Paramecia thrive, there can be no reasonable doubt. That it did not continue to furnish conditions favorable to the continued life of Paramecium has been demonstrated. The question arises as to whether the high rate of metabolism of the animals in M during this period was due to a preponderance of favorable types of bacteria which prevented the increase of unfavorable types. Since it is well known that bacteria vary in their resistance to drying, is it not possible that some of the more resistant forms are unsuitable food for Paramecium?

The foregoing observations may be summarized as follows: The data dealing with the behavior of Paramecium when fed in a standard 0.1 per cent timothy hay infusion upon a diet consisting of either a pure culture of C', or its mixtures A'C' and A'B'C', and of the chance mixture M, demonstrate that it is possible for this animal to live upon a single article of diet, although the metabolism under such conditions is not so high as when a mixed diet is used. The chance mixture M was very much more satisfactory during the first period than either of the artificial mixtures, but such a chance mixture is not easily kept constant, and under the conditions of these experiments failed to maintain a continued normal rate of metabolism in the animals fed upon it. Of the artificial mixtures, A'C' proved to be the most satisfactory and

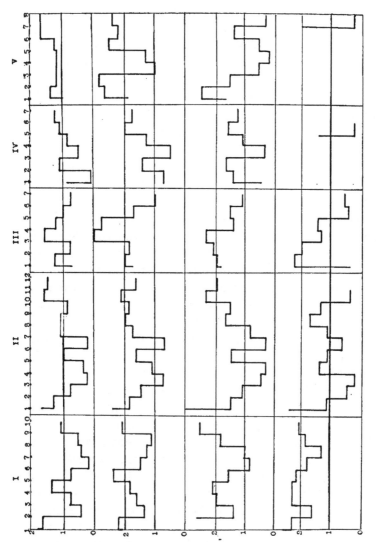

Fig. 1 Graphs showing behavior of Paramecium when fed with bacteria C, A'C', A'B' C, and M, August 28, 20, to Feb 7, 1921. Roman numerals for to periods and and as follows: I, Aug. 28 to Sept. 28; II, Oct. 20 to Nov. 24; III, Dec. 5 to Dec. 28; IV, Jan. 12 to Feb. 5; V, Feb. 13 to March 7. age division rates ted on ordinates number o three- ly intervals on abcissas.

appears to be capable of maintaining a normal rate of metabolism over an extended period.

2. *Change of food of Paramecium without change of medium.* This section of the work consisted of two parts. First, Paramecia which had been fed for a long time upon the bacteria C', A'C', A'B'C', and M were shifted from one such food to another, as, for example, animals fed upon C' were washed and then placed in the A'C' mixture; the presumably unfavorable foods, A' and B', were added to M, and other such shifts made as described below. The second part of the work consisted in an entire change of food. Animals which had been fed upon one of these combinations were washed and placed in pure cultures of bacteria used in the earlier part of the work. Table 4 gives the data for the first type of food change, and in table 5 are the probable errors and significance factors computed from the data of table 4. The original cultures were used as controls, and the change of food is indicated by changing the order of the letters used, e.g., cultures containing animals changed from C' to a combination of A' with this bacterium are designated as C' A', to distinguish them from A'C', the control of such a mixture. The culture, C', serves as a control for all the mixtures used.

It is evident from the data in table 4 that the change from C' to C'B' resulted in a serious decrease in the vitality of the animals, for they could be kept alive but nine days in this mixture.

Animals changed from C' to C'A' appeared to experience an increase in metabolism. This increase was too slight to be due to the change in food as is suggested by the significance factor, 1.985. The disturbing effect of the change in environment was not overcome in the time during which the animals were under observation. The division rate failed to reach that of A'C', and it is evident that in the change from C' to a food presumably more satisfactory, the mere fact of the change tended to counteract the effect of the more favorable food to such an extent that the animals failed to show a marked reaction to this type of variation in the environment. The significance factors for all the other changes save MA'B' tend to show the same thing. In the case of MA'B', however, the significance factor, 3.965, indicates that the addition of A′ and B′ to the chance mixture had a stimulating effect.

TABLE 4

Data for the growth of Paramecium aurelia in the foods C′, A′ C′, A′ B′ C′, and M, with the following changes in food; C′, to C′ A′, C′ B′, C′ A′ B′; A′ C′, to A′ C′ B′; M, to MA′, MB′, and MA′ B′, during the time from December 5, 1920, to December 28, 1920

FOOD	AVERAGE NUMBER DAILY DIVISIONS	PER CENT HIGH DIVISIONS	PER CENT LOW DIVISIONS	PER CENT OF DEATH	NUMBER OF DAYS
C′	1.13	21.6	33.6	4.8	21
A′ C′	2.108	59.8	16.3	7.7	21
A′ B′ C′	1.883	50.8	18.2	11.3	21
M	1.098	23.4	36.3	7.3	20
C′ A′	1.535	50.6	18	10	20
C′ B′	0.794	10.4	44 8	27.6	9
C′ A′ B′	1.880	52.6	15.9	7.9	20
A′ C′ B′	1.936	57.1	14.3	7.2	18
MA′	1.177	29	29	10.1	19
MB′	1.233	30	30	7.5	20
MA′ B′	1.504	38.6	26.3	7	16

TABLE 5

Mean division rates with their probable errors, differences in mean division rates with probable errors of the differences, and significance factors for the data summarized

FOOD	MEAN DIVISION RATE	PROBABLE ERROR	DIFFERENCE IN MEAN DIVISION RATE			PROBABLE ERROR OF DIFFERENCE	SIGNIFICANCE FACTOR
C′	1.13	0.0911	C′ and	A′ C′	0.978	0.1072	9.140
A′ C′	2.108	0.0180	C′ "	A′ B′ C′	0.755	0.1453	5.193
A′ B′ C′	1.883	0.1132	C′ "	C′ A′	0.405	0.2045	1.985
M	1.098	0.1735	C′ "	C′ A′ B′	0.653	0 2417	2.709
C′ A′	1.535	0.1831	A′ C′ "	A′ C′ B′	0.172	0.1667	1 036
C′ A′ B′	1.880	0.2239	M "	MA′	0.230	0.2771	0.830
A′ C′ B′	1.936	0.1570	M "	MB′	0.427	0.2431	1.757
MA′	1.177	0.2162	M "	MA′B′	1.150	0 2902	3.965
MB′	1.233	0.1705					
MA′ B′	1.504	0.2327					

This experiment shows that changes of the sort attempted are usually accompanied by so much disturbance that no marked increase in metabolism results and the animals fail to reach the metabolic level of their respective controls. Although these results indicate that it is desirable to maintain a constancy of satisfactory food, one would not be justified in maintaining that this is always the case. Further investigation of such phenomena as have been described is needed.

After testing the effect of such changes in food, it was decided to make a more radical change and determine how this particular line of Paramecium would react toward certain of the bacteria which had previously been found to be unsatisfactory. The bacterial cultures J, K, and L were used separately and in combination with one another. They proved to be as unsatisfactory as when previously used, the change from a favorable food like A′C′ being immediately followed by a marked depression of the division rate and a speedy death of the animals.

Under the conditions of the above experiments, change in food of Paramecium had one of two effects dependent on the nature of the food. When the change was from one satisfactory type to another, the evidence was in favor of a constancy of food under the conditions obtaining in this experiment. If, however, the change was from a markedly satisfactory to a decidedly unsatisfactory type of bacteria, a sudden lowering of the metabolic rate occurred, resulting in the speedy death of the animals.

3. Changing the medium without varying the food. Since so much emphasis has been laid by workers with Paramecium upon the nature of the medium in which the animals were cultivated, it was considered desirable to try the effect of a change in medium without varying the food, in order to throw more light, if possible, on the relation of this animal to the medium when it was furnished with a food which long trial had shown to be satisfactory. The media used were made according to the same formula as the standard hay infusion, and used in the same way for growing the bacteria used for feeding. The first infusion tested was made from dry, uncured swamp hay, and the results of using this are summarized in table 6.

It will be seen from a study of this table that any variation in the division rate of the animals in the new medium from that of those in the standard hay infusion was so slight as to be of no importance. The fact that M failed during the last period of this experiment when used in the new medium is not necessarily

TABLE 6a

Showing the results of feeding Paramecium with the bacteria C', A' C', A' B' C', and M, in the 0.1 per cent standard timothy hay infusion and in 0.1 per cent uncured swamp-hay infusion, during the time from November 6, 1920, to November 26, 1920. Data for the swamp-hay medium in the second row in each case

FOOD	AVERAGE NUMBER DAILY DIVISIONS	PER CENT HIGH DIVISIONS	PER CENT LOW DIVISIONS	PER CENT OF DEATH	NUMBER OF DAYS
C'	1.184	18.4	32.5	6.6	18
	1.144	18.8	37.1	4	18
A' C'	1.677	42.3	30.6	2.4	18
	1.536	40.9	31.3	4.5	18
A' B' C'	1.506	38.8	29.1	2.4	18
	1.546	41.5	36.3	2.3	18
M	1.315	21.1	37.4	14.3	16
	1.386	11.8	32.7	13.7	16

TABLE 6b

Showing results of feeding Paramecium with the same bacteria in the same media as in 6a, during the time from December 5, 1920, to December 28, 1920

FOOD	AVERAGE NUMBER DAILY DIVISIONS	PER CENT HIGH DIVISIONS	PER CENT LOW DIVISIONS	PER CENT OF DEATH	NUMBER OF DAYS
C'	1.138	28.6	33.6	4.8	21
	1.186	26.7	20.4	7.9	21
A' C'	1.877	59.8	16.3	7.7	21
	1.861	58	21.3	3.6	21
A' B' C'	1.868	50.8	18.2	11.3	21
	1.771	46.5	19.4	10	16
M	1.386	23.4	36.3	7.3	20
	1.45	32.1	8.9	26.8	7

TABLE 6c

Summarizing the data contained in tables 6a, and 6b, for the time from November 6, 1920, to December 28, 1920

FOOD	AVERAGE NUMBER DAILY DIVISIONS	PER CENT HIGH DIVISIONS	PER CENT LOW DIVISIONS	PER CENT OF DEATH	NUMBER OF DAYS
C′	1.162	19.8	32.9	5.9	39
	1.113	22.9	28.4	6.2	39
A′ C′	1.756	49.4	24.7	4.6	39
	1.705	49.9	26	4.1	39
A′ B′ C′	1.695	45.5	23.2	7.3	39
	1.652	43.9	28	6.1	34
M	1.348	22.1	36.9	11.1	36
	1.478	21.9	20.8	20.2	23

due to a depressing effect of the medium. It is just as probable that the change in environment in a culture which was becoming unsatisfactory was enough to cause the death of the animals. Had the food in the mixture M been as satisfactory as was the case when this was first used, it is not likely that the death rate in the new medium would have been greater.

The second infusion tested was made of entirely different material, the dry, uncured, succulent moneywort, Lysimachia nummularia L. It presumably had a different chemical composition than either of the other infusions used. As in the case of the uncured swamp hay, it was tested in connection with controls in the standard 0.1 per cent hay infusion. The effect of using this infusion is shown in table 7.

The first impression one gains from a study of this table is that the change from hay infusion to moneywort produced a marked increase in metabolism of the animals thus treated. The significance factors for these changes do not indicate that this was the case save for the mixture A′B′C′. However, since there was a stimulating effect of the medium in this one case, we are justified in concluding that this particular change in medium tended to be stimulating rather than depressing or without effect.

During the second period of observation the animals passed through an endomixis. This was accompanied by a lowering of the division rate, as was to be expected, and after recovery the rate for the animals in the new medium remained slightly

TABLE 7a

Showing results of feeding Paramecium with the bacteria C', A' C', A' B' C', and M in the 0.1 per cent standard timothy hay infusion, and in 0.1 per cent dry uncured moneywort infusion, during the time from January 12, 1921, to February 5, 1921. Data for the moneywort infusion in the second line in each case

FOOD	AVERAGE NUMBER DAILY DIVISIONS	PER CENT HIGH DIVISIONS	PER CENT LOW DIVISIONS	PER CENT OF DEATH	NUMBER OF DAYS	SIGNIFICANCE FACTOR
C'	0.906	15.5	34.8	7	20	2.453
	1.485	52	16.7	2.5	20	
A' C'	1.303	38	29.3	3.8	20	2.647
	2.023	54.5	15	9	20	
A' B' C'	1.168	28.3	28.8	4.9	21	4.080
	2.184	58.5	23.5	5.5	21	
M	0.764	9.4	20.8	37.7	6	
	0.643	3.5	24.6	42.1	6	

TABLE 7b

Showing the results of feeding Paramecium with the same bacteria in the same media, as in table 7a, during the time from February 13, 1921, to March 7, 1921

FOOD	AVERAGE NUMBER DAILY DIVISIONS	PER CENT HIGH DIVISIONS	PER CENT LOW DIVISIONS	PER CENT OF DEATH	NUMBER OF DAYS	SIGNIFICANCE FACTOR
C'	1.390	10.4	51.3	4.3	23	
	1.296	6.1	58.3	0.8	23	
A' C'	2.084	47.8	18.2	1	23	2.168
	1.505	27	24.3	5.2	23	
A' B' C'	1.390	15.7	25.9	10.2	23	
	1.162	14.6	36.6	19.5	23	
M	0.727	0	15.4	15.4	3	
	0.455	0	38.5	15.3	3	

TABLE 7c

Summarizing the data contained in table 7a and b, for the time from January 12, 1921, to March 7, 1921

FOOD	AVERAGE NUMBER DAILY DIVISIONS	PER CENT HIGH DIVISIONS	PER CENT LOW DIVISIONS	PER CENT OF DEATH	NUMBER OF DAYS
C′ {	1.094	13.5	41.7	6	43
	1.399	32.4	34.3	2	43
A′ C′ {	1.591	41.5	25.1	3.1	43
	1.832	44.4	18.5	7.6	43
A′ B′ C′ {	1.234	23.9	27.5	6.7	44
	1.916	45.7	27.3	9.6	44
M {	0.755	9.1	19.7	31.8	9
	0.595	3	27	37	9

below that for those in the hay infusion. That this depression was not due to the changed medium is shown by the significance factor for A′C′, the only mixture in which the depression appeared to be at all marked.

When the data for the two observation periods are combined, it is seen that on the whole this change in medium was without significant effect. In general it may be said that changes in medium which were tried had no appreciable effect over a considerable time. The only result of change in medium of any importance was that of a slight transitory stimulation of the metabolism in animals changed to the moneywort infusion.

4. Change in medium accompanied by change in food. In order to determine whether a change in medium would produce an effect upon the metabolism of Paramecium if the food was changed at the same time, animals were washed and placed in 0.1 per cent standard timothy hay infusion and in 0.1 per cent moneywort infusion with the bacteria J, K, L and the mixtures JK, JL, KL, and JKL as food. The culture C′ was used as control. It was found that the animals in the moneywort infusion did not show any significant increase in rate of metabolism, but the duration of life in the moneywort was in most cases slightly longer. In the hay infusion none of the animals fed with the bacteria J, K, L, or their combinations, lived over six days, whereas many of those in the moneywort with the same food lived seven days. There was very little effect which could be ascribed to the change in medium, but the increased length of life mentioned may indicate a slight tendency of the moneywort to stimulate. However, in case this was so, the effect was not great enough to overcome the unfavorable nature of the bacterial food.

This experiment indicates that a change of medium in the presence of a food known to be unsatisfactory does not so affect the animals that they are able to utilize such food for any length of time. It would seem, therefore, that whereas the stimulating effect of a changed medium may be an aid in reviving cultures of Paramecium which are not in a thriving condition, the nature of the food is at least as important, and if this be unsuited to the requirements of the animal, no amount of artificial stimulation by altering the medium will prevent the death of such cultures.

5. *The effect of sterile media.* In order to determine whether the results obtained by feeding certain bacteria which did not support active metabolism in Paramecium were due to the toxic action of such bacteria or merely to the fact that they were for some reason not utilizable for food, certain individuals were thoroughly washed and placed in sterile media. The average length of life in hay infusion was 2.15 days, in uncured swamp-hay infusion, 2.54 days, and in moneywort infusion, 1.72 days. The maximum duration of life for animals in the first infusion was 4 days, for those in the second, 5, and for those in the moneywort, 4 days. The minimum duration of life in all three infusions was the same—less than twenty-four hours. Five lines were tested in each case, and controls carried in C' during this time showed the normal rate of metabolism for animals fed upon this organism.

These tests show that Paramecium aurelia will not live for any length of time in sterile media of the type ordinarily used in the laboratory. It is, therefore, to be assumed that the failure of the animals in the preliminary experiments to live in pure or mixed cultures of the bacteria used, was due in most cases to starvation, rather than to any toxicity of the bacteria used for food. In such cases as the cultures L′ and K′L′, however, it is very probable that the bacteria were toxic, since all Paramecia died within twenty-four hours.

HISTORICAL AND DISCUSSION

Whatever the title of the article or scope of the investigation, all who have worked with Paramecium have been concerned directly or indirectly with some phase of the life-history of this animal. Earlier workers did not conceive the complexity of. function and behavior revealed by later studies of its metabolic processes. The sum total of these results reveals a striking similarity to the vital processes of many-celled organisms. The food of an animal is not the least important factor in determining its behavior, yet this has sometimes been ignored, sometimes obscured by too great emphasis of other factors by workers with Paramecium, and still remains to be clearly understood.

The reason for this neglect lies in the fact that since the food of Paramecium consists mainly of bacteria, any study of this factor must necessarily overlap the field of bacteriology. Accordingly, the zoologist has left it alone almost completely, either because of lack of technical skill or because of hesitation to enter unknown territory. Hargitt and Fray (′17) have devised a technic which makes possible the study of the relationship of Paramecium to food. The technic of getting pure cultures of bacteria is the only part of the work essentially new to the protozoologist, and this can easily be learned.

One of the most striking things met with in reading the literature is that the earliest worker to apply experimental methods to the study of Paramecium (Maupas, ′88) had possibly a clearer idea of the importance of food than any who followed him until a few years ago. He says: "Possibly the use of Pasteur's methods for culturing bacteria would prove to be more suited to the needs of the Ciliates, but I have never made any attempt in this direction." It seems strange that this very pertinent suggestion should have been so long overlooked, but again it may be that the technical difficulties of obtaining such cultures discouraged workers from entering this field and forced them to be content with more or less general methods of food control.

Meissner ('88) was one of the first to investigate the food habits of Protozoa. His experiments dealt with the digestibility of the various types of food, such as starch grains, oil drops, and particles of albumen. His work is of importance in demonstrating that the Protozoa are not essentially unlike the Metazoa in their power to utilize food. Moreover, Meissner showed that not all types of food are digested with equal ease, and some not at all. These findings are of significance in suggesting that variations in digestibility may exist in the normal food of these animals which are of importance when one is attempting to study their behavior.

Environmental factors other than food have received the greater amount of attention of investigators of the Infusoria. Of such factors, the nature of the medium has been perhaps too much emphasized. That the medium is important, there can be no doubt, but if it is of such a nature as to fulfill the osmotic requirements of animals living in it, and to furnish the proper salt content, it would seem that its importance, as far as Paramecium is concerned, lies in its ability to furnish a proper food for the bacteria upon which this animal feeds. Calkins at first ('02 a) thought of the medium as being of main importance, but he soon came to appreciate the relation of bacteria ('02 b). His statement that B. subtilis was the chief food of his Paramecia is probably not justified. We now know that the variety of bacteria living in an ordinary hay infusion is not confined to a single group. Moreover, bacteriologists have subdivided the group of bacteria known as B. subtilis until it includes a large number of forms. Not all of these will support the life of Paramecium, as is shown by a comparison of my results with those of Hargitt and Fray ('17). Calkins was careful to sterilize his media by heating to 90°, but we now know that this temperature is not sufficient to kill many spores. No attempt was made to exclude air-borne bacteria which he assumed furnished the greater part of the bacterial food. He may, therefore, have been using an infusion in which deleterious bacteria came to predominate.

Woodruff ('08) was led to believe that the death of Calkins'
animals might have been due to too constant a medium, and
accordingly began the cultivation of Paramecium in what he
termed a 'varied medium.'   He believed that by collecting material at random, he would provide infusions which would more
nearly approximate the natural environment of this animal
than a constant medium of hay infusion.   The conclusion of the
procedure was truly remarkable, resulting as it did in the continuing of a single line for several years without a conjugation
and a very complete analysis of the life-history.

Later, Woodruff and Baitsell ('11 a) were successful in cultivating Paramecium in a constant medium of beef extract.   This
work established that mere constancy of medium, provided the
fluid used was suitable, did not interfere with the normal course
of the life-history of the animal.   A constant medium did as
well as one which was continually varied.

None of these investigators were working with a closely
controlled food.   They were, therefore, continually dealing with
an important unknown factor.   With a known food it is possible
to test further the effect of medium upon Paramecium.   Two
types other than the 0.1 per cent standard timothy hay infusion
were used in this work.   The infusion made with uncured swamp
hay differed from that made with cured timothy hay in that it
was prepared from a mixture of grasses which had grown in a
swampy lowland district, and was presumably different chemically from the standard infusion.   The moneywort differed even
more.   It was uncured when used, was extremely succulent to
start with, and much richer in cell sap than either type of grass.
The method used in preparing these infusions was the same, so
that gram for gram, equal concentrations were used.   The result
showed that there was no significant difference in effect of the
two grass infusions.   The moneywort tended to have a slight
stimulating effect for a short time, as shown in the case of A'B'C',
and in the fact that when used with such unfavorable foods as
J, K, and L, the duration of life of the animals was somewhat
greater than in the standard hay infusion.

Since the animals fed with C′, A′C′, and A′B′C′ showed no tendency toward diminished vigor for a period of more than six months in the standard hay infusion, it is possible that the diverse results obtained by Woodruff and Calkins may have been due to fundamental differences in the bacterial content of their infusions rather than to the nature of the media or to the old age of the Paramecia.   May it not be that the hay infusion is not so well suited to the maintenance of a favorable mixture of bacteria as is beef extract, and that this accounts for Woodruff's results with this medium?   Granted that substances do occur in media which are capable of stimulating Paramecia for a time, is it not more likely that the most important factor in the environment is food?

The full importance of food in the behavior of Paramecium was first recognized by Jennings ('08).   He not only recognized that cultural conditions should be identical for different series of animals, but he took special precautions to make them so. He realized that the nature of the bacteria in a given culture was very important, for he says: "It is not sufficient to attend merely to the basic fluid, the bacteria in the culture must be the same."   His methods were not directed toward a determination of the exact nature of this food, and he was probably correct in maintaining that if precautions were taken as to the method of making cultures, frequent changes of the animals, and the like, a reasonably constant mixture of bacteria would be obtained, in which enough favorable types would be present to effect a normal metabolism in the animals observed.

The first account of which we have record of feeding Paramecium with a particular bacterium is that of Popoff ('10).   He speaks of feeding Paramecium caudatum with cultures of B. proteus mirabilis grown upon potato.   Popoff's attention was centered upon the effect of various media on cells.   In this case the medium used was ammonia-rich water.   He supposed that the food was uniform, but makes no mention of having freed the animals of other bacteria.   These Paramecia lived but a few days, and their death was attributed by Popoff to the nature of the medium.   Since my work has shown that Paramecium aurelia will not live upon B. proteus, it would seem that Popoff's results may have been due to the effect of unfavorable bacterial food rather than the result of an unfavorable medium.

Any experiments dealing with feeding Protozoa pure cultures of bacteria necessarily involve a special technique. It is, however, not difficult for the protozoologist to master. Hargitt and Fray ('17) were the first to devise a method for rendering Paramecium bacteria-free, which is very simple and involves no new method of handling the animals. All that is necessary is to have absolutely sterile apparatus and media. Their method consisted in a thorough washing of the animals in several changes of sterile water. The bacteriological tests for controlling cultures in the progress of the work are of the simplest, and with the prepared media now on the market, take little more time than does the making of ordinary hay infusions. In my experiments I have so shortened the method described by Hargitt and Fray that it is possible to carry a very large number of cultures, and my work has shown that with care as to the sterility of all apparatus, one need not fear contamination for some time. This reduces the amount of washing necessary to a minimum—an important factor in economy of time. It would seem, therefore, that there should be no technical obstacles in the way of further investigations concerning food, and that much of the work now under way regarding the effect of glandular extracts and vitamines on the Protozoa might be made more exact by the use of these methods.

The method of making control cultures containing chance mixtures of bacteria is important. Jennings ('08) describes his method for keeping the mixture of bacteria constant. He took Paramecia from vigorous stock cultures and washed them in a large amount of culture fluid. Since the animals were from different infusions, a very representative mixture of bacteria was obtained in the washing fluid. This was used to seed the cultures in which he wished to grow pure lines of Paramecium. By doing this every few days he was able to keep the bacterial content of his cultures at an optimum efficiency as to food. It seemed to me that a chance mixture even more reliable than this might be had if sterile media were inoculated every day with fresh hay, and the cultures obtained used when twenty-four hours old. Experience demonstrated that this method is not reliable. Cultures prepared in this way failed to support the life of Paramecium after a period of four months. Since bacteria are known to vary considerably in their resistance to drying, it is very probable that in the method used, I was actually dealing with a progressively changing bacterial content rather than a fairly constant one, as I at the time supposed. It is, therefore, probable that some such method as described by Jennings is to be preferred for keeping Paramecium in vigorous condition upon a chance mixture of bacteria. It is possible that the determination and selection of the predominant bacteria in an infusion by pouring agar plates every day or two, and seeding infusions with predominating colonies thus obtained, without resorting to first growing them in pure culture, might give satisfactory mixtures. This method is, however, open to objection. It is at the best tedious and more time-consuming than the familiar way. Moreover, one is not sure that all types of bacteria found in infusions grow equally well on the ordinary media. Considerable study would be necessary to determine this, and additional time consumed in finding a suitable medium for the growth of all types of bacteria.

The rate of fission is our main index of the metabolic condition of Paramecium. Calkins ('02 a) introduced the method now in use of taking the average division rate for a given number of days as the basis for conclusions regarding the behavior of this infusorian. The number of days chosen in computing the averages is arbitrary and varied with different investigators. Calkins preferred the ten-day interval. Woodruff used the five-day interval in his investigations. Since, in dealing with the effect of known food, it has seemed desirable to have a record of all fluctuations possible and avoid undue error, a three-day interval was chosen for averaging division rates in this study.

Although it is true that the division rate is the only visible index of the progress of metabolism in Paramecium, and it must serve as the basis for all data bearing on the subject, the phenomena indicated by the actual rate of division may be expressed in other ways, such as the percentage of high divisions above the mean rate for a given period, percentage of divisions below this mean, and the percentage of deaths. It is believed that such expressions of the activities of the animals are helpful in interpreting their reaction to food. These percentages have been computed and are included in the tables.

One of the newer methods in biometry is the determination of what I have elsewhere termed the 'significance factor.' I have used this term as being less clumsy than any phrase describing the mathematical processes involved, believing that its meaning is as clear as a longer expression. Gross ('20) and MacDowell ('21) have used this method in interpreting physiological results, but, so far as I am aware, it has not previously been used in experimental work with the Protozoa. The use of the significance factor has proved to be the most helpful of any instrument employed in interpreting the facts. It has been possible by its use to decide as to the relative value of different foods or the effects of changes in medium. Division rate alone, even when aided by percentages of high divisions and the like, fails to reveal the entire truth in such cases. For instance, in dealing with changes in medium, the conclusion from an inspection of these data would be that a moneywort infusion was stimulating when first used in all the lines tested. Determination of the significance factor reveals that the only real evidence of acceleration of metabolism was in the case of the mixture $A'B'C'$. We may state, therefore, that although the division rate is the truest index of the metabolism we have yet found for the Protozoa, the interpretation of its meaning is greatly helped by other indices, especially by the use of the 'significance factor,' which at once enables one to settle any doubt as to the trend of the metabolism as expressed by the rate of division.

Under normal conditions Paramecium feeds mainly upon bacteria. The enormous variety of bacteria in infusions makes it possible for this animal to obtain adequate food under natural conditions. However, since Paramecium has no choice as to the type of food, but must ingest whatever comes in its way, it is entirely unprotected against deleterious bacteria if these predominate in infusions in which Paramecium is living. The animals may die from starvation in the midst of an abundance of unavailable food, or from poisoning should these bacteria be toxic.

The lack of exact knowledge concerning the characteristics of saprophytic bacteria is a great handicap in undertaking the study of any bacterial mixture serving as food for Paramecium. Hargitt and Fray ('17) found it impossible to identify adequately the organisms with which they worked. Miss Watt and I found the same difficulty in the course of this investigation. I have included cultural characteristics of the organisms used in these experiments with the idea that they might be of some slight help to anyone desiring to undertake similar work, but there is great need for a thorough investigation of the saprophytic bacteria. Until such a study shall have been made, it is desirable that any satisfactory cultures which may be discovered shall be carefully kept and made available to any investigators who wish to use them. The cultures A', B', and C' are being maintained, and subcultures will be given to those who desire them.

For ordinary routine work, a satisfactory chance mixture maintained at a high state of efficiency is to be preferred to an artificial mixture such as used in this investigation. The behavior of animals fed with M during the first month amply demonstrates this, for this mixture was superior to the artificial ones during this period. However, if one is investigating the behavior of Paramecium with great care, a known bacterial content is necessary in infusions. Deleterious bacteria certainly lower the rate of metabolism or cause an undue percentage of deaths and so modify normal metabolism as to greatly influence the interpretation of results.

A varied diet is natural for most animals. Although a given environment may be normal for a particular organism, it is not necessarily one in which the optimum of metabolism can be maintained. Variety of food seems to produce an optimum metabolic rate in all cases investigated, largely because no one type of food has been found which combines all the factors necessary to maintain this rate. The question arises, does this generalization hold true for the Protozoa? Will Paramecium thrive upon a single type of food, as illustrated by a pure culture of a bacterium, or is a mixed food better for this animal? The fact that Paramecia were able to live upon the streptothrix C' for a period of nine months, when the line was discontinued, would seem to indicate that it is possible for them to live upon a single article of diet. Moreover, in the case described, the metabolic rate was more uniform than with any food save the chance mixture during the first month. The division rate was not so high among animals fed with C' as with the unknown mixture before it failed, or the mixtures A'C' and A'B'C'. That such organisms as C', which are capable of supporting the life of Paramecium when fed in pure culture, are not numerous is also indicated. Of the nine cultures of bacteria isolated from infusions and tested, only C' could be so used. It is also true that satisfactory mixtures of known bacteria are hard to find. Two only were discovered out of twelve tested. There is no question but that the method used in this work, and by Hargitt and Fray, is not suitable for providing a thoroughly efficient food, but on the other hand it is the only one yet devised whereby the food can be adequately controlled and known types of bacteria included. These bacteria can be determined only by long and tedious tests, and they must first be obtained in pure culture if the food is to be adequately controlled.

Pure cultures of bacteria usually fail to support the life of Paramecium. This failure may be due to one of two reasons. Paramecium may not be able to digest them and so utilize their contained energy, or else the bacteria may contain or excrete substances which are toxic to this animal. In the first instance the bacteria are ingested, but are not digested. The animal starves as it would in a sterile medium. In the second case death comes more quickly.

Khainsky ('10) undertook to study the relationship between the structure and the physiological state in Paramecium caudatum. He studied the course of the food vacuole and the changes occurring within it by the use of vital stains. It is possible that this technique might prove useful in testing the food value of particular bacteria for Paramecium.

The question of toxicity is more complex. In the account of the preliminary experiments of this study, 100 per cent of deaths occurred the first twenty-four hours among animals fed with L' and K'L'. A wholesale mortality such as this would lead one to suspect that the bacterium L' was toxic for Paramecium. Yet, when it was combined with J', itself not a satisfactory food, immediate death did not result. Phenomena such as these need further investigation.

The evidence presented in this paper seems to show that mixtures of bacteria furnish the most satisfactory food for Paramecium. Of all the mixtures tried, the chance mixture M seems to have been the best, could it have been maintained at an optimum of efficiency. The objection to the use of such a mixture in certain types of experimental work is that its exact content is unknown and is subject to daily variation. A mixture such as A'C', when studied with regard to the division rate and significance factor, is found to be so nearly the equal of the chance mixture during a long period of time, that it may be said to be on the whole as advantageous. It possesses the advantage, moreover, of being known and subject to control.

Mention has been made of the stimulating action of mixed cultures of bacteria as contrasted with pure cultures. The word stimulating is not used here in the same sense as in the account of the effect of change in medium. In this latter instance, any acceleration of metabolism noted is probably due to the action of some chemical constituent of the medium less complex than a food. Mixing foods may cause an acceleration of metabolism simply because more energy becomes available from an outside source, whereas the effect of a chemical stimulant is to release energy locked up within the organism. The increase in metabolic rate as a result of combining pure cultures of bacteria has been the usual experience during this work. Frequent as it has been, however, but two instances, A'C' and A'B'C', were observed where this acceleration of metabolism continued for any great time.

Woodruff and Erdmann ('14) showed that what they describe as rhythms in Paramecium are due to the internal reorganization of endomixis. The study of Paramecium with reference to food seems to show that minor fluctuations occur in the course of rhythms which, like these latter, are largely independent of environmental conditions. Such fluctuations have been noted by others, but have been attributed to such unknown factors as variation in food or medium. In my experiments food and medium were as exactly controlled as possible. All the organisms were subjected to the laboratory temperature, but, as has been pointed out, since these fluctuations took place when there was little change in the temperature, it would appear that this factor had little to do with their occurrence. More carefully controlled experiments are necessary to determine this. There appear, however, to be fluctuations in the metabolism not directly connected with the rhythms of endomixis, but, like them, due to some intrinsic characteristic of the protoplasm. These fluctuations occur in all cultures irrespective of changes in food or medium and do not appear to be directly due to environmental factors, but rather to be manifestations of protoplasmic changes. This subject needs careful investigation before definite conclusions can be drawn.

Having determined that mixtures of bacteria are more satisfactory than pure cultures, one is led to inquire if such a mixture should be kept constant. Such experiments as were performed in testing the effect of change of food showed that under the conditions obtaining, change of food was usually accompanied by so much disturbance in the metabolism that no marked increase in division rate resulted, and that with these mixtures at least constancy of diet was preferable to change. This result with a known food supports the contention of workers with chance mixtures, namely, that every effort should be made toward maintaining a uniform bacterial content in media used for growing Paramecium.

The behavior of Paramecium in sterile media is of interest for two reasons. It demonstrates that these animals are incapable of utilizing for food any substances which may be dissolved in such a medium. For this reason, it is valuable in testing the availability of various bacteria for food. If Paramecia live no longer, or but little longer than in sterile medium, we are justified in assuming that these organisms are unsuitable food; that their energy content is unavailable. Since it has been demonstrated that Paramecium does not utilize the substances dissolved in sterile media of the sort ordinarily used in the laboratory, it would seem that Peters' ('20) contention that Protozoa are capable of saprophytic existence needs further investigation. His experiments dealt with the growth of Colpidium in a sterile synthetic medium very different from hay infusion. Colpidium encysts, and so has a means of becoming adjusted to marked environmental change. It would seem, therefore, that Paramecium or some other non-encysting protozoan should be tested in this synthetic medium before generalizations can be made regarding the ability of Protozoa to live as saprophytes.

SUMMARY

The study of the behavior of Paramecium aurelia when fed with known bacteria shows that it is perfectly feasible to control the bacterial content of a medium and that the technique required is not too laborious. In studying such conditions, it has been found that, in addition to the rate of fission, a consideration of the percentages of high and low divisions and the death rate is of value. The application of the significance factor to data of this type has proved exceedingly useful, making possible decisions with regard to effects of food and media which could not otherwise have been reached.

The contention of Hargitt and Fray that pure cultures of bacteria are as a rule unsatisfactory food for Paramecium has been sustained in this work. Moreover, in the single instance in which a pure culture could be used over a long period, the metabolic rate was consistently lower than that of any mixture employed. Mixtures of bacteria would then appear to be the most satisfactory food for Paramecium. Of all the artificial mixtures tested, but two were found which furnished adequate food over a long period of time. These two artificial mixtures were nearly as satisfactory as the usual chance mixture.

When Paramecia are fed with known mixtures of bacteria, it is found that minor fluctuations in division rate occur which are independent of endomixis, and like it do not seem to be greatly influenced by environment. The evidence gathered tends to favor constancy of food rather than frequent change. The influence of ordinary media is practically without effect. Paramecium is unable to utilize food substance dissolved in such media.

LITERATURE CITED

CALKINS, GARY N. 1902 a Studies on the life history of the Protozoa. I. The life history of Paramecium caudatum. Arch. f. Ent-Mech., Bd. 15, S. 139–186.
1902 b Studies on the life history of the Protozoa. III. The six hundred and twentieth generation of Paramecium caudatum. Biol. Bull., vol. 3, pp. 192–205.

CHESTER, F. D. 1914 A manual of determinative bacteriology. The Macmillan Company.

GILLESPIE, L. J. 1920 Colorimetric determination of the H-ion concentration without buffer mixtures, with special reference to soils. Soil Sci., vol. 9, pp. 115–136.

GROSS, A. O. 1920 The feeding habits and chemical sense of Nereis virens Sars. Jour. Exp. Zool., vol. 32, pp. 427–442.

HARGITT, G. T., AND FRAY, W. W. 1917 The growth of Paramecium in pure cultures of bacteria. Jour. Exp. Zool., vol. 22, pp. 421–455.

HUNTINGTON, E. J. 1918 Handbook of mathematics for engineers. McGraw Hill Book Company.

JENNINGS, H. S. 1908 Heredity, variation and evolution in Protozoa. III. Proc. Amer. Phil. Soc., vol. 47, pp. 393–546.
1910 What conditions induce conjugation in Paramecium? Jour. Exp. Zool., vol. 9, pp. 279–299.

KHAINSKY, A. 1910 Physiologische Untersuchungen über Paramecium caudatum. Biol. Centr., Bd. 30, S. 267–278.

MACDOWELL, E. C., AND VICARI, E. M. 1921 Alcoholism and the behavior of white rats. I. The influence of alcoholic grandparents upon maze behavior. Jour. Exp. Zool., vol. 33, pp. 209–291.

MAUPAS, E. 1888 Recherches expérimentales sur la multiplication des infusoires ciliés. Arch. de Zool. Exp. et Gén., Ser. 2, T. 6, pp. 165–277.

MEISSNER, M. 1888 Beitrage zur Ernahrungsphysiologie der Protozoen. Zeit. f. Wiss. Zool., Bd. 46, S. 498–516.

PETERS, R. A. 1920 a Nutrition of the Protozoa. The growth of Paramecium in sterile culture medium. Jour. Physiol., vol. 53; Proc. Physiol. Soc., Feb. 21, 1920, cviii.

1920 b Nutrition of the Protozoa. 2. Carbon and nitrogen compounds needed for the growth of Paramecium. Jour. Physiol., vol. 54; Proc. Physiol. Soc., Oct. 16, 1920, i.

POPOFF, M. 1910 Experimentelle Zellstudien. III. Ueber einige Ursachen der physiologischen Depression der Zelle. Arch. f. Zellf., Bd. 4, S. 1–43.

WINSLOW, C.-E. A., AND OTHERS 1920 The families and genera of the bacteria. Final report of the committee of the Society of American Bacteriologists on characterization and classification of bacterial types. Jour. Bact., vol. 5, pp. 191–229.

WOODRUFF, L. L. 1908 The life cycle of Paramecium when subjected to a varied environment. Am. Nat., vol. 42, pp. 520–526.

1911 Evidence on the adaptation of Paramecium to different environments. Biol. Bull., vol. 22, pp. 60–65.

WOODRUFF. L. L., AND BAITSELL, G. A. 1911 The reproduction of Paramecium aurelia in a 'constant culture medium' of beef extract. Jour. Exp. Zool., vol. 11, pp. 135–142.

WOODRUFF, L. L., AND ERDMANN, R. 1914 A normal periodic reorganization process without cell fusion in Paramecium. Jour. Exp. Zoöl., vol. 17, pp. 425–518.

Resumen por los autores, Carl G. Hartman y W. F. Hamilton.

Un caso de hermafroditismo verdadero en la gallina, con observaciones sobre los caracteres sexuales secundarios.

El presente trabajo es una descripción de un individuo de raza Rhode Island Red, de nueve años de edad, el cual poseía un testículo espermatogenético en el lado derecho, y un ovotestículo y un oviducto normal en el lado izquierdo. El ovotestículo contiene ovocitos de un tamaño máximo de 20 mm. y túbulos testiculares activos. El animal poseía barbillas enormes y un solo espolón; el plumaje era por completo el de una gallina. Puso un huevo, cuidaba de los pollos y cacareaba como una gallina; también cantaba como un gallo de un modo excesivo, perseguia a las gallinas y era objeto de peleas con los gallos. Los resultados de los estudios citológicos llevados a cabo en este ejemplar son los siguientes: El ovotestículo contenía células "luteínicas" y también las llamadas células "intersticiales." El autor considera a las primeras como las células endocrinas del ovario, pero su homología es dudosa. Las células "intersticiales" constituyen un estado basofílico de los leucocitos eosinófilos. Las células endocrinas del testículo del gallo se desconocen por completo.

El ejemplar descrito es semejante a una docena descrita en la literatura sobre este sujeto, los cuales son objeto de tabulación y revisión. Los autores no encuentran razón alguna en contra de la interpretación en perfecta armonía con los resultados de los experimentos de castración y transplantación en la gallina. Presentan un análisis de las pruebas relativas a los caracteres sexuales secundarios de la gallina, incluyendo las que pueden aducirse como resultado del estudio de los hermafroditas. Los puntos que han sido objeto de mayor controversia son objeto de discusión y los autores sugieren nuevos experimentos. Como conclusión un cuadro tentativo de ciertos caracteres sexuales secundarios de la gallina y sus relaciones causales es presentado como base de nueva discusión.

Translation by José F. Nonidez
Cornell Medical College, New York

A CASE OF TRUE HERMAPHRODITISM IN THE FOWL WITH REMARKS UPON SECONDARY SEX CHARACTERS[1]

CARL G. HARTMAN AND WILLIAM F. HAMILTON

Department of Zoölogy, The University of Texas, Austin

TWO PLATES (TEN FIGURES)

Since less than a score of instances of true hermaphroditism in birds have been described, the case herein presented is of interest in relation to the problem of secondary sex characters in birds. The fact that the writers have in hand the entire history of their hermaphrodite and the further fact that it exhibits some unique feures are additional reasons for the publication of this study. We shall first describe the behavior and the anatomy of the specimen and then discuss some of the theoretical aspects of the subject. For the reader's convenience the other cases described in the literature have been compiled in table 1.

BEHAVIOR

The observations under this head have been put together after a careful sifting of the testimony given by a dozen or more persons. Unfortunately, however, while in our possession and for some months preceding, the bird exhibited only an indifferent sex behavior, perhaps on account of its advanced age, which was nearly nine years.

The bird was hatched as a robust chick and developed into an apparently normal Rhode Island Red pullet.[2] The following spring the comb and wattles began to enlarge and the bird, after

[1] Contribution from the Department of Zoology, The University of Texas, no. 152.

[2] For details as to the life-history and habits of the specimen our thanks are due our fellow townsmen, Mr. Joe Amstead, Jr., who presented the bird, and Mr. and Mrs. E. Raven, who raised it.

TABLE

Cases of hermaphroditic birds described in the literature

CASE[1]	GONADS[2]	COMB	WATTLES	SPURS	EGG-LAYING	REMARKS
1429	Ov. with tur ('parovarium')	♂	♂	Large	0	Indifferent sex behavior
1428	Ov. with tur ('many ... tu')	♂	♂	Large	0	Indifferent sex behavior
1427	Ov. with ... ('oe l cords')	♂	♂	Large	0	Indifferent sex ... bv or
1425	Ov. ... (t ...')	♂	♂	?	0	Indifferent sex ... bv or
1426	Active Ov. (left), Active T.	♂	♂	Large	0	Indifferent sex bv or
1422	Ov. (... Small T. right)	Small	?	Large	25 eggs	Male carriage; fought by ... les and ...
1349	O. ... ide T.	♂	♀	Large	0	Hemophroditic ... ehor
1616	O. , ta e T.)	♂	♂	None	12 eggs	N e d; ... d to tread hens; ought by ♂ ... to ... ; fought by ♂
Pearl and Curtis ('09)	Ov. (left), T. right	♂	♂	?	1 egg(?)	
Shattuck and Seligmann ('06)	Ot. (left), T. right	♂	♂	Large	0	Indifferent sex behavior
Tichomiroff ('88)	Ot. (?)	♂	?	Large	?	Fought by ♂ ... ed
Hartman and Hamilton	Ot. (left; active T. (r ght; active)	Small	♂	Left	0 1 egg	Hemophroditic bv or

To this list ... be added Poll's bullfinch ('09) and ... chaffinch ('99), ... with ... male feathering and Ov. on left side, male feathering and ... on right side; also Bond's pheasant (14) with male plumage and ovary on le t side and fe male plumage and ... on right ..., spur on left leg only.

[1] The numbered specimens are ... o Boring and Pearl ('18).

[2] *Ov.*, ...; *T.*, ...; ... *O* , ovotestis.

a few abortive attempts, learned to give the genuine crow of a rooster. The pitch of the voice deepened, the change taking place while the animal was still in the growing age (cf. Brandt, '89). Its nightly crow was easily recognizable on account of the deep voice, and by day the animal was often observed to join the chorus of roosters in the barnyard, crowing excessively like the hen-feathered but large-combed bird of Tichomiroff (Brandt, '89) and specimen no. 1616 of Boring and Pearl ('18). It was often seen scratching in the ground and calling the flock to an alleged morsel of food, and though it was never seen to tread hens it would strut and make advances to them after the manner of cocks. Finally, it was fought by roosters, though itself quite non-combative.

The female behavior of the bird was as follows. For years it would 'sing' like a laying hen. On two occasions it adopted broods of incubator chicks, caring for them day and night and clucking like a normal hen; but it was never seen sitting on a nest. On one occasion, while the owner was stroking the 'pet' on its ventral surface, it dropped an egg, which, though small and elongate, showed the bird to be in possession of functional ovary and oviduct.

ANATOMY

The external appearance (fig. 1) marked the bird as an hermaphrodite even to the casual observation of the layman, and this conclusion was fully borne out on dissection.

Male secondary sex characters consist of the upright character of the comb, the enormous wattles, the spur on the right leg (fig. 3), and the general carriage. Of these characters the unmistakably male wattles were regarded as decisive evidence of male testicular hormones.

Female external characters include the hen-feathering, infiltration of fat in the core of the comb (Smith, '11a), and a spur rudiment on the left leg (fig. 3). The comb is not unusually large for a female comb and the fatty core is in correlation with the generally fat condition of the animal. However, the hen-feathering seemed to us conclusive evidence of the presence of ovarian tissue.

A study of the internal anatomy proved the animal to be a true hermaphrodite, unique in that up to the time when it was killed, at the age of nine years, it possessed active testicular and ovarian tissue containing both fully formed spermatozoa and large oocytes. A partial dissection is shown in figure 5. On the left side is an ovotestis and a coiled oviduct, but the former was not recognizable as a hermaphroditic gonad until a microscopic examination of sections had been made.[3] The surface of the gonad is studded with oocytes of every size up to a diameter of 20 mm. and looks not unlike the ovary of a normal hen approaching the laying season. Several cysts, mostly about the size of the larger oocytes (fig. 5), are found not only upon the surface of the ovotestis, but also in the very plentiful fat which fills the abdominal cavity. As may be seen from figure 6, the ovotestis is an intimate mixture of ovarian and testicular tissue. Some tubules have degenerated, but most of them are normal and contain ripe spermatozoa. There are follicles and ova in various stages of growth and degeneration.

On the right side is seen a testis (fig. 5), an elongate body that widens anteriorly where it attaches to the mesentery. There is a thin and straight vas deferens on each side. The testis consists of tubules containing ripe spermatozoa (figs. 7 to 10) with here and there a more or less completely degenerated tubule (fig. 4). In one portion of the organ there is a group of smaller and more compact tubules which contain epithelial cells within the lumina. This is probably the epididymis.

[3] The microscopical anatomy of the organs was observed from pieces fixed in Bouin's fluid, Flemming's fluid, and 10 per cent neutral formalin. The Bouin material was stained with iron haematoxylin or haematoxylin picrofuchsin; some of the formalin material with Mann's eosin-methyl blue stain. The Flemming material was treated twenty-four hours with a mixture of pyroligneous acid and chromic acid, with a view of rendering the stained lipoid insoluble in alcohol and oil. Frozen sections, cut from some of the formalin fixed material, were subjected to the action of sudan III and osmic acid according to the usual procedures.

For comparison, the same technique was applied to the gonads of a nearly grown normal pullet, a cockerel of about the same age, a mature rooster, and a two-year-old hen that had the habit of treading other hens.

THE ENDOCRINE CELLS OF OVARY AND TESTIS

Two types of endocrine cells have been described for the ovary of fowl: the 'luteal' cells of various authors and the 'interstitial' cells of Boring and Pearl ('17).

The latter type of cell has now been shown by Nonidez ('20) to be the basophilic stage in the development of the eosinophilic leucocytes, as first pointed out by Goodale ('18). We find these cells in large numbers in our preparations of normal ovaries, and they are conspicuous structures in the ovotestis of our hermaphrodite. The hematopoietic rôle of these cells was emphasized to us by our finding one in the blood stream of a normal ovary. There is little wonder, therefore, that Boring and Pearl ('17) found no correlation between these cells and the secondary sex characters.

The 'luteal' cells are easily demonstrable in normal ovaries and in the ovotestis of the hermaphrodite. They occur in well-defined groups and singly in the thecae of the follicles as well as in the stroma about the follicles. Fat droplets of spherical shape but of variable size could be demonstrated in the cytoplasm of these cells, but we looked in vain for eosinophilic granules in them. There appears no reason, however, for doubting the endocrine rôle of these 'luteal' cells.[4]

The endocrine cells of the bird testis are, however, in doubt (cf. Goodale, '18). The basophilic 'interstitial' cells of Boring and Pearl are certainly absent from the adult testis, a point on which we satisfied ourselves. We have further studied the large fat-containing cells which des Cilleuls ('12) and Reeves ('15) have seen in the interspaces between the seminiferous tubules. These cells are found in the testes both of normal birds and of our hermaphrodite. Whether they are ordinary adipose-tissue cells similar to those found elsewhere in the body and the fat con-

---

[4] It is our opinion, however, that their homology is by no means certain. They are not confined to the thecae internae and may not even all arise therefrom (cf. Pearl and Boring, '17). The use of the term 'luteal' or 'lutein' as applied to these cells seems to us to be most premature, particularly since the corpus luteum of mammals is conceded by all who have studied complete series to originate for the most part from the granulosa membrane.

tained in them ordinary fat of metabolism, as Pearl and Boring ('12) contend, further study must determine. Massaglia ('21) tacitly homologizes these cells by calling them the 'cells of Leydig.'

On the mammalian side there is more definite information on these details, for here it is possible to destroy the germinal portion of the testis (leaving the interstitial not only intact, but even hypertrophied) by taking advantage of the differential susceptibility of these two elements to certain harmful stimuli: to radiation (Steinach and Holzknecht, '16), to alcohol (Kostitch, '21), to the pressure due to ligation of the vas deferens (Steinach, '12, '21). Similar experiments done on fowl might lead to illuminating results. Several workers have ligated the vasa deferentia of roosters, but with conflicting results. Shattock and Seligmann ('04) find that ligation of the vas does not interfere with the spermatogenesis nor does it produce any germinal degeneration. This is, of course, rather to be expected, in view of the fact that healthy testicular transplants in fowl, unlike those in mammals, always contain spermatozoa (Guthrie, Goodale). The present writers repeated the experiment in one case with the identical results obtained by Shattock and Seligmann. Massaglia's recent experiments ('21), however, yielded different results, for in those animals in which the vas remained occluded the testis gradually atrophied in all of its elements except the 'cells of Leydig,' which in themselves, according to the author, sufficed to maintain male sex characters (comb and wattles) and behavior.

To summarize, then, it seems to us, first, that the 'luteal' cells of the ovary are very likely the ones responsible for a part of the female hormones of that organ and, second, that we cannot with certainty identify the endocrine cells of the testis. In the next section some of the theoretical bearings of these conclusions will be discussed.

DISCUSSION

We desire now to review briefly some of the evidence on the relation of the gonads to secondary sex characters with reference particularly to partial or complete avian hermaphroditism.

Cock-feathering in hens has long been known to be associated with degeneration of the ovaries (Yarrell, '27; Brandt, '89). This conclusion has been put on an experimental basis by the castration experiment of Pezard, Guthrie, Goodale, and Morgan and falls in line with the observations of Poll ('11) on sterile hybrid ducks and of Smith and Thomas ('13) on sterile hybrid pheasants. Evidence on the positive side is furnished by Goodale ('10, '18), who succeeded in producing hen-feathering in cockerels by means of ovarian transplantation. The ovary is, therefore, shown to be responsible for hen-feathering. But Morgan ('19, '21) carried the analysis a step further, making it reasonably certain that the luteal cells are the endocrine cells involved. For after castration of his hen-feathered Sebright and Campine males, these birds became male-feathered; and in complete harmony with the theory, Boring and Morgan ('18) found characteristic luteal cells in the testis of these birds.

Such are the facts relative to the internal secretion of the bird ovary. The subject cannot be pursued further here. In table 2 the reader will find a summary of the facts in condensed form. The table emphasizes the fertile suggestion of Goodale, that each sex character acts like a unit—a view which it is well to bear in mind.

Passing now to a consideration of the avian testis, we find the view expressed (Boring and Pearl, '18) that there is "only a very general correspondence such as in body shape and carriage between the secondary sex characters and the primary sex organs, . and that spurs, comb and wattles vary regardless of primary sex organs." Certain well-known facts seem to substantiate this view; for example, laying hens occasionally have spurs, capons and hens may exhibit certain phases of male behavior, and hens often crow. Such aberrant individuals have been described as 'normal'; but until we know more about the functional cytology of the fowl testis, it is idle to apply the term 'normal' to these specimens. There may be cited, further, the obscure and disconcerting gynandromorphs described by Weber ('90), Poll ('11), and Bond ('13). Nevertheless, a consideration of all the facts force us to take the contrary view to that proposed by Boring and Pearl; for if the internal secretion of the testis be denied, the sex characters of the hermaphroditic birds listed in table 1 would have to be explained simply on the basis of the degeneration of ovaries (cf. Morgan, '19, p. 39). We believe this explanation to be inadequate for the following reasons:

First, if the three cases of gynandromorphism cited above are invoked against the internal secretion of the testis, the internal secretion of the ovary must also be denied.

Second, while degeneration of the ovaries of zygotic females is correlated with the assumption of male plumage, such male-feathered females have atrophic, capon-like combs and wattles but the hermaphrodites have almost invariably *cock's combs and wattles associated with hen feathering*.

Third, capons possess mere rudimentary (not female) combs and wattles; but regeneration of incompletely extirpated testes (Bond, '13), transplantation of testes into capons (Bernard, '49), injection of testicular extract, even of mammalian testes (Pezard, '11, '18, '21 a; Loewy, '03)—all result in a resumption of growth of head furnishings. Experimental evidence of this character is usually considered conclusive with reference to any other endocrine organ (Biedl, '13).

Fourth, Pezard ('18) was able to cause growth to comb and wattles in an ovariotomized pullet by successfully engrafting testicular tissue, which is quite in accord with Steinach's results of heterosexual transplantation of gonads and the production of hermaphrodites in mammals.

Fifth, if it is urged that nos. 1429, 1428, 1427, and 1425 of table 1 are females which exhibit hermaphroditic tendencies simply because of the presence of degenerating ovaries, we would answer with the well-known fact that the presence of germ cells is no criterion of the endocrine activity of the gonad.[5]   Further-

---

[5] Compare Pearl and Curtis ('09) and various vasectomy experiments on mammals (Steinach, etc.).

more, there are structures in these birds and in those described by Tichomiroff which are highly suggestive of testicular tissue,[6] and there is no reason, in view of the present state of our knowledge of the testis, for denying the presence of cells that may give rise to a testicular hormone.[7]

Sixth, when cocks are feminized (Goodale, '18), the comb and wattles and plumage "become indistinguishable from the female form." This shows that male head furnishings are not a function of zygotic maleness, like behavior for example (table 2), but may be changed to the female type only upon stimulation by female hormone. When we compare the feminized cocks with hen-feathered Sebrights or Campines, the contrast between the male head furnishings of the latter and the female head furnishings of the former is striking. Both classes have luteal cells and hen feathers; the male Sebright or Campine has male comb and wattles associated with testicular tissues. The hermaphrodites listed in table 1 are like the hen-feathered males: they are hen-feathered, they have large combs and wattles, they possess luteal cells and testicular tissue.

We believe, therefore, that the hermaphrodites that have been described are to be interpreted quite in accord with the experimental evidence that has accumulated in this field.

CONCLUSION

The evidence reviewed in the foregoing discussion appears to the writers ample to establish a definite endocrine function for both testis and ovary in relation to secondary sex characters of birds. Yet it is also clear that our knowledge is lacking in many details. The endocrine cells of the testis are an unknown quantity, those of the ovary have been insufficiently interpreted. On

---

[6] Attention is again called to the findings of Massaglia ('21), according to which small, hard, atrophic testes which are hardly recognizable as such are still capable of maintaining the cock's combs and wattles in the vasectomized animals.

[7] It should be further pointed out in this connection that Boring and Pearl's no. 1349 is most male-like in body shape (upon which the authors lay much stress) although its gonads contain the most luteal cells, while their no. 1427 is described as possessing no luteal cells, although the animal was very female in appearance, with complete hen-feathering.

the experimental side, the work on mammals is in a much more satisfactory state. For example, experiments have hardly been begun on the masculination of pullets and hens and on the production of avian hermaphrodites by transplantation of the gonads.

It would be interesting, moreover, to note the result of castration of a spontaneously occurring hermaphrodite such as the one here described—a manifestly difficult operation. Such an animal should also be studied cytologically in order to determine whether the ZY or the ZZ condition prevails in the soma of the specimen.[8] Cytological studies of gynandromorphs among the vertebrates would also go far toward differentiating genetic from physiological factors in the problem. Indeed, it is of prime importance for the experimenter to know the genetic substratum of the material upon which he works (cf. Morgan, '19). For example, in some races of fowl the females have large combs, in others small ones; in some strains a large proportion of the females are normally spurred, in others a spurred female occurs never or very rarely. Again the sexual activity of the fowl may influence certain secondary sex characters, for example, the size of the comb, which may increase as much as 130 per cent in area before the egg-laying period (Smith, '11 a). Lack of information like this can only lead to confusion.

In conclusion, we bring before the reader table 2, pages 196 and 197, in which the secondary sex characters are classified according to the physiological or genetic causes underlying them. The classification is to be considered strictly tentative, and is presented as a possibly useful basis for further discussion.

---

[8] We attempted this investigation, but, due to the well-known difficulties with the fixation of chicken material, we were unable to arrive at a satisfactory conclusion. Counts of eight haploid plates gave either seven or eight chromosomes. The spermatogonial plates were not very clear and we were unable to be sure of any sex elements.

SUMMARY

1. There is described in this paper a Rhode Island Red fowl, nine years of age, which possessed an ovotestis and a testis, both active. The hermaphrodite displayed external characters and behavior of both sexes.

2. Other similar cases previously described in the literature are summarized in table 1 and reviewed.

3. The ovotestis contains both the 'interstitial' cells of Boring and Pearl and the 'luteal' cells of various authors. The former are regarded as the basophilic stage in the life-history of eosinophilic leucocytes. The latter are probably the endocrine cells of the ovary responsible for hen-feathering.

4. Although we are in almost total ignorance concerning the endocrine cells of the testis, yet the fact appears to be established that this organ does secrete a hormone which influences the behavior and certain secondary sex characters, notably the comb and wattles. The evidence, including that furnished by the hermaphrodites, is reviewed.

5. Experiments are suggested which might be expected to clear up some of the mooted and obscure points in this field.

6. Genetic and cytological data must supplement physiological facts in the elucidation of secondary sex characters in fowl. In the writers' opinion, a mere beginning has been made towards this end.

TABLE 2

Summary of certain sex characters of fowl in their causal relations to or independence of gonadial tissues. The table is to be considered purely tentative. The plus sign (+) signifies stimulating, the minus sign (−) inhibiting

CHARACTER	AS INDICATED BY:
1. Action of testicular hormones	
♂ Comb and wattles +.......	Atrophic in capons; enlarge after transplantation or regeneration of testes; hermaphrodites have large combs and wattles
♂ Behavior +.............. ⎫	Shown by castration and transplantation
♂ Crowing[1] +.............. ⎭	experiments and study of hermaphrodites
♂ Body size −..............	Capons grow larger than cocks and their
♂ Feather size −............	wing coverts and sickle feathers become excessively large
♀ Behavior −..............	Capons sometimes brood
♀ Fat distribution[2] −........	Capons are fat like hens
2. Action of luteal cells	
♂ Plumage −.................	Castrated hens of all races and hen-feathered male Sebrights become ♂ feathered[3]
♀ Comb and wattles +.......	Feminized ZZ has ♀ comb and wattles
♀ Plumage +................	(?)[4]
3. Action of ovary not due to luteal cells	
♂ Sex behavior + (ZZ only)...	Feminized ZZ behaves like ♂
♀ Sex behavior + (ZW only)..	Castrated ♀ is indifferent; hermaphrodite often exhibits hermaphroditic behavior; Sebright ♂ behaves only as a ♂
♂ Spurs − (ZW only)	Castrated ♀ has spurs, but luteal cells of Sebright testis does not inhibit spurs in ♂
4. Genetic factors *a. ZZ*	
♂ Size (and carriage ?)	Body size cannot be feminized by heterosexual transplantation of ovaries or by castration.
♂ Behavior—in presence of *either* ovary or testis !	
b. ZW	
♀ Size (and carriage ?)....... ⎫	? (There are no experiments to test effect of
♀ Behavior............... ⎭	masculination)

TABLE 2—*Continued*

CHARACTER	AS INDICATED BY:
c. ZZ or ZW	
♂ Spurs....................	Castrated individuals of both sexes have spurs
♂ Plumage..................	Male plumage develops on castrates of both sexes
Certain phases of behavior seen, e.g., in brooding capons	

[1] Exceptions are seen in crowing hens; but these have been only superficially studied.

[2] Too little is known concerning the deposition of fat in the body of fowl to draw definite conclusions. The altered metabolism of castrates complicates the situation (Meisenheimer, '11, and Pezard, '18).

[3] Sebright males, possessing large combs and wattles, have luteal cells in the ovary; but here the female combs and wattles may be covered up by the larger male combs and wattles, due to testicular hormones.

[4] There are some genetic differences here, for after castration the primary coverts enlarge in the male, not at all in the female; feminized males do not always develop female feather colors.

LITERATURE CITED

(All of the following papers have been consulted by the writers)

BERTHOLD, PROF. 1849 Transplantation der Hoden. Muller's Archiv, S. 42.

BIEDL, A. 1913 Innere Sekretion. Berlin.

BOND, C. J. 1913 Some points of genetic interest in regeneration of the testes after experimental orchectomy in birds. Journ. of Genetics, vol. 3, pp. 131–139.
1914 On a case of unilateral development of secondary male characters in a pheasant, with remarks on the influence of hormones in the production of secondary sex characters. Journ. of Genetics, vol. 3, pp. 205–216. (See also brief account in Nature, 1913, p. 338.)

BRANDT, ALEX. 1889 Anatomisches und Allegemeines über die sogenannte Hahnenfedrigkeit und über anderweitige Geschlechtsanomalien bei Vogeln. Zeitsch. f. wiss. Zool., Bd. 48, S. 101–190.

BORING, ALICE M. 1912 Sex studies. III. The interstitial cells and the supposed internal secretion of the chicken testes. Biol. Bull., vol. 23; pp. 141–153.

BORING, A. M., AND MORGAN, T. H. 1918 Lutear cells and hen-feathering. Journ. of Gen. Physiol., vol. 1, pp. 127-131.

BORING, A. M., AND PEARL, R. 1917 Sex studies. IX. Interstitial cells in the reproductive organs of the chicken. Anat. Rec., vol. 13, pp. 233–252.
1918 Sex studies. XI. Hermaphroditic birds. Jour. Exp. Zool., vol. 25, pp. 1-48.

198   CARL G. HARTMAN AND WILLIAM F. HAMILTON

Des Cilleuls, F. 1912 A propos du déterminisme des charactéres secondaires chez les Oiseaux. C. R. Soc. Biol., T. 73; pp. 371-2 (reviewed in Journ. Roy. Mic. Soc., Dec., 1912).

Goodale, H. D. 1913 Castration in relation to the secondary sex characters in Brown Leghorns. Am. Nat., vol. 47, pp. 159–169.
1916 a A feminized cockerel. Jour. Exp. Zoöl., vol. 20, pp. 421-428.
1916 b Gonadectomy. Publication no. 243 of the Carnegie Inst. of Wash.
1916 c Further developments in the ovariotomized fowls. Biol. Bull., vol. 30, pp. 286-293.
1918 Feminized male birds. Genetics, vol. 3, pp. 276–299. (See also Anat. Rec., vol. 11, pp. 512–514, 1917.)
1919 Interstitial cells in the gonads of domestic fowl. Anat. Rec., 16: 247-250.

Guthrie, C. C. 1910 Survival of engrafted tissues. Journ. of Exp. Med., vol. 12, pp. 269–277.

Kostitch, A. 1921 Sur la dissociation de la glande seminale et de la glande interstitielle déterminee par l'alcoölisme expérimentale. Stérilité sans impuissance. C. R. Soc. Biol., pp. 569–571.

Lowey, A. 1903 Neuere Untersuchungen zur Physiologie der Geschlecht-sorgane. Erg. d. Physiol., Bd. 2, S. 130–158.

Loisel, G. 1902 Sur le lieu d'origine, la nature et le rôle de la secrétion interne du testicule. C. R. Soc. Biol., p. 1034.

Massaglia, A. C. 1920 The internal secretion of the testis. Endocrinology, vol. 4, pp. 547-566.

Meisenheimer, J. 1911 Ueber die Wirkung von Hoden- und Ovarial-substanz auf die sekundären Geschlechtsmerkmale der Frösche. Zool. Anz., Bd. 38, S. 53–60.

Morgan, T. H. 1919 The genetic and operative evidence relating to secondary sex characters. Publication no. 285 of the Carnegie Inst. of Wash.
1920 a The effects of castration of hen-feathered Campines. Biol. Bull., vol. 39, pp. 231–247.
1920 b The endocrine secretion of hen-feathered fowls. Endocrinology, 4, 381–385.

Nonidez, Jose F. 1920 Studies on the gonads of the fowl. I. Hematopoietic processes in the gonads of embryos and mature birds. Am. Jour. Anat., vol. 28, pp. 81–113.

Pearl, R., and Boring, A. M. 1912 Sex studies. IV. Fat deposition in the testis of the domestic fowl. Science, n. s. vol. 36, pp. 833–835.
1918 Sex studies. X. The corpus luteum in the ovary of the domestic fowl. Am. Jour. Anat., vol. 23, pp. 1–16.

Pearl, R., and Curtis, M. R. 1909 A case of incomplete hermaphroditism. Biol. Bull., vol. 17, pp. 271–286.

Pezard, A. 1911 Sur la détermination des charactéres sexuels secondaires chez les Gallinacès. C. R. Acad. de Sci., T. 153, p. 1027. (See also further papers by this author in the same journal: 1912, T. 154, p. 1183; 1914, T. 158, p. 513; 1915, T. 160, p. 260.)
1918 Le conditionnement physiologique des charactéres sexuels secondaires chez les oiseaux. Bull. Biol. France et Belg., T. 52, p. 1–176.

PEZARD, A. 1919 Castration alimentaire chez les coqs soumis au régime carné exclusif. C. R. Acad. de Sci., T. 169, pp. 1177-9.

1920 Secondary sexual characters. Endocrinology, vol. 4, pp. 527-540.

1921 Numerical law of regression of certain secondary sex characters. Jour. of Gen. Physiol., vol. 3, pp. 271-283. (See also C. R. Acad. de Sci., 1920, T. 171, pp. 1081-3; 1921, T. 172, pp. 89-92 and 176-178.)

POLL, H. 1909 Zur Lehre von den sekundären Sexualcharakteren. Sitzungsber. d. Ges. naturforsch. Freunde zu Berlin, S. 331-358.

1911 Mischlingsstudien. VI. Eierstock und Ei bei fruchtbaren und unfruchtbaren Mischlingen. Arch. f. mikr. Anat., Abt. II, Bd. 78, S. 63-127.

REEVES, T. P. 1915 On the presence of interstitial cells in the chicken testis. Anat. Rec., vol. 9, pp. 383-386.

SHATTOCK, S. G., AND SELIGMANN, C. G. 1904 Observations upon the acquirement of secondary sexual characters, indicating the formation of an internal secretion by the testicle. Trans. Path. Soc., London, vol. 56, pp. 57-80.

1906 An example of true hermaphroditism in the domestic fowl with remarks on the phenomenon of allopterotism. Trans. Path. Soc., London, vol. 57, pp. 69-109.

1907 An example of incomplete glandular hermaphroditism in the domestic fowl. Proc. Roy. Soc. Med., vol. 1, pp. 3-7.

SMITH, GEOFFREY 1911 a Studies in the experimental analysis of sex. Part 6. On the cause of fluctuations in growth of the fowl's comb. Quart. Journ. Mic. Sci., vol. 57, pp. 45-51.

1911 b Studies in the experimental analysis of sex. Part 5. On the effects of testes extract injections upon fowls. Quart. Jour. Mic. Sci., vol. 56, pp. 591-612.

SMITH, GEOFFREY, AND MRS. HAIG THOMAS 1913 On sterile and hybrid pheasants. Jour. of Genetics, vol. 3, pp. 39-53.

STEINACH, E. 1912 Willkürliche Umwandlung von Säugetiermännchen in Tiere mit ausgepragt weiblichen Geschlechtscharakteren u. weiblicher Psyche. Pflüger's Archiv., Bd. 144, S. 71-108.

1920 Verjüngung durch experimentelle Neubildung der alternden Pubertätsdrüse. Julius Springer, Berlin. (See also Vienna letter, Restoration of Youth. Journ. Am. Med. Ass'n, vol. 75, p. 490, Aug. 14, 1920, and vol. 75, p. 617, Aug. 28, 1920.)

STEINACH, E., AND HOLZKNECHT, G. 1916 Erhöhte Wirkungen der inneren Sekretion bei Hypertrophie der Pubertätsdrüsen. Arch. f. Entwicklungsmech., Bd. 42, S. 490-507.

TICHOMIROFF, A. 1887 Russian; see Brandt, 1889.

1888 Androgynie bei den Vögeln. Anat. Anz., Bd. 3, S. 221-228.

WALKER, C. E. 1908 The influence of the testis upon the secondary sexual characters of fowls. Proc. Roy. Soc. Med., Path. Sec., vol. 1, pp. 153-156.

WEBER, M. 1890 Ueber einen fall von Hermaphroditismus bei Fringilla coelebs. Zool. Anz., Bd. 13, S. 508-512.

YARRELL, WM. 1827 On the change in the plumage of some hen pheasants. Phil. Trans. Roy. Soc., pp. 268-275. (Abstracted in Proc. Roy. Soc. 1827, p. 317.)

PLATE 1

1 Hermaphroditic Rhode Island Red chicken, nine years old. It is seen to be hen-feathered, has spur on left leg and has male wattles.

2 Head of same. Because the head was laid on a glass plate in order to photograph it the wattles appear foreshortened.

3 The feet of the hermaphrodite, showing spur on left and spur rudiment on right.

4 Section of testis; the light area consists of degenerated tubules containing cellular debris.

5 Partial dissection of abdominal organs photographed fresh. C, cyst; CL, cloaca; OD, oviduct; E, oocyte; T, testis.

200

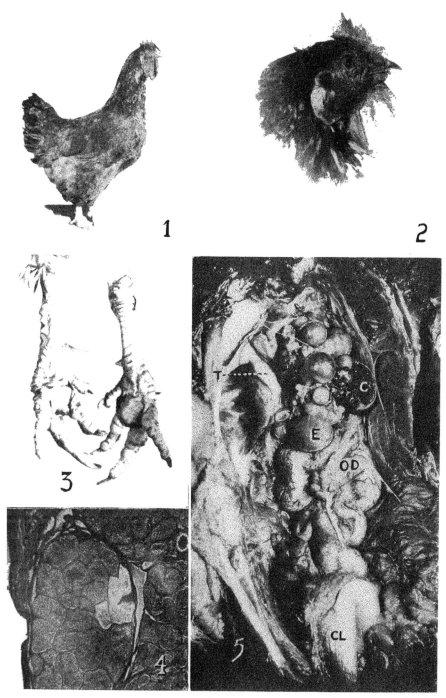

6   Section through ovotestis, showing atresic follicle (*E*) and atrophic tubules (*T*) side by side.

7   Section through testis, showing at *A* seminiferous tubules, at *B*, epididymal (?) tubules.

8   Portion of seminiferous tubule, showing spermatozoa.

9   A portion of tubular epithelium of testis, to show active cell division.

10   A portion of figure 8, still more highly magnified.

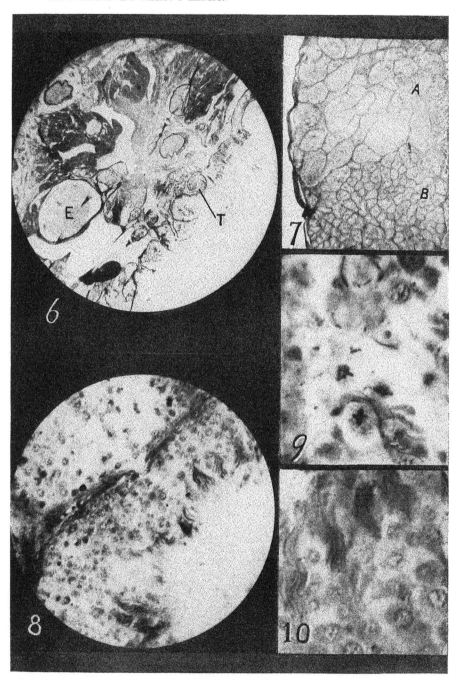

Resumen por el autor, George H. Parker.

El salto del caracol marino (Strombus gigas Linn.).

Strombus gigas se mueve de un sitio a otro saltando, en vez de deslizarse del modo característico de la mayor parte de los caracoles. El salto tiene lugar mediante la extensión anterior del pié, su fijación al substrato mediante sus extremos anterior y posterior, y una vigorosa contracción muscular, a consecuencia de la cual la concha es arrojada hacia delante, viniendo a situarse a una distancia a veces equivalente a la mitad de la longitud de dicha estructura.

Translation by José F. Nonídez
Cornell Medical College, New York

BY THE BIBLIOGRAPHIC SERVICE, JUNE 12

THE LEAPING OF THE STROMB (STROMBUS GIGAS LINN.)

G. H. PARKER

Zoological Laboratory, Harvard University

TWO FIGURES

It has long been known that strombs differ from other gastropods in their method of locomotion. These conchs progress over the substrate by sudden leaps and not by the slow gliding movement so characteristic of most snails. Adams ('48, p. 493), in describing the living Strombus, says that "it is, in fact, a most sprightly and energetic animal, and often served to amuse me by its extraordinary leaps and endeavors to escape, planting firmly its powerful narrow operculum against any resisting surface, insinuating it under the edge of its shell, and by a vigorous effort throwing itself forward, carrying its great heavy shell with it, and rolling along in a series of jumps in a most singular and grotesque manner." ˙ This description portrays fairly well the movements of this giant conch.

While I was at the Miami Aquarium I had the opportunity of studying the locomotion of Strombus gigas Linn., which, at least in immature specimens, was common in the neighborhood of Miami. I am under obligations to the Miami Aquarium Association for the privilege of carrying out this work at the laboratory of the Aquarium.

Immature but large specimens of Strombus gigas are to be found creeping about on the weed-covered flats in Biscayne Bay, Florida. One large animal whose shell measured 15 cm. long would progress half the length of its shell at a single bound and in doing so it lifted its shell off the substrate at least 4 cm. Ordinarily these snails would progress 4 to 5 cm. at a single leap.

When such specimens were brought into the laboratory and studied closely in a glass aquarium, the details of their locomotion could be readily made out. In this operation the foot and body musculature is the active part. As compared with other mollusks, the foot of Strombus is rather peculiarly shaped. Anteriorly it has the form of a broad, flattened finger which can be applied very closely to the substrate (fig. 1). Behind this comes a second portion which is smooth and rounded from side

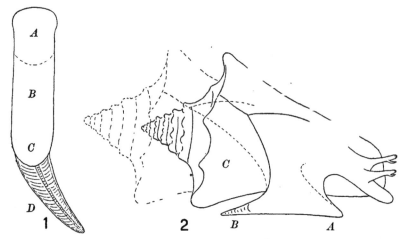

Fig. 1 Ventral view of the foot of Strombus gigas. *A*, anterior flattened end; *B*, middle rounded portion; *C*, posterior or metapodial portion carrying the operculum, *D*.

Fig. 2 Outline of a lateral view of a stromb immediately after a leap. The dotted outline indicates the general position of the parts before the leap, the full outline after the leap. *A*, anterior end of the foot; *B*, operculum; *C*, shell.

to side; this part is seldom in close contact with the surface over which the snail is moving. Finally, in the posterior region the foot tapers off, carrying at its hind end the long, dark-brown, pointed operculum. This metapodial portion is thrust vigorously into the ground at each spring of the animal.

When an active stromb is put on its side in a glass aquarium, it soon begins to protrude its foot and eye-stalks. At the least movement on the part of the observer it is likely to withdraw into its shell and with a suddenness quite surprising for a snail.

As compared with other gastropods, Strombus is remarkably alert and active, and in the quickness of its movements it reminds one more of a vertebrate than of a mollusk. Its eyes, too, are highly developed and are moved and directed in such a way as to give it the appearance of no small degree of intelligence.

After the stromb has protruded its foot a few times in a tentative way, it will gradually extend this organ till it reaches the substrate. The anterior finger-like end of the foot is then pressed vigorously against the ground, the middle section of the foot arching over to the metapodium, which together with the operculum is moved in under the shell and thrust vigorously backward. At the same moment the general musculature of the foot and of the body contracts in such a way as to raise the shell over the support given by the foot and throw the shell vigorously forward, as though the animal as a whole made a spring (fig. 2). After such a leap, which, as already stated, may be half the length of the shell, the animal usually withdraws a little, then thrusts out the foot far forward, regains a hold on the substrate, and leaps again. Thus step by step it progresses at a rate quite surprising for a mollusk.

The principle upon which the locomotion of Strombus rests appears to be the forward extension of the body and especially the foot, the attachment of the latter by its two ends to the substrate, and the lifting and throwing of the shell forward to the advanced location occupied by the foot. In preparing for a leap the anterior finger-like end of the foot is closely applied to the substrate and appears to attach itself by suction. Such, however, is not the case. When a stromb is about to spring, its shell may be laid hold of by the experimenter without interrupting the action of the animal, and under such circumstances the anterior end of the foot may be freely lifted from the substrate, showing that it is not exerting suction, for this end leaves a mud, wood, or glass surface at once and without the least sign of being especially attached. The posterior end of the foot, which at the moment of the spring is drawn well under the shell, is fixed by having the point of the operculum energetically driven into the substrate. So vigorous is this backward thrust of the operculum that divers who are collecting strombs are said often to be cut on the breast by these conchs when, with an armful held tightly to the body, the diver is swimming to the surface. Strombus, then, attaches its foot by pressing the two ends of that organ vigorously against the substrate and by relying more or less upon the weight of the shell to hold the foot in place while the shell itself is being thrown forward.

The weight of the shell is considerable and no one can watch the locomotion of Strombus without being impressed by its strength. In an immature specimen whose living body weighed 49 grams the shell weighed 173 grams, making a total of 222 grams, yet the relatively small amount of musculature in this animal was sufficient to enable it to make considerable leaps even in the air. In the sea-water the shell is, of course, relatively lighter, and in consequence the animal can leap rather farther than in air. The shell that weighed 173 grams in the air, weighed only 105 grams in sea-water, the whole animal, shell and soft parts, weighing 110 grams in water as against 222 grams in air. Thus in its more usual habitat, in sea-water, the stromb has less work to do in moving its shell than it has in air, but such work even when the animal is in the water is by no means inconsiderable.

Thus the locomotion of Strombus is accomplished by means radically different from those used by other gastropods, for the muscular action involved in the leap of this snail is in no obvious way related to the muscular waves that pass over the foot of most moving gastropods and is absolutely distinct from ciliary activity which, contrary to my former view (Parker, '11, p. 157), has been recently shown by Copeland ('19) to be the means of locomotion in several gastropods.

SUMMARY

Pedal locomotion in gastropods is accomplished either by gliding (an operation dependent upon muscular waves passing over the foot or by ciliary action) or, in Strombus, by leaping, an act that involves the forward extension of the foot, its fixation in the substrate by its anterior and posterior ends, and a vigorous muscular contraction whereby the animal's shell is thrown well forward, to the extent even of half the length of that structure.

LITERATURE CITED

ADAMS, A.   1848   Notes from a journal of research into the natural history of the countries visited during the voyage of H. M. S. Samarang. In E. Belcher: Narrative of the Voyage of H. M. S. Samarang during the years 1843–46, vol. 2, pp. 223–532.

COPELAND, M.   1919   Locomotion in two species of the gastropod genus Alectrion with observations on the behavior of pedal cilia. Biol. Bull., vol. 37, pp. 126–138.

PARKER, G. H.   1911   The mechanism of locomotion in gastropods. Jour. Morph., vol. 22, pp. 155–170.

PROMPT PUBLICATION

The Author **can** greatly assist the Publishers of this Journal **in** attaining prompt publication of his paper by following these four suggestions:

1. *Abstract.* Send with the manuscript an Abstract containing not more than 250 words, in the precise form of The Bibliographic Service Card, so that the paper when accepted can be scheduled for a definite issue as soon as received by the Publisher from the Editor.

2. *Manuscript.* Send the Manuscript to the Editor prepared as described in the Notice to Contributors, to conform to the style of the Journal (see third page of cover).

3. *Illustrations.* Send the Illustrations in complete and finished form for engraving, drawings and photographs being protected from bending or breaking when shipped by mail or express.

4. *Proofs.* Send the Publisher early notice of any change in your address, to obviate delay. Carefully correct and mail proofs to the Editor as soon as possible after their arrival.

By assuming and meeting these responsibilities, the author avoids loss **of** time, correspondence that may be required to get **the** Abstract, Manuscript and Illustrations in proper form, and does all in his power to obtain prompt publication.

THE JOURNAL OF EXPERIMENTAL ZOOLOGY, VOL. 36, NO 3
OCTOBER, 1922

Resumen por el autor, Edgardo Baldi

Investigaciones sobre la fisiología del sistema nervioso de los insectos.

II. Movimientos circulares de los coleopteros.

El autor estudia en algunos géneros de coleópteros (Blaps, Pimelia, Carabus y otros) aquellos movimientos circulares causados por lesiones de la región lateral de los ganglios supraesofágicos. Después de haber investigado la naturaleza de la alteración necesaria para producir tales movimientos y de haber descrito el mecanismo de la locomoción normal de los coleópteros, afirma que los movimientos circulares están causados, sobre todo, por un mayor grado de flexión en todos los grupos musculares del lado del organismo opuesto 'al de la herida cerebral, especialmente por el mayor grado permanente de la contracción de los músculos adductores de las patas (articulación tibio-femoral). También discute basándose en este punto de vista más general, las conclusiones de otros autores que le han precedido, especialmente las de Bethe. Indica como una herida cerebral produce una serie de alteraciones generales de la musculatura, las cuales afectan a todo el cuerpo del insecto y no están localizadas, conforme han creído otros muchos autores, en una mitad separada del organismo. Basándose en estas observaciones y en un estudio cuantitativo de la disimetría muscular (empleo de los reogramas) propone una teoría de los movimientos circulares de los coleópteros que se diferencia de todas las precedentes en el hecho de fundarse solamente en datos experimentales sin recurrir a las expresiones vagas y ambiguas de "inhibición" y "tono."

Translation by José F. Nonidez
Cornell Medical College, New York

STUDI SULLA FISIOLOGIA DEL SISTEMA NERVOSO NEGLI INSETTI

II. RICERCHE SUI MOVIMENTI DI MANEGGIO PROVOCATI NEI COLEOTTERI

EDGARDO BALDI

Istituto di Zoologia, Università di Pavia

QUARANTA FIGURE

INDICE

1. INTRODUZIONE. MOTI DI ROTAZIONE E MOTI DI MANEGGIO

Prima di accingerci allo studio descrittivo ed interpretativo dei fenomeni di maneggio, e pure attendendo da questo studio luce che ci possa permettere di afferrarne il significato e quindi di trovare loro un posto tra le altre manifestazioni dell'attivitá fisiologica degli insetti studiati,[1] converrá tentare di circoscrivere il concetto dei movimenti di maneggio, cosí come li studieremo, trovando per essi un approssimativo collocamento tra gli altri fenomeni consimili, di andamento normale od anormale, che ci sono presentati dagli organismi animali, dagli insetti in ispecie.

---

[1] Delle specie che mi hanno interessato ho fatto cenno in una precedente nota, alla quale rimando come ad una informazione preliminare ed in certo modo necessaria alla comprensione di talune pagine della presente: "Ricerche sulla fisiologia del sistema nervoso negli insetti. I. L'influenza dei gangli cefalici sulla locomozione dei coleotteri." Atti Soc. ital. Scienze Naturali, vol. ~~40~~, Milano, 1921.

I moti di rotazione compiuti da un organismo non sedentario e che interessino l'intero suo corpo possono effettuarsi attorno ad un asse che passi per una regione del corpo, oppure attorno ad un asse che cada fuor d'esso corpo. I primi non sono evidentemente legati ad una trasposizione spaziale dell'organismo in regione sufficientemente lontana dall'iniziale: l'organismo, cioé, vi puó obbedire rimanendo *in situ*, mentre i secondi causano un suo trasporto attraverso lo spazio. I due moti possono essere abbinati nella locomozione ad elica od a cavaturaccioli, consueta ad alcuni animali inferiori e provocabile in altri mercé opportune lesioni ai gangli cefalici.

Anche moti di rotazione attorno ad assi cadenti fuori del corpo possono presentarsi non anormalmente. Cosě dell'*Agromyza* scrive lo Zetterstedt che sia *"motu valde agilis, inquieta, per folia ʃere in circulo currens."*[2] Né é raro osservare, anche in insetti comunissimi—io l'ho vista in piú generi di ditteri—l'apparizione sporadica di moti in circolo, in ambiente naturale e probabilmente in particolari condizioni di illuminazione. Tali moti peró sono occasionali e di breve durata; né molto difficile sarebbe il riunire qui altri esempi di moti rotatorii in senso lato, genericamente presentati da organismi illesi, in ambiente normale. Ma non é di essi—fenomeni fuggevoli e di vario ed impreciso significato—che noi intendiamo di occuparci.

Noi ci riferiremo infatti, in queste nostre ricerche, all'individuo singolo e terremo sopratutto conto, nello studiare i fenomeni di giro in circolo, in esso artificialmente provocati, come delprincipale loro fattore, del grado e della natura della dissimmetria sensoria e locomotoria ch'esso individuo presenta. E, per chiarirci il concetto del moto in circolo, considerato come conseguenza di una alterazione nella simmetria delle condizioni che regolano l'attivitá normale dell'organismo dell'insetto, considereremo quest'ultimo come segue.

Nell'animale normale i gruppi muscolari che presiedono, con fenomeni di contrazione attiva e di tono, alla dirittezza della

[2] Citata da E. Corti "Contributo alla teratologia dei ditteri" (Lavori dell'Istituo Zoologico dell'Universitá di Pavia, No. 14, 1913) secondo i "Diptera Scandinaviae disposita et descripta" dello Zetterstedt (VII. 2729).

deambulazione ed alla conservazione di un certo orientamento nello spazio, sia dell'intero organismo, sia di singole sue regioni, sono simmetricamente distribuiti ai due lati del piano sagittale. Cosí avviene ad esempio, dei muscoli che si inseriscono a parti chitinizzate mutuamente mobili, come i segmenti del corpo ed i diversi tratti degli arti ed in genere alle appendici pari dell'organismo. Per riassumere in una espressione sola questo tipo di disposizioni, potremo dire che sussista, nell'organismo dell'insetto, uno "scheletro motorio" simmetrico. Ma questa simmetria delle disposizioni e delle attivitá muscolari é sorretta e guidata in certo senso e dentro certi limiti, da un'analoga disposizione ed attivitá delle regioni sensorie alla superficie dell'organismo. Quelle regioni infatti della superficie esterna dell'organismo che sono devolute all'ufficio di zone concentratrici e trasmettitrici delle azioni del mondo ambiente, le regioni sensibili, cioé—e, fra esse, gli organi di senso ed i sensilli—sono del pari e ad un dipresso distribuiti in maniera simmetrica sui due antimeri dell'animale. I piú vistosi organi di senso, gli occhi, le antenne, ι palpi, sono pari e simmetrici; gli eventuali organi impari, taluni ocelli, ad esempio, sono contenuti nel piano sagittale, e per i nostri fini, possono venir considerati come due organi pari infinitamente vicini. Volendo indicare con una nuova espressione questa disposizione di cose, potremo dire che, accanto ad uno scheletro motorio esiste, nell'organismo di molti insetti uno "scheletro sensorio." Fra questi due scheletri simmetrici, come li abbiamo chiamati, per amore di schematicitá, intercedono definite relazioni.

Si puó ammettere che, in maggiore od in minore grado, l'insetto venga guidato, nella sua deambulazione, dalla azione di agenti esterni, dalla luce, dagli effluvii chimici, dalla temperatura, da particolari direzioni di spostamento del fluido ambiente e cosí, via.[3]

Le percezioni di tali diverse qualitá di stimoli, *aventi valore motorio*, avvengono mercé quelle giá nominate regioni sensorie

---

[3] Fenomeni che nella teoria dei tropismi assumono ordinatamente il nome di fototropismo, di chimiotropismo, di termotropismo, di reotropismo ed anemotropismo ecc.

della superficie dell'organismo e vengono incanalate come stimoli attraverso vie di conduzione che talora é comodo supporre
—agli effetti fisiologici—mantengano un andamento antimerico
connesso con la disposizione simmetrica degli organi percipienti.

Ecco quindi come siano legate in certo modo la simmetria sensoria e la simmetria motrice dell'organismo. Comunque si
immaginino costrutte e disposte le relazioni fra le due, *quest'é
certo, che in determinati casi un disturbo di quella prima simmetria
si traduce in un'alterazione della seconda.*

Occorre qui segnare quali note differenziino questo nostro
modo di esprimere la determinazione dell'orientamento del
piano sagittale dell'organismo in un ambiente stimolatore, dalla
nota concezione del Loeb.[4]   Per il Loeb, l'organismo é pure una
superficie simmetricamente sensibile; l'ambiente che la circuisce
é considerato come un campo di forze nel senso rigidamente
fisico, anzi, faradayano, dell'espressione.   L'incontro di un tubo
di flusso con la detta superficie vi provoca mutamenti prevalentemente e presumibilmente di ordine chimico, di entitá proporzionale al valore dell'intensitá del flusso.   É in tali mutamenti
che risiede la ragione dell'eccitazione che la regione sensibile
trasmette, ad esempio, ad un gruppo muscolare, il quale verrá
posto cosí in attivitá con una intensitá proporzionale ancora
all'intensitá del flusso. E l'organismo, in forza di questa destata
attivitá muscolare, si orienta rispetto alla direzione delle linee
di forza dell'agente stimolatore in modo che l'eccitazione si
uguagli sulle due regioni simmetricamente sensibili.   Tende cioé,
in definitiva, a disporre il proprio asse longitudinale nella direzione delle linee di forza del campo od in quella delle risultanti
dalla composizione delle linee di forza di due campi che si com-
penetrino.[5]

Cosí si puó formulare in breve la teoria dell'orientamento quale
si può trarre dalle interpretazioni loebiane dei fenomeni di tro-

Veggasi il cenno bibliografico e per una moderna e pur fedele esposizione del
pensiero di Loeb, veggasi Bouvier. La vie psychique des insectes-Flammarion.
Parıs 1920 e Kuhn.   Die Orientierung der Tiere im Raum.   Fischer, Jena, 1919
ove, con alquanto diversa forma, sono esposte vedute che si accostano. a quelle
discusse nella presente nota.

[5] Particolare sul quale ha insistito dettagliatamente il Bohn.

pismo. La differenza fra di essa ed il nostro punto di vista é di carattere sopratutto logico; in realtá il biologo, nel considerare un organismo che risponda a variazioni delle condizioni di ambiente (stimoli) alterando comunque il proprio comportamento, non saprebbe risolvere se—nel decidere del senso e del modo di quelle alterazioni—prevalga il fattore puramente fisico e chimico, se cioé l'organismo sia in piena balia del mondo stimolante, oppure se tale decisione dipenda da una destata connessione di attivitá proprie all'organismo e vincolate agli effetti fisiologici della sua sensorietá, se cioé vi sia in esso ed in quale misura, una spontaneitá psicologica. La fisiologia non ci da modo di risolvere il problema e l'aderire all'una piuttosto che all'altra spiegazione, alla psicologistica piuttosto che alla chimicofisica, non é giá conclusione, ma presupposto piú o meno necessario, della ricerca. Alla stregua quindi dei dati di fatto, l'una concezione non é piú infondata dell'altra.

Ora quel modo, che abbiamo detto, di rappresentarci le connessioni fra l'azione degli stimoli del mondo fisico e la reazione motoria dell'organismo, evita entrambi quegli scogli, in quanto non si riferisce tanto alla distribuzione degli stimoli nello spazio ambiente, non tanto, cioé, allo stimolo considerato a sé, quanto alla distribuzione della loro azione sulla superficie sensoria dell'organismo. Cioé a quello che—per estensione—si potrebbe chiamare la percezione dello stimolo da parte dell'insetto, ove della percezione si tenga sopratutto presente l'effetto fisiologico.

L'ambiente viene cosí, strettamente e continuamente, riferito all'organismo e quasi percepito attraverso ad esso; posizione del ricercatore, cui sono strumento, beninteso, le convenzioni che porge la fisiologia comparata dei sensi, nonché le acquisite nozioni sull'anatomia dell'organismo.

Questo punto di vista permette senz'altro una analogizzazione ed una generalizzazione di certo interesse. Si identificano per esso, ad esempio, le alterazioni della simmetria in quello che abbiamo chiamato lo scheletro sensorio, provocate per alterazione della distribuzione degli stimoli fisici nello spazio ambiente e quelle provocate per alterazione della distribuzione delle zone sensorie alla superficie dell'organismo. Cosí possono apparire intimamente connessi i fenomeni descritti dal Bohn, di giro in circolo, in taluni decapodi, per diseguale illuminazione delle superfici oculari e quelli descritti, ad esempio, da Dewitz, dal Parker, dal Dolley, di giro in circolo in un campo luminoso uniforme, per opacamento di una cornea dell'insetto cimentato. Né meno interessante é la connessione che si puó istituire fra queste alterazioni e quelle che il Dubois provocava nel *Pyrophorus noctilucus*, otturando con cera annerita un degli organi luminosi laterali. L'animale deviava in circolo verso il lato illuminato. Il Dubois anzi supponeva che l'illuminamento effettivamente servisse all'animale per l'orientamento. Nelle tre alternative: esperimenti con ischermi opachi, opacamento delle cornee, soppressione della luminescenza laterale, vi é sempre un tratto comune, che é rappresentato dalla soppressione della distribuzione simmetrica delle "percezioni luminose."

Infine, passando dalla considerazione della pura sensorietá dell'organismo a quella delle vie anatomiche di essa sensorietá, dovremo comprendere entro gli elementi di quel primo nostro "scheletro," lo stesso sistema nervoso centrale il quale é appunto costruito, generalmente, con un'architettura simmetrica.

Altro puntosul quale é utile soffermarci qualche poco, e che sembra militare in favore del concetto dei moti in circolo, provocati per generica alterazione della simmetria sensoria dell'animale, si é quello della *specificitá delle reazioni dell'organismo rispetto agli stimoli in azione*. L'aver mutilato un animale di un dato organo di senso, l'averne cioé lesa la simmetria sensoria, non é sufficiente a provocare sempre e comunque nello scheletro motorio tale dissimmetria, che ne sia determinato un moto in circolo. Perché questo compaia, occorre effettivamente che la sensibilitá dell'animale venga eccitata, ma non una qualsiasi forma di sensibilitá, bensí, ed eventualmente, quella di cui è strnmento l'organo destinato all'amputazione unilaterale. Ad esempio, i maschi del bombice del gelso cui sia stata amputata una antenna, compiono moti in circolo solamente allorché siano stati eccitati dagli effluvii odorosi che emanano dal corpo della femmina.[6] Analogamente gli individui della *Drosophila ampelo-*

[6] Kafka. op. cit. Vo. 6. (Confermato da Kellogg nel 1907).

phila, amputati dell'articolo terminale di una antenna, compiono moti in circolo solamente alloché vengano eccitati dalle emanazioni degli alcoli, degli eteri, degli acidi risultanti dalla fermentazione delle frutta.[7] Il Ràdl ha osservata una Dexia girare in circolo attorno ad un lume solamente allorché le era stata resa opaca una cornea.[8] Dolley e Parker hanno riferito di fatti simili vella *Vanessa antiopa*, né sarebbe difficile trovare, nella bibliografia, altri esempi di connessione specifica fra la alterazione motoria e l'alterazione di qualche elemento dello scheletro sensorio.

I risultati di esperienze consimili non sono quindi facilmente generalizzabili ed occorre in essi tenere rigoroso conto delle condizioni dell'ambiente fisico, degli stimoli che cioé possono avere agito sull'organismo. L'aver constatato, ad esempio, che l'ablazione di una antenna in un carabo non produce alcuno squilibrio della locomozione in condizioni normali (condizioni che generalmente si riassumono nella presenza di un campo luminoso uniforme), non implica che tale ablazione debba per sempre ed assolutamente rimanere senza influenza alcuna sulle attivitá dell'animale, ad esempio, sulla sua deambulazione. Se siesperimentasse infatti con uno stimolo cui l'antenna dell'animale fosse sensibile, probabilmente anche in esso si dimostrerebbero squilibrii di qualche sorta nel comportamento. Nelle nostre specie non abbiamo mai osservati fenomeni consimili, che potessero sovrapporsi a quelli di maneggio genuino. Riservandoci di esaminare in seguito quanto é stato visto da altri autori in proposito e di discutere la portata delle loro interpretazioni, prenderemo in esame quei fenomeni di rotazione che si accompagnano a lesioni unilaterali dei gangli sopraesofagei e che—come abbiamo indicato—si possono considerare come interessanti ad un tempo lo scheletro motorio e lo scheletro sensorio.

[7] Kafka. ibidem. Bouvier op. cit. II. Barrows, op. cit. 1907.
[8] Loeb. Die Tropismen, cit.

2. SUPPOSIZIONI SULLA DETERMINAZIONE DEL MOTO DI MANEGGIO

Esporremo con maggior dettaglio altrove i procedimenti che abbiamo posti in atto per gli interventi sperimentali sui nostri coleotteri, analizzandone le particolari modalitá. Le speciali condizioni di esperimento non permettendoci altra tecnica piú delicata, abbiamo aggrediti i gangli sopraesofagei indirettamente, attraverso la cuticola chitinosa della volta cranica, servendoci di aghi a manico. Poiché il procedimento puó sembrare grossolano e dare appiglio a giustificati dubbi circa la portata della lesione, ci preoccuperemo ora di eliminare questi dubbi, allo scopo di appianare la via alla ulteriore esposizione.

Il dubbio puó essere formulato cosǐ i moti di maneggio osservati nelle nostre esperienze dipendono effettivamente dalla lesione di una regione nervosa?

E cominceremo da un'obiezione, da taluno mossami, che si puó tradurre nel concetto che il moto di maneggio dipenda esclusivamente da una sorta di "fuga continuativa" del dolore, causata dalla lesione. Non discuteró ora dell'opportunitá di introdurre in fisiologia quest'ambigua nozione del dolore, accontentandomi di ricordare qui la nota esperienza del lombrico dimezzato, riportata dal Loeb nel giá citato libro;[9] ma ritenendo pure che ad una simile sensazione sia dovuta, ad esempio, quella retrazione degli arti che segue ad un toccamento dei tarsi e che in definitiva porta ad un incurvamento nell'opposto senso della deambulazione dell'animale, oppure quella serie di moti degli arti che per toccamento di un'antenna sorte medesimo effetto e cosí via, potrebbe sembrare che una persistenza di stimolo, quale deve essere legata ad una grave lesione di una metá dei cerebroidi sarebbe capace di determinare una persistente analoga reazione.

---

[9] La quale esperienza é peró di W. Normann (che il Loeb non cita) e fu compiuta sull'Allobophora caligınosa. Pfluger's Archiv, 1897. Un'idea analoga ci sembra aver visto espressa, in altra forma, dal Mast a proposito dei moti di rotazione presentati dalla *Planaria torva*. Riferisco le stesse parole dell'Autore: *"Planaria with one eye removed turns contınuously from the wounded sıde for some tıme, evidently owing to the stimulations of the wound"* (Mast. Preliminary report in the reactions to lıght in marine turbellaria. Carnegie Inst. Yearbook. Vol. 9. Cit. sec. Talıaferro Reactions to light in *Planarıa maculata*. Journal of exp. Zool. luglio 1920.)

Ma non si comprende allora perché la sensazione dolorifica dovrebbe essere esclusivamente legata ai gangli cerebroidi; qualsiasi
altra grave lesione, infatti, di qualunque altra regione sensibile e
laterale, dovrebbe trarre seco analoghe sensazioni dolorose ed
analogamente produrre moti in circolo, il che non avviene.

Puó sorgere il dubbio che il moto di maneggio dipenda dalla
lesione di qualche altro apparato o sistema che non sia il nervoso,
contenuto nel capo e nei pressi dei cerebroidi.   Sollevando infatti la chitina dalla fronte all'occipite, accanto ai ciuffi muscolari
laterali e sopra il ganglio, viene messa allo scoperto una porzione
di corpo adiposo e, sotto questo, un brillante plesso di ramuscoli
tracheali.   Fini diramazioni tracheali si estendono anche attorno
e sopra ai gangli, avvolgendoli come in una lassa rete, in genere
superficiale e talora qualche poco approfondita nello strato córti-
cale del ganglio, che, per la sua trasparenza, la lascia intravve-
dere al fuocheggiamento.   In taluni casi tale rete si fa tanto
cospicua da attirare vivamente l'attenzione e da affacciare con
essa un sospetto.   I processi di ossidazione della biomolecola
nervosa infatti non ne debbono essere indipendenti, poiché la tra-
chea é il veicolo dell'aria respirata.   La recisione di qualche ramo
collettore tracheale, operata dall'ago introdotto nel capo potrebbe
aver compromessa la normale ossigenazione di qualche regione
del cervello ed aver causati con ció in esso disturbi funzionali
tali da tradursi in alterati comportamenti dell'intero organismo.
Ma poiché lo squilibrio dinamico del corpo qual'é espresso nei
moti di maneggio, deve riposare sul turbamento di una simmetria
funzionale, occorrerá, per render responsabile la lesa tracheazione,
che questa normalmente si effettui secondo dispositivi simmetrici.
Ora, da apposite ricerche sul cervello delle *Blaps* e delle *Pimelia*
tale simmetria risulta confermata (fig. 1); né il decorso delle
diramazioni tracheali segue, nella specie, vie costanti per tutti gli
individui, bensí individualmente alquanto variabili, nella rete
cerebrale.   Solamente nella *Pimelia* é accennata una distribuzione grossolanamente simmetrica (fig. 2): da un grosso nodo
tracheale postcerebroidale, ove per breve tratto si fondono due
cospicui tronchi dando origine a. tasche tracheali ed a nuove
diramazioni, si dipartono tre ramuscoli distintamente, i quali

Fig. 1   Tracheazione dei gangli sopraesofagei di *Blaps mortisaga*.
Fig. 2   Tracheazione dei gangli sopraesofagei di *Pimelia undulata*. Si osservi la distribuzione grossolanamente simmetrica dei 3 rami tracheali.

mandano diramazioni rispettivamente alla metá interna della superficie superiore della porzione sinistra del sopraesofageo—all'avvallamento centrale che segna nel ganglio una sorta di strozzatura mediana, con fini diramazioni laterali—alla metá interna della superficie superiore della parte destra del ganglio. Vi sono poi altri plessi di minor entitá, che discendono lungo le commissure con il sottoesofageo.

Ma checché si pensi di questa dubbia simmetria, il fatto che alla superficie del cervello, i ramuscoli tracheali, per distinte frequentissime anastomosi sono in mutua comuncazione, é sufficiente a far porre da parte anche questa interpretazione dei

Fig. 3   Distribuzione dei ramuscoli tracheali nella cortica dei gangli sopraesofagei di *Pimelia*.

moti di maneggio (fig. 3).   Perché la recisione di un ramo tracheale, anche se laterale, possa infatti dar luogo agli effetti sup posti, é d'uopo che le sue diramazioni conservino un'autonomia ed una mutua indipendenza che mantengano valore alla sua disposizione simmetrica, la quale altrimenti diviene un puro fatto morfologico senza grande significato fisiologico.

Vi é, infine, un'ultima possibilitá degna di venire presa in considerazione: un coleottero leso che descriva un moto durevole di maneggio presenta il capo volto da quella banda da cui gira. Non potrebbe darsi che l'introduzione dell'ago nel capo avesse lesa la muscolatura del collo cosí da obliterarne la funzionalitá da un lato e da sottoporre il capo all'unica azione dei muscoli dal lato opposto che lo costringano in quella posizione?   E non potrebbe indi essere che—locomovendosi un organismo, ad un dipresso secondo le idee del Loeb, in un campo di forze stimolatrici nella direzione del piano sagittale del capo (del piano mediano, cioé, fra le superfici fotosensibili recettive) tale deviazione puramente meccanica del capo fosse sufficiente a determinare il moto di maneggio?   Ma tale abnorme disposizione del capo puó venire provocata ad arte mercé la resezione della muscolatura del collo dall'un dei lati, previa asportazione di un tratto di chitina dal corsaletto.   Orbene, un animale cosí operato ed in cui il capo presenta una costante piegatura da un lato senza che i gangli cefalici siano stati tocchi—si muove sicuramente e coordinatamente secondo una traiettoria contenuta nel piano sagittale, non del capo, ma del rimanente corpo.   Risultato che non solamente esclude che ad un tale obliquamento del piano interoculare sia dovuto il moto in circolo, ma che é tale da infirmare forse lo stesso concetto del Loeb, nella sua applicazione almeno a questo organismo.   Altre lesioni inferte ad altre regioni della muscolatura degli arti e praticate lungo le pareti laterali del torace, allo scopo di disturbare la motilitá degli arti da un lato non hanno avuto altro effetto che una parziale soppressione dell'attivitá motoria degli arti interessati, la quale non ha mai condotto a moti in circolo, ma all'assunzione di atteggiamenti anormali, per la soppressıone dei fattori meccanici dell'equilibrio.

E' lecito concludere, da quanto precede, che il maneggio non é vincolato ad una alterazione statica della simmetria muscolare, cioé alla inattivitá di determinati gruppi di muscoli.   Dipende esso forse, in prima analisi, da una alterazione dinamica di essa simmetria—da una alterazione, cioé, della funzionalitá simmetrica degli arti?

3. LA LOCOMOZIONE NORMALE DEI COLEOTTERI

Per poterne giudicare, é necessario ricordare come avvenga, nei coleotteri delle nostre esperienze, la locomozione normale. La deambulazione delle imagini degli insetti, vincolata alla presenza in essi di sei arti contemporaneamente attivi, si attua in uno specialissimo modo, che, nelle sue linee generali ed in modo sufficientemente preciso, era noto giá al Weiss e che indi é stato accuratamente studiato da Paul Bert, dal Graber, dal Plateau, dal Dahl. Tale deambulazione implica un complicato meccanismo di regolazione dei moti, meccanismo accuratamente studiato, nei suoi particolari statici e cinematici, da Jean Demoor. Secondo la precisa ed espressiva designazione del Graber, l'insetto in cammino si puó considerare come un doppio treppiedi ambulante e gli arti in moto possono venire raggruppati in due terne, ciascuna costituita dall'anteriore e dal posteriore degli arti di un antimero e dall'arto medio dell'antimero opposto, terne le quali si alternano nelle fasi di moto e di apparente riposo.   Mentre una terna d'arti, quella ad esempio, rappresentata dal primo sinistro, dal secondo destro e dal terzo sinistro, é sollevata da terra e sta descrivendo un'arcata per prender terra poco piú avanti, la seconda terna, rappresentata dal primo e dal terzo destro e dal secondo sinistro posa a terra e costituisce supporto al corpo dell' animale durante il "passo." Essa a sua volta si porrá in moto e descriverá un altra arcata all'avanti, quando la prima si sará posata a terra.   L'arcata viene descritta separatemente da ciascuno degli arti che costituiscono la terna ed il cui moto é contemporaneo o quasi.   In linea generale non si da peró contemporaneo moto degli arti appartenenti alle *due* terne.   Al loro turno, gli arti simmetrici compiono regolarmente escursioni di uguale ampiezza, cosicché, uguagliandosi gli effetti di trazione e di propulsione da entrambi i lati dell'animale in moto, viene determinata la marcia rettilinea.

Designeró come "coordinazione normale" dei moti locomotorii questo ritmico alternarsi di condizioni di attivitá e di quiete fra i sei arti avendo sopratutto riguardo al criterio cronologico, cioé al tempestivo e reiterato intervento di ogni arto nella locomozione e non tenendo conto delle particolari modalitá di impiego di ogni arto, modalitá connesse alla loro morfologia.   La coordinazione dei moti non interessa quindi la velocitá dei moti medesimi, né puó influire sulla forma della traiettoria descritta dall'animale o sulla velocitá del moto lungh'essa.

Il discorrere del particolare impiego di ciascun arto nella locomozione, come di fenomeno intimamente legato alla morfologia dell'arto stesso—mentre le osservazioni precedenti valgono ad un dipresso per la locomozione degli insetti in genere—ci conduce a limitare la nostra esposizone ai soli coleotteri, riferendoci a quanto direttamente abbiamo potuto osservare sugli esemplari avuti in esame.

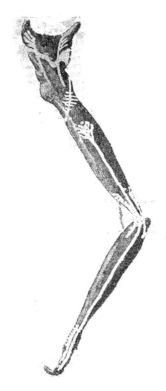

Fig. 4   Schema di un arto di insetto (modificato dal Dahl).

Un arto di coleottero adulto é un insieme di parecchie leve (fig. 4) le une vincolate alle altre in modo ben determinato, cosí da limitarne le possibilitá di spostamento nello spazio.   In modo approssimativo e schematico possiamo riferire questi movimenti a due piani perpendicolari (fig. 5) nei quali tipicamente hanno luogo i movimenti delle due leve principali dell'arto: il femore e la tibia.   Porremo l'un piano orizzontale e tangente alla superficie sternale del segmento nel suo punto mediano, l'altro, normale al

primo, orientato verticalmente ed individuato dal suddetto
punto sternale e dall'asse della tibia. Nel primo piano si compiono i moti del femore, nel secondo quelli della tibia; generalmente, gli spostamenti di questa sono compresi entro un angolo
massimo, determinato in valore dalla morfologia dell'articolazione
tibiofemorale.

In realtá le cose non istanno cosí semplicemente: il piano dei
femori non é orizzontale che in particolari casi ed in genere é inclinato cosí da riuscire (mediante opportuni trasporti paralleli)
tangente, non alla linea mediana degli sterniti, ma ad un punto
della superficie del corpo piú o meno presso le pleure. Il piano
tibiale non mantiene nello spazio un'orientazione costantemente

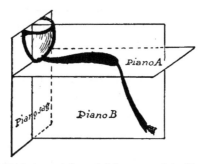

Fig. 5   Piani di riferimento dei moti del segmenti degli arti medi nella deambulazione (vedi testo).

verticale, bensí spesso ruota leggermente e ritmicamente intorno
ad un immaginario asse orizzontale, contenuto nello stesso piano.
Infine la serie di leve ch'é rappresentata dalla linea dei tarsi ha
spesso una meccanica propria e variabile da specie a specie.   Ma,
attenendoci a quello schema generale, caratterizzeremo l'impiego
delle tre paia d'arti, constatando che i moti delle leve del primo
paio avvengono tipicamente in un piano B e solo secondariamente
in un piano A (fig. 5).   Occorrerá assegnare un verso al moto con
cui sono descritti gli angoli tibiali nel piano B; e chiameremo peró
positivo il verso del moto descritto per adduzione della tibia sul
femore e negativo quello descritto per abduzione del primo segmento sul secondo.   Nella locomozione, il moto dei primi arti é
tipicamente efficiente per essere descritto nel piano B con verso

positivo; il moto degli arti medii é invece caratterizzato dal pre-
valere di spostamenti nel piano A e dalla subordinata importanza
(almeno nella marcia rettilinea) degli spostamenti tibiali nel
piano B. Il moto degli arti posteriori si avvicina a quello degli
arti anteriori: sono cioé particolarmente efficienti in esso gli
spostamenti nel piano B, ma descritti con verso negativo.

Supposto, per comoditá, che tutto il sistema di leve dell'arto
sia contenuto in un unico piano, nel piano B, ad esempio, é facile
constatare come, per le singole paia di arti, questi piani non siano
ugualmente inclinati sul piano sagittale; ad un dipresso ortogonale
gli é quello dei medii, mentre quello dei primi gli é obliquo
e diretto all'avanti; parimenti obliquo, ma diretto all'indietro é
quello dei terzi arti. Riferendoci ad un asse orientato, rappresentato dall'asse sagittale dell'animale, diretto dall'addome al
capo, l'angolo compreso fra di esso e la traccia del piano B nel
primo paio di arti é minore di un retto, é prossimo ad un retto per
il piano mediano, é molto maggiore di un retto ed in taluni casi
prossimo ad un piatto per il terzo paio.

Possiamo ora in altre parole esprimere la diversa funzione di
ogni arto nella locomozione; gli arti medii hanno un'azione emi-
nentemente propulsiva, sospingono cioé il corpo all'avanti, descrivendo una doppia serie di arcate: arcate sollevate ed all'avanti
per "compiere il passo" e prendere indi terra—arcate in posizione
di appoggio al terreno e dirette all'indietro, per sospingere il
corpo. I terzi arti pure compiono un'azione schiettamente propulsiva[10] contraendo la tibia sul femore dapprima perché i tarsi
e l'estremo tibiale prendano terra all'avanti ed estendendo indi
la tibia sul femore, cosí da impellere il côrpo verso l'innanzi. Vi
é dunque fra gli arti del secondo e del terzo paio una parziale
somiglianza d'impiego, ottenuta con diversi mezzi, poiché in
quelli il principale lavoro é sostenuto dall'articolazione coxo-
femorale ed in questi dalla femorotibiale. Una funzione ben caratteristica spetta agli arti del primo paio; essi, alquanto rivolti
verso l'avanti, hanno un'azione prevalentemente attrattiva sul
corpo, si stendono, di afferano al substrato con i tarsi, indi con-

---

[10] Poiché i secondi in condizioni che non siano quelle della marcia rettilinea
entrano pure in gioco con azioni attrattive nel piano B.

traggono la tibia sul femore e avvicinano il corpo al punto di aggrappamento dell'ultimo articolo tarsale. A questo moto che avviene nel piano B con verso positivo e che caratterizza l'impiego degli arti del primo paio si accompagna generalmente un moto ad arcata, operato nel piano A, il quale peró non propelle il corpo, ma modifica il valore del moto di pura trazione esercitato dalla contrazione tibiofemorale suddescritta. I primi arti godono, rispetto ai successivi, di un certo grado di libertá e di movimenti, che fa veramente di essi—come giá da tempo si é detto—i timoni dell'animale. Potendosi essi aggrappare nell'atto della loro estensione, in posizioni alquanto laterali, possono causare oscillazioni e deviazioni della deambulazione dalla linea retta. Gioverá notare peró che anche per gli stessi coleotteri, l'impiego dei primi arti puó mostrarsi nel dettaglio, abbastanza vario, cosí come varia la loro morfologia, dalla snellezza e dalla elegante sagomatura degli arti dei carabici alla tozzezza della tibie appiattite e dentate dei *Copris*, dei *Gymnopleurus*, degli scarabei in genere, in cui la linea dei tarsi si riduce e puó anche mancare completamente come negli *Ateuchus*, l'insetto servendosi allora per la locomozione delle estremitá distali dei femori.

Se queste sono le linee generali della locomozione dei coleotteri, é d'uopo convenire che la meccanica ne é assai plastica e che l'animale, di fronte ad amputazioni gravi dei tarsi ed anche di parte delle tibie od a svariati disturbi meccanici della morfologia dell'arto, reagisce con fenomeni che potrebbero essere detti di autoregolazione, mercé i quali viene consentita la prosecuzione della marcia. Noi non abbiamo tracciato quindi—per quanto in modo molto sommario—che lo schema della locomozione di un coleottero morfologicamente integro in condizioni normali di funzionalitá, procedente in linea retta. Intervengono fatti nuovi, allorché il coleottero in marcia cambia di rotta. Il che agevolmente si puó ottenere premendo leggermente con una cannuccia ed anche sfiorando leggermente con un pennello uno dei tarsi, in ispecie delle due prime paia di arti; allo stimólo l'animale risponde deviando dal lato opposto da quello stimolato, allontanandosi cioé dallo stimolo. L'arto stimolato si é vivamente retratto (come accade sempre, anche allorché l'insetto sia in quiete

od impossibilitato per particolari condizioni fisiologiche, alla locomozione) posandosi indi a terra in un punto diverso da quello
che avrebbe occupato nello stabilirsi normale della terna di riposo.
Passando da questa nuova posizione alla terna di moto, poiché il
punto d'appoggio dell'arto al terreno é piú presso al piano sagittale di quello che sarebbe stato il normale, l'impulsione fornita
dalla distensione dell'arto stimolato é piú intensa di quella fornita in una terna precedente dall'arto simmetrico, oltrecché
diversamente applicata. L'equilibrio degli sforzi propulsivi dai
due lati del corpo viene turbato ed il piano sagittale ruota cosí
da indirizzare la marcia verso quel lato da cui la somma di detti
sforzi é rimasta inalterata e quindi minore, cioé dalla parte non
stimolata.

Ma l'insetto puó girare anche "volontariamente" su sé stesso,
cioé in seguito all'azione di stimoli non direttamente applicati
dall'osservatore. La rotazione puó avvenire da fermo ed essere
operata da una serie di piccoli spostamenti eseguiti principalmente dagli arti del primo paio cui secondano i successivi con
piccoli moti di aggiustamento; tali spostamenti consistono, nel
caso che l'animale volga, ad esempio, a destra, nell'esagerarsi
dell'attivitá delle leve femorotibiali, le quali compiono dal lato
destro e lateralmente, moti di aggrappamento e di adduzione, dal
lato sinistro e pure lateralmente, moti di abduzione, cosí che il
corpo, per l'azione combinata di queste attivitá antagoniste
viene ruotato verso destra, finché l'animale non riprenda la
deambulazione rettilinea. La nuova rotta, fenomeno caratteristico di questo tipo di rotazione, forma un angolo ben deciso con
la vecchia direzione (fig. 6 e 7). Avviene talora, invece, che la
curva abbia piú grande raggio e che non si palesi fra le due direzioni un brusco mutamento (fig. 8). Ma lo studio del meccanismo di questo secondo tipo di svolta richiede che ci rifac-
ciano ai procedimenti di inscrizione grafica degli atti ambulatorü
dei coleotteri. Noteremo peró ancora che la coordinazione dei
moti durante la locomozione non tralascia mai di dimostrarsi
nella marcia normale; il meccanismo testé descritto, e che im-
pegna pochi arti dell'insetto, prevalementę due arti simmetrici,
potrebbe sembrare una contraddizione. Ma le svolte ad angolo

retto si producono solamente allorche' l'animale abbia sospesa la locomozione progressiva e sia quindi in quiete.   Non é peró il caso di parlare di coordinazione, poiché ogni arto—salvo le de

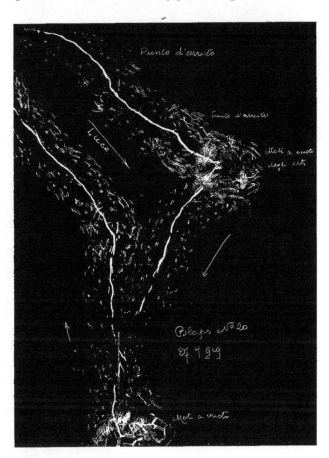

Fig. 6   Reogramma di *Blaps con* svolte ad angolo (rid. 1 /2).

bite eccezioni—puo venire usato indipendentemente dagli altri (nei moti di pulizia, ad esempio) in tutti quei movimenti che non siano di locomozione.

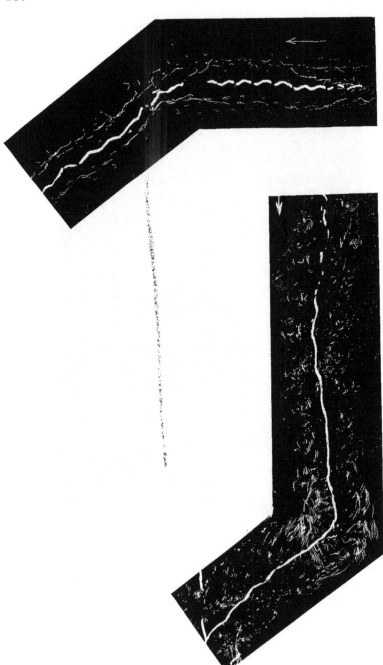

Fig. 8 Reogramma di *Cetonia floricola* on svolta **loc** (**id. 2 /3**).

Fig. 7 Reogramma di *Bups* on svolta ad ı ago mo uto be nella fig. 6 (rid. 2 /3).

4. L'INSCRIZIONE GRAFICA DEI FATTI LOCOMOTORII

Onde poter praticamente applicare il criterio di fare dell'arto uno strumento di segnalazione, che con la sua attività normale od alterata designi l'andamento dei fenomeni che si svolgono nell'intimità dell'apparato neuromuscolare interessato, occorre trovar modo di registrare continuativamente i fatti della marcia dell'insetto. L'inchiostratura diretta degli arti non sorte buon effetto e per l'untuosità caratteristica della cuticola dei coleotteri che impedisce un regolare deflusso della vena d'inchiostro sul sottostante foglio di registrazione, e per l'inomogeneità del tratto e per la troppo breve durata. Demoor afferma di aver usato un simile procedimento, ma non da particolari di tecnica. È preferibile trasportare dal tamburo di Marey sul tavolo registratore il procedimento tanto usato in fisiologia, di inscrivere i moti su di un foglio di carta affumicata. Di questo procedimento già aveva usato il Dubois, il quale pubblicava nelle sue "Lezioni di fisiologia generale" alcuni clichés di grafici consimili, da lui però eseguiti occasionalmente e non istudiati sistematicamente ed in dettaglio. I coleotteri in particolare si prestano bene a simili inscrizioni. La scarsa reattività delle nostre specie a stimoli fotici faceva sí che esse non rimanessero gran che turbate dalla presenza di una superficie nera sotto il loro corpo.

Molteplici sono i vantaggi offerti da siffatti reogrammi. Riconosciute le relazioni fra le tracce ed i moti corrispondenti degli arti che le hanno descritte, l'osservatore può studiare, con tutto suo comodo a tavolino, una serie di manifestazioni che mal si lasciano cogliere a volo nell'affaccendato susseguirsi degli arti nella locomozione ed i dati ricavati dalle due forme di esperienza mutuamente si possono integrare. Un reogramma permette confronti precisi tra due fasi successive di una medesima locomozione e consente l'effettuazione di misurazioni, l'introduzione, cioé, di un criterio quantitativo nell'indagine dei fenomeni.

La comparazione di reogrammi ottenuti da diverse specie, permette l'immediato riconoscimento di una caratteristica, almeno specifica, della locomozione; 'ogni specie fornisce un suo tipo di reogramma, dotato di certa costanza. La successione di tracce brevi e distanziate di una *Blaps* (fig. 9, 10), in cui i graffiti

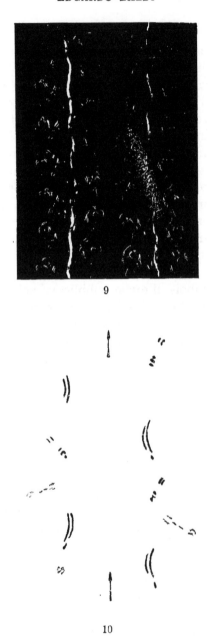

9

10

Fig. 9 Reogrammi di deambulazione rettilinea di *Blaps* (rid. 2/3).
    Fig. 10 Schema della disposizione delle tracce deambulatorie per l'inter-
pretazione dei reogrammi di *Blaps*. Le tracce esterne ed oblique all'indietro sono
degli arti medi. Le interne doppie a semiluna, degli arti posteriori.

dovuti alle estremitá dei tre arti si trovano ravvicinati in grup-
petti ben distinti, mentre rivela la leggerezza e la sveltezza della
marcia dell'animale, nettamente la distingue, ad esempio, dalle
tracce pesanti, spesseggianti, continuative di un *Osmoderma* (fig.
11), le quali sembrano giá rivelare la tozzezza dell'animale e la
lentezza impacciata dei suoi moti.

Né, per quanto superficialmente si rassomiglino, si potranno
confondere le tracce di un *Carabus* (fig. 12–13–14) con quelle di una
Aromia (fig. 15). Le strisciature falcate del primo, dovute al
rapido moto delle leve scriventi, sono ben diverse dalle tracce
bene impresse dell'*Aromia*, le quali talora permettono di distin-
guere il numero degli articoli tarsali appoggiati a terra e che
denotano un posamento lento, preciso, durevole, degli arti sul
terreno. Né sarebbe difficile continuare per altri tipi in interpretazioni del medesimo genere. Giá dunque di primo acchito il
reogramma dice qualcosa, circa la locomozione e la motilitá in
genere del coleottero (fig. 16–17–18–19).

Anche piú dice un suo esame dettagliato. Ben evidente é il
trascinamento dell'estremo dell'addome lungo una linea sinuosa,
trascinamento gia' notato con precisione dal Demoor e di cui
l'Uexküll ha indicate le ragioni meccaniche. Riprendiamo sui
reogrammi lo studio dei mutamenti di rotta ad incurvamento
dolce che abbiamo distinti dalle svolte da fermo e riferiamoci ad
un simile grafico desritto da una *Blaps mortisaga* normale; i
gruppi di tracce nella deambulazione normale, lasciati da tutti
gli arti da una banda, sono disposti alternatamente dalle due
parti del piano mediano, cosicché ognuno di essi gruppi corrisponde all'intervallo fra due gruppi attigui, segnati dagli arti
dell'antimero opposto. Le due fascie di tracce, destra e sinistra,
lasciano fra di loro una zona vuota di segni, e sinuosamente ed
interrottamente rigata dalle tracce dell' addome che a tratti
viene a strisciare sul terreno; zona la quale corrisponde alla
proiezione sul piano della zona di spazio occupata dal corpo
dell'animale. Il margine esterno delle fascie laterali é segnato
dalla successione dei colpi di unghia degli estremi tarsali dei
medii arti, punti facilménte reperibili. Misurando le distanze che
li separano per i due lati del tratto di marcia che presenta curvatura, si ottengono, in decimi di millimetro, i dati che seguono:

158

190

160

180

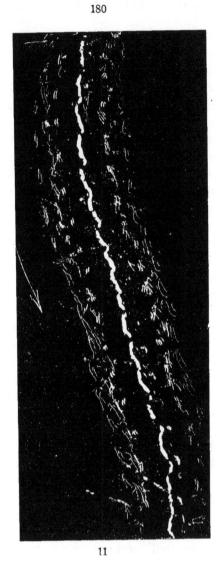

11   12

Fig. 11   Reogramma di *Osmoderma heremita* (rid. 2/3).
Fig. 12   Reogramma di *Carabus morbillosus* (rid. 2/3).

Fig. 13 Schema della disposizione delle tracce nei reogrammi di *Carabus*.
Le tracce marginali trasverse sono degli arti medi, quelle falcate dei posteriori.
Le tracce interposte alle falcate sono dei primi arti.

Allorché si tenga presente che la distanza fra due punti omotopi dei gruppi di tracce corrisponde all'ampiezza del colpo d'arco
che l'arto relativo ha data al proprio moto propulsivo—e si può
quindi ritenere proporzionale all'intensitá stessa dell'impulso che
l'arto ha trasmesso al corpo nell'atto locomotorio—appare evidente come l'animale, per isvoltare a destra abbia raccorciate le
ampiezze dei colpi d'arco dal lato destro, abbia cioé diminuita
l'intensitá degli impulsi motorii degli arti impressi al corpo da

236   EDGARDO BALDI

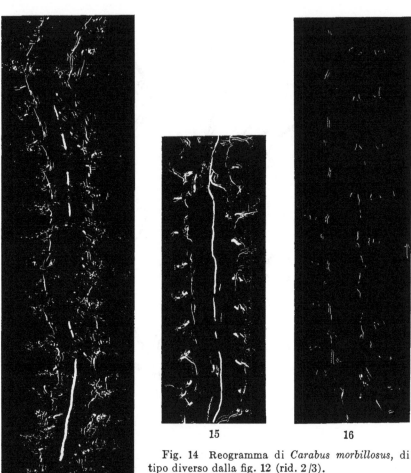

15   16

Fig. 14 Reogramma di *Carabus morbillosus*, di
tipo diverso dalla fig. 12 (rid. 2/3).
    Fig. 15 Reogramma di *Aromia moschata* (rid. 2/3).
    Fig. 16 Reogramma di *Pimelia undulata* (rid. 2/3).

14

quella banda, mentre ha esagerate le ampiezze dei moti e le
intensitá degli impulsi dal lato sinistro. L'intervallo normale,
infatti, tra due posizioni omotope nella serie locomotoria, rileva-
bile da altri tracciati di *Blaps*, é di quindici millimetri.

L'analisi che compiremo in seguito con i medesimi mezzi sui
tracciati dei movimenti di maneggio ci dimostrerá una certa

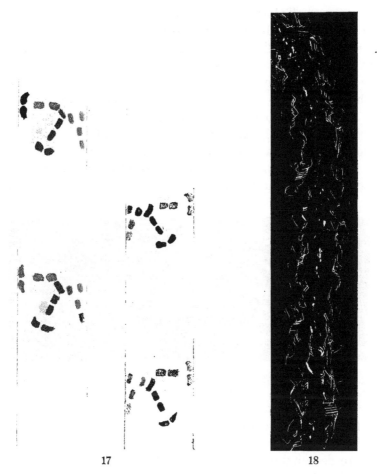

17
18

Fig. 17   Schema della disposizione delle tracce degli arti nella deambulazione di *Lamia*.

Fig. 18   Reogramma di *Oryctes griphus* (rid. 2/3).

somiglianza qualitativa fra di essi ed i fenomeni di svolta "volontaria," somiglianza sulla quale ritorneremo.

L'esame diretto dei fenomeni della deambulazione può chiarire qualche dettaglio dei tracciati reografici.   Cosí, nella Blaps mortisaga normale tutti e tre gli arti poggiano sul terreno con la spina della estremitá distale della tibia e con le unghie dell'ultimo

anello tarseo. Talora anche, per un meno pronunciato incurvamento della linea dei tarsi, il substrato è sfiorato anche dalle
apofisi dei rimanenti articoli tarsali, col che viene inscritta sulla
carta affumicata la serie di piccoli tratti che qualche volta si
scorgono dietro il graffio delle unghie. Graffio che nei primi
arti ha spesso una caratteristica forma lunata. Infatti l'azione
degli arti del primo paio si può scin dere in tre momenti successivi:

1. I tarsi unghiati, aggrappandosi al substrato, esercitano una
trazione dovuta alla tibia che viene addotta verso il femore;

2. la tibia avvicinatasi maggiormente al corpo fa forza sulla
sua spina distale puntellandosi molto obliquamente all'indietro:

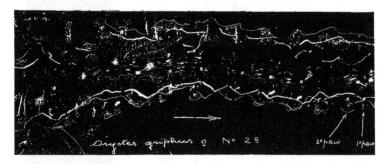

Fig. 19 Reogramma di *Oryotes griphus*. Le tracce dei terzi arti sono più
strisciate che nella fig. 18   (rid. 2/3).

3. frattanto la linea dei tarsi, facendo perno su detta spina,
descrive un piccolo angolo all'indietro tracciando sul terreno le
arcature dell'unghia ed i trattini delle apofisi tarsee, che si rilevano nel reogramma.

La meccanica dei secondi e terzi arti nella Blaps si conforma
ad un dipresso alle regole generali che abbiamo date più su. Nè
molto se ne discosta la locomozione della *Pimelia undulata*, in
cui i moti dei primi arti nel piano B sono meno evidenti che il
consueto ed in cui tali arti descrivono con la linea tibiotarsale
brevi arcate quasi rettilinee ed orientate parallelamente all'asse
sagittale dell'animale. Nella *Pimelia* è invece spiccata la caratteristica meccanica di propulsione all'indietro secondo lo schema
descritto, degli arti dell'ultimo paio.

É caratteristica delle specie, buone e rapidi camminatrici,—carabi e blapsidi, ad esempio,—una molto precisa coordinazione dei moti ambulatori, la quale spesso si esprime nel raggruppamento e nel distanziamento delle traccie locomotorie sui reogrammi. Con minor precisione essa si effettua nelle specie che camminano malamente; in esse, come nelle *Cetonia* e negli *Osmoderma* la contemporaneitá nel moto degli arti di una terna è rotta da un regolare ritardo di fase degli arti posteriori, i quali iniziano il loro moto allorché i primi sono giá presso al termine della loro escursione.   In tutte le specie che ho cimentate, vi é, del resto, assai piú assieme; sussistono, cioé, connessioni funzionali assai piú rigide fra gli arti delle prime due paia che fra questi ed i terzi, i quali, nei riflessi complessivi dell'organismo, godono di una certa autonomia.

Accennato appena all'esistenza di una deambulazione "sollevata" non ci dilungheremo oltre sulle modalitá della locomozione in tutte le specie che ci hanno servito nelle nostre ricerche, poiché nell'esporle, ci fonderemo sopratutto sui dati che ci hanno fornito le *Blaps*, i *Carcbus* e le *Pimelia*.

5. LE ALTERAZIONI DELLA DEAMBULAZIONE NORMALE NEL MANEGGIO

Lasciando per ora da parte il problema delle cause intrinseche del movimento di maneggio nei coleotteri—in qual misura debba cioé esso moto venir riportato all'alterazione di un'attivitá del sistema nervoso centrale—non occupiamoci che della questione del determinismo suo immediato.   Con quali modificazioni dell'-attivitá locomotoria normale viene effettuato un continuativo giro in circolo?

Scindiamo senz'altro il procedimento di risoluzione del problema in due parti.   Esamineremo dapprima lo spostamento delle relazioni dinamiche fra l'attivitá dei singoli arti, rispetto alla traslazione complessiva del corpo, esamineremo cioé, la proiezione di detti moti sul piano di deambulazione, servendoci dei reogrammi.   In un secondo tempo passeremo dal piano nello spazio esaminando—per cosí dire—la morfologia del movimento—,

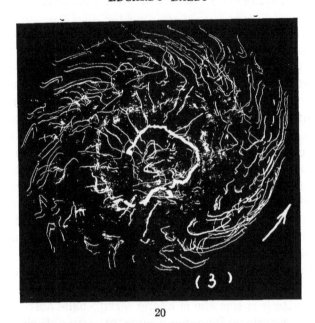

20

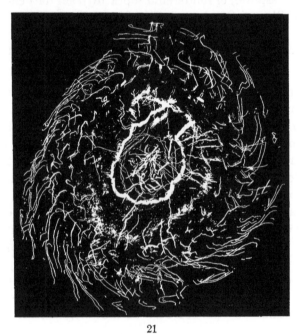

21

Fig. 20   Maneggio di *Osmoderma heremita* (grandezza naturale).
Fig. 21   Maneggio di *Osmoderma heremita* (grandezza naturale).

esaminando cioé come si scomponga l'attivitá globale di ogni arto in quella delle singole leve che lo compongono.

Un primo evidente segno di una dissimmetria locomotoria nel movimento di maneggio si rileva nei relativi reogrammi, dall'essere il tracciato relativo nettamente distinguibile in due metá, l'una concentrica all'altra, talora visibilissimamente separate dallo striscio lasciato sulla carta affumicata dall'estremitá posteriore dell'addome (fig. 20–21–22). Ciascuna di queste due zone concentriche, descritte rispettivamente dagli arti di ogni antimero, ha un suo proprio carattere; qualunque sia il senso in cui il giro è descritto la fascia esterna del reogramma corrisponde agli arti dalla banda lesa, l'interna a quelli della banda sana. Ora, la zona esterna ha sempre una larghezza maggiore dell'interna, particolare che gioverá ritenere poiché ne troveremo altrove la spiegazione. Tale differenza puó agevolmente venire tradotta in cifre.

Reogramma fig. 22

	LARGHEZZE MINIME IN MILL			LARGHEZZE MÁSSIME IN MILL		
	Largh totale	Larghezza fascia int	Larghezza fascia est.	Larghezza totale	Larghezza fascia int	Larghezza fascia est.
a	22	7,0	15,0	a' 25	8,5	16,5
b	21	6,5	14,5	b' 22	8,0	14,0
		7,5				
	23	7,5	15,5	c' 23	9,0	14,0

Riferendo tali cifre alla larghezza totale della fascia, assunta come unitá, si ottengono i seguenti rapporti, che dimostrano la relativa estensione della fascia interna.

$$a = 1/3,14 \qquad a' = 1/2,04$$
$$b = 1/3,25 \qquad b' = 1/2,75$$
$$c = 1/3,06 \qquad c' = 1/2,55$$

Cioé, nel citato reogramma, la fascia interna ha approssimativamente l'estensione di un terzo della fascia totale; supera tale valore nei punti di strozzamento del reogramma, gli é inferiore in quelli di allargamento.

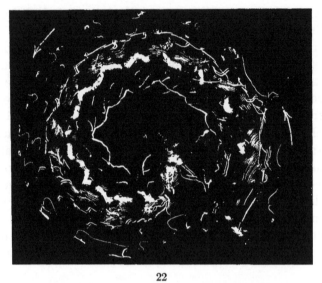

22

23

Fig. 22   Maneggio di *Cetonia aurata* (grand. natur.).
Fig. 23   Maneggio di *Blaps mortisaga* (grandezza natur.).

Reogramma fig. 23

PUNTI	LARGHEZZA TOTALE	FASCIA INTERNA	FASCIA ESTERNA
a	28	10,5	17,5
b	27	10,5	16,5
c	26	9,5	16,5
d	25	8,0	17,0
e	25	8,5	16,5

Donde, calcolando come dianzi i rapporti della larghezza della fascia interna alla larghezza totale:

$$a = 1/2,66; b = 1/2,57; c = 1/2,73; d = 1/2,12; e = 1/2,94$$

Anche qui, quindi, l'ampiezza della zona interna si approssima ad un terzo dell'ampiezza totale. Altri reogrammi di *Blaps* (fig. 23) mostrano ad un dipresso le medesime relazioni. Un valore pure prossimo ad un terzo della larghezza totale della fascia assume la larghezza della zone interna del reogramma di un carabo (fig. 24)

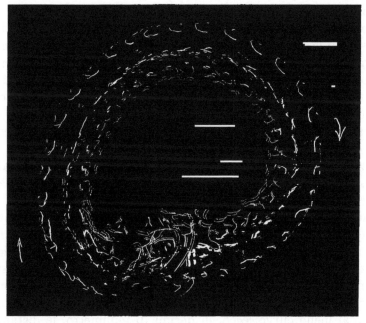

Fig. 24 Maneggio di *Blaps mortisaga*, con amputazione delle tibie sinistre (grand. natur.).

laddove, almeno essa é misurabile. Tale larghezza della fascia ha il suo significato; infatti i margini che la determinano segnano i punti in cui i tarsi, o comunque, la parti distali degli arti si sono posati a terra, per sospingere od attirare il corpo. Nella marcia normale, eguagliandosi l'ampiezza delle due fascie dai due lati, le impulsioni trasmesse attraverso due bracci di leva di uguale lunghezza, cioé attraverso agli arti in normale estensione, parimenti si uguagliano. Una diminuzione della lunghezza di un braccio di leva da un lato, dato che l'intensitá degli sforzi motorii si mantenga simmetrica, dovrá tradursi in una diminuzione di impulso ivi, l'organismo dovrá volgersi da quella banda. Vedremo quanto vi sia di attendibile in questa interpretazione suggerita dalla minore ampiezza della fascia interna del reogramma. Inoltre tale fascia interna si distingue talora dall'esterna anche per il carattere dei tratti che la costituiscono. Mentre la fascia esterna presenta una successione di gruppi di tracce ben distanziati, non dissimile da quella che si osserva nei reogrammi della deambulazione normale, l'interna si mostra costituita da tracce continue, da ghirigori allungati e continuantisi, quasi che le arcate descritte dagli arti corrispondenti, anziché elevarsi ritmicamente dal suolo, per ritornarvi in punti decisi, si fossero schiacciate, appesantite e l'arto talora si trascinasse sul terreno. Tale sintomo di abbassamento dell'attivitá locomotoria degli arti é specialmente manifesto nei posteriori. La misurazione delle distanze fra i punti omotopi di gruppi successivi di tracce da altresí modo di poter misurare le differenze di attivitá fra gli arti dei due antimeri, una quantitá proporzionale, cioé, allo squilibrio dinamico fra di essi. Un esame superficiale del reogramma (fig. 23) descritto da una *Blaps* in movimento di maneggio, mostra come la curvatura del moto non sia costante, ma come la traiettoria sia approssimativamente una ellissi. La curvatura ne é minima (cioé il moto é piú rettilineo) nelle vicinanze della piccola freccia, massimo nella regione che ne dista di un arco di circa 90° gradi, il che é bene osservabile nella zona interna descritta dagli arti sinistri. Passando dal tratto piú curvo al meno curvo, ecco, in decimi di millimetro, le distanze fra i colpi d'unghia degli arti medii:

Arti sinistri   *Arti destri*

30

                              170

50

                              170

50

                              180

60

                              180

70

                              190

Fig. 25   Schema per la interpretazione delle tracce periferiche del reogramma fig. 24 (grigio = primi arti—bianco = secondi arti —nero = terzi arti).

Da esse risulta non solamente l'altissima differenza di attivitá fra gli arti dei due antimeri, ma anche che, essendo le due curve su di esse costruibili diversamente inclinate sull'asse dei tempi tale differenza é minima per il tratto iniziale, meno curvo, del reogramma ed é massima per il tratto che nel reogramma si presenta come piú curvo. E poiché la curva dei destri é molto meno inclinata di quella dei sinistri, ne segue che l'essersi fatto piú stretto il maneggio é effetto di una ulteriore diminuzione di attivitá dei sinistri, piú che di un aumento di attivitá dei destri, *allorché beninteso, si considerino le componenti delle prestazioni locomotorie degli arti nel piano di deambulazione e non nello spazio.*

Il reogramma (fig. 24) si presta abbastanza bene alla misura per essere stato descritto da una *Blaps,* cui, per altri fini, erano stati recisi i tarsi degli arti sinistri, il che ha contribuito ad aumentare la schematicitá e la nettezza delle tracce. Il maneggio é destrorso; le tracce della fascia interna sono facilmente interpretabili grazie allo schema di cui alla fig. 10 (deambulazione normale). Le tracce periferiche alterate si scindono come mostra lo schizzo a fig. 25.

Riferiamoci quindi alle tracce dei primi arti, facilmente individuabili e misuriamo le distanze da gruppo a gruppo dalle due parti in decimi di millimetro.

Arti sinistri	Arti destri
110	
	90
100	
	90
120	
	90
120	
	90
120	
	90
100	
	90
105	
	80
100	
	70
105	
	80
110	
	85
125	
	90
120	
	95
115	
	100
130	
	95
120	

Sussiste fra le arcate dei singoli arti la costante differenza che giá altrove abbiamo notata. Analogo computo possiamo compiere sulle arcate dei secondi arti.

Arti sinistri	Arti destri
110	
	90
105	
	70
120	
	70

Arti sinistri	Arti destri
125	
	70
120	
	85
125	
	70
105	
	80
100	
	55
95	
	50
115	
110	
	55
120	
	65
130	
122	
	70
120	
	70
120	

Ripetiamo infine la misura per le tracce dei terzi arti

Arti sinistri	Arti destri
100	
	80
110	
	80
115	
	100
115	
	80
105	
	85
95	
	75
85	
	60
105	
	80
95	
	80

Arti sinistri	Arti destri
110	
	80
120	
	80
120	
	90
125	
	90
115	
	80
115	
	70
125	
	90

Gioverá notare che l'ampiezza delle arcate destre e sinistre é anche attenuata dal fatto dell'amputazione delle tibie sinistre, le quali, cosí mozze, diminuiscono la lunghezza della leva arto e quindi l'ampiezza della arcate sinistre. A tale causa sono probabilmente dovuti taluni lievi scarti rilevabili nella tabella. Le cifre relative all'attivitá degli arti destri sono peró costantemente inferiori a quelle misurate per i sinistri. Su tali dati é costruito il diagramma fig. 26.

Ad una prima ispezione le curve del diagramma rivelano nelle frequenti irregolaritá di ciascuna la presenza di qualche fattore secondario disturbatore del loro regolare andamento, che crediamo in parte almeno poter riferire alla inevitabile imprecisione di una misurazione su di un tracciato eseguito da arti inegualmente mozzi, indi al numero non sufficiente di misurazioni, indi a condizioni interne della locomozione, di apprezzamento non immediato. Comunque, appare netta la distinzione fra il gruppo superiore, che traduce l'attivitá degli arti sinistri e l'inferiore, relativo ai destri. Il gruppo superiore mostra indi un certo parallelismo nell'andamento delle sue tre curve (e segnatamente nella seconda metá del diagramma) che é buon segno della normalitá di funzionamento degli arti relativi. Vi sono in esso oscillazioni d'assieme, cui le tre curve contemporaneamente obbediscono e che provano, al piú, una obbedienza complessiva degli arti dell'antimero alle condizioni che hanno determinate quelle oscillazioni. Verosimilmente, se il reogramma fosse stato descritto da un animale dotato di arti illesi, dalla banda sinistra esso gruppo di curve si sarebbe spostato anche piú in alto, segnando piu nettamente la sua distinzione dall'inferiore.   Il gruppo inferiore di curve, relativo agli arti destri, é meno unito nelle sue oscillazioni, come

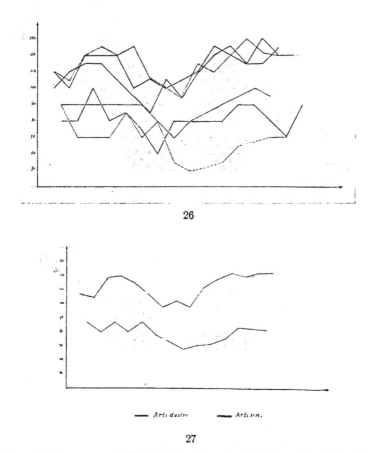

Fig. 26 Diagramma dell'attivitá degli arti nella Blaps di cul alla fig. 24. Vedi testo. (dall'alto in basso: 2° sinistro, 1° sinistro, 3° sinistro = 1° destro, 2° destro, 3° destro.)

Fig. 27 Diagramma delle medie per i due fasci del precedente (Cfr. testo).

risulta dal diagramma fig. 27, nel quale sono prese per ordinate le medie fra le ordinate di punti di uguale ascissa nel primo diagramma, per ciascuno dei due fasci di curve. Consideriamo, per chiarircene l'andamento, le relazioni fra le singole coppie di curve nei due fasci, dovute al medesimo paio di arto. Appare manifesto un parallelismo fra le curve relative all'attivitá dei terzi arti, curve che mantengono ad un dipresso lo stesso andamento, maggiormente scostandosi verso la fine del diagramma. Ció significa che la differenza di attivitá fra i terzi arti si mantiene ad un dipresso costante, accentuandosi un poco nei punti di minima curvatura del reogramma. Non é dai terzi arti, quindi, come insegna anche l'osservazione immediata, che principalmente dipende il meccanismo del maneggio. Una constatazione analoga suggeriscono le curve relative all'attivitá dei primi arti, le quali mantengono un andamento grossolanamente parallelo e sprovvisto persino di quella divergenza terminale che abbiamo rilevata nelle curve per i terzi arti. Ma nella meccanica dei primi arti intervengono altri fenomeni non rilevabili da un reogramma (in quanto prevalentemente si svolgono nello spazio) e che loro assegnano una parte principale nei moti di maneggio. E' da notare, peró, nel reogramma, l'avvicinamento delle tracce relative alla linea mediana del tracciato, nella quale talora esse penetrano.

Il piú cospicuo grado di dissimmetria é porto dalle curve relative agli arti del secondo paio, le quali, con la loro reciprocitá di andamento, ricordano diagrammi precedentemente costruiti—e precisamente su dati di arti mediani. Conviene riconoscere che, per quanto riguarda l'entitá degli impulsi propulsivi *in un piano parallelo a quello di marcia*, la parte principale del moto di maneggio, *in quanto squilibrio dinamico*, viene assunta dagli arti mediani. Fatto che si puó osservare evidentemente trascritto nel diagramma fig. 28, nel quale sono riavvicinate le curve del secondo arto sinistro e la curva delle medie degli arti sinistri, esprimente cioé la globale attivitá dell'antimero sinistro. Ora, tra le singole curve, le cui medie hanno servito alla costruzione del predetto diagramma (fig. 27) quella che piú si accosta alla forma della curva di media, quella che, cioé, in certo senso, le conferisce il suo proprio carattere, é precisamente la curva dell'isolato secondo sinistro, come il nuovo diagramma appunto mostra. Dal primo diagramma si possono inoltre ricavare le differenze fra ordinate di punti corrispondenti appartenenti alle due curve degli arti di una coppia simmetrica: dei due primi, del due secondi, dei due terzi arti.   Tali differenze specificatamente e continuativamente indicano i valori della dissimmetria dinamica esistente

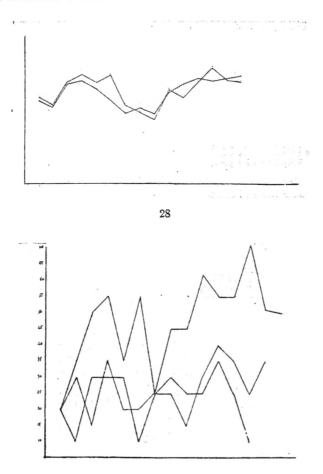

Fig. 28   Vedi testo.   Fig. 29   Vedi testo.

fra gli arti della coppia e, prese complessivamente, il valore della dissimmetria dinamica esistente fra i due antimeri.   Con tali differenze, ordinatamente calcolate per ogni coppia di arti, é stato costruito il diagramma fig. 29, dal quale evidentissimamente risulta il maggior valore che tale differenza assume per gli arti del secondo paio, rispetto a quelli del primo e dell'ultimo paio. La curva dei secondi arti abbraccia pressoché completamente le altre due.

Sarebbe ardito pretendere di voler ricavare altre conclusioni dalla analisi dei reogrammi. Un'ultima occhiata al secondo—diagramma di medie—mostra la netta distinzione delle curve relative ai due antimeri ed il loro divergere nei punti di maggiore curvature del reogramma. In questo la disposizione alternata dei gruppi di tracce ancora ben riconoscibile, prova che la coordinazione non é andata perduta, durante il maneggio. In altri reogrammi, questo particolare non é nettamente visibile, la fascia interna peró presenta una sorta di festonatura, in cui gli apici ed altri punti di facile riferimento cadono fra i corrispondenti gruppi della fascia esterna.

Altri reogrammi, sui quali, piú o meno agevolmente, si possono istituire computi analoghi, mostrano diversi tipi del moto di maneggio, sui quali ritorneremo dicendo delle particolari condizioni in cui, di volta in volta, si attua quella meccanica tipica del maneggio, che ora schematizziamo.

Con la convinzione che ulteriori misure di reogrammi netti e significativi piú di quelli che noi possiamo considerare qui possano condurre ad interessanti constatazioni sulla dinamica comparativa degli arti nel maneggio, riassumeremo le conclusioni che si possono trarre dall'esame di quelli che abbiamo sott'occhio.

1. L'attivitá degli arti dei due antimeri si mostra dissimmetrica, per essere la fascia descritta dagli arti del lato sano piú ristretta dell'opposta di circa un terzo dell'ampiezza totale.

2. E per essere ivi le tracce talora altrimenti disposte che nei reogrammi normali e nella fascia opposta.

3. Per essere talora la fascia interna costituita di tratti continuativi e non bene differenziati in gruppi.

4. Per essere sopratutto le arcate descritte dai singoli arti assai meno ampie dal lato integro che da quello leso.

5. Tutti segni—questi—di una minore attivitá degli arti dal lato sano, rispetto a quella degli arti dal lato leso.

6. Parte preponderante in questo disquilibrio dinamico della locomozione hanno gli arti medii.

7. V'é fra gli arti degli antimeri una certa dipendenza funzionale, per cui quella proporzionalitá fra le attivitá di ciascuno si mantiene grossolanamente costante, oscillando l'inferioritá degli arti dal lato sano rispetto all'attivitá degli arti opposti da un terzo ad un mezzo di quest ultima.

Gioverá soffermarci un istante sul punto segnato con il numero 5. Nel caso della *Blaps* abbiamo giá visto come l'intervallo normale fra due gruppi di tracce locomotorie sia di 150 decimillimetri in media. Poiché un moto in circolo é genericamente provocato da nua *differenza* di attivitá, vediamo se nei nostri reogrammi esso sia dovuto ad una eccedenza sul normale dell'attivita' degli arti esterni oppure ad una deficienza degli interni. Scegliamo in ultimo un esempio nel reogramma fig. 30 descritto da un *Carabus* (veggasi anche fig. 31). La media distanza fra i gruppi di tracce in esso non supera i 27 o 28 millimetri dal lato leso, valore proprio alla deambulazione normale retta e non lenta degli individui congeneri sani. Un confronto fra i valori delle distanze in punti di diversa curvatura si puó istituire sul medesimo reogramma, laddove il tracciato momentaneamente si allarga, assumendo curvatura molto minore. I seguenti dati sono misurati in esso lungo il margine esterno e relativamente alle tracce degli arti medii.

a.	mm. 25		h.	mm. 25
b.	mm. 23		i.	mm. 25
c.	mm. 23		l.	mm. 24
d.	mm. 24		m.	mm. 24
e.	mm. 24		n.	mm. 24
f.	mm. 25		o.	mm. 23
g.	mm. 27			

Íl divario é pressoché insensibile. Nell'ultima parte del tracciato, laddove ne comincia la diminuzione di curvatura, é possibile una sommaria misurazione tra le insenature manifeste nella curva lasciata dal terzo arto sinistro, la cui sinuositá continuativa permette di seguirla per un tratto entro il reogramma stesso. Eccone i dati paragonati a quelli per il corrispondente arto destro.

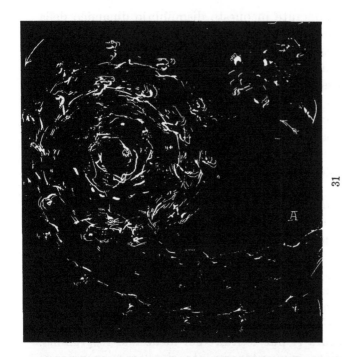

31

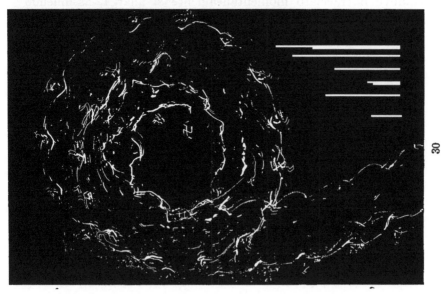

30

Fig. 30 Maneggio di Carabus morbill   tura variabile.
Fig. 31   gio di   bus morbillosus a   tura variabile.

Arti destri	Arti sinistri
	110
210	
	130
225	
	150
230	
	150
230	
	175
235	
	170
220	
	175
220	
	180

La curvatura della traiettoria va continuamente diminuendo dall'alto al basso della tabella.   Il grafico fig. 32 traduce in curve

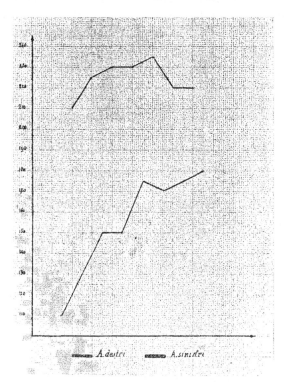

Fig.32   Vedi testo.

quèsti valori. Da esso risulta evidente la differenza di anda
mento delle due curve. Mentre quella degli arti destri non di-
scende che leggermente verso sinistra, cioé verso il punto di
maggiore curvature della marcia, la curva degli arti sinistri nella
seconda metá precipita rapidamente, dimostrando come la maggiore strettezza del giro sia dovuta piú ad una diminuzione di
attivitá nei sinistri, che ad una esaltazione di essa nei destri.
Si noti, infine, come le curve si riferiscano ad una coppia di terzi
arti che giá nella *Blaps* abbiamo visto non avere la parte principale nello squilibrio dinamico del maneggio.   Passiamo ora all' e-
same del moto degli arti nello spazio.

Descriveró la meccanica degli arti nel maneggio tipico, riferen-
domi specialmente alle *Blaps* che ne hanno fornito i migliori casi.
Supporro', per comoditá, nelle considerazioni che seguono, che
l'animale descriva un maneggio sinistrorso; parlando cioé di arti
sinistri e destri, intenderó riferirmi rispettivamente agli arti dal
lato illeso ed a quelli dal lato leso.   Le differenze fra l'attivitá degli arti destri e quella dei sinistri risulta a prima vista dall'osservazione che, mentre nel descrivere le relative arcate, gli arti
destri si raggiungono o quasi, nelle posizioni estreme, i sinistri
rimangono costantemente lontani gli uni dagli altri.   Gli arti
destri, in forza delle piú ampie traiettorie descritte linearmente,
sembrano animati da maggiore velocita'.   In realtá, la coordinazione non viene gran che menomata; persistono, cioé, le relazioni
fra i tempi di inizio dei moti per ciascuno degli arti, in valori
uguali o proporzionali a quelli normali.

Inoltre gli arti sinistri sono tutti—e specialmente i primi—piu'
flessi dei corrispondenti arti destri. Ossevando l'animale dal-
l'alto si veggono cosí gli arti destri sporgere dal corpo di un tratto
maggiore che i sinistri; infine il corpo presenta verso sinistra una
inclinazione piú o meno accentuata.   Nell'arto medio si puó
talora osservare chiaramente la disposizione che causa tale abbassamento in pari tempo che tale flessione (fig. 33).   Rispetto
ai destri, i femori sinistri sono maggiormente flessi sulle coxae
cioé piú aderenti alle pareti del corpo; in pari tempo le tibie sono
maggiormente flesse che nel normale, rispetto ai femori ossia si
sono ad un tempo fatti piú acuti i due angoli cox ofemorale e

femorotibiale. Poiché, comunque il punto d'inserzione della tibia all'estremo distale del femore viene sollevato nello spazio rispetto a quella che é l'orientazione normale del corpo (esprimentesi nella verticalitá del piano sagittale) e poiché, d'altro canto, é forza che l'estremo distale della tibia o taluno fra i primi articoli tarsei tocchi terra, tale orientazione viene menomata ed il corpo é obbligato ad obliquare verso sinistra. Il piano sagittale non é piú coincidente con un piano verticale, ma fa con esso un angolo sulla sinistra, piú o meno acuto. Fenomeni analoghi, se forse non cosí schematici, si osservano nel primo e nel terzo arto.

La meccanica dei primi arti conserva loro ancora un compito direttivo nella marcia ed é perció particolarmente interessante.

Fig. 33 Meccanismo dell'inclinazione del corpo nel maneggio.

Nel primo arto destro é sopravvenuta una modificazione nell'impiego delle singole sue leve che lo raccosta all'arto destro del secondo paio. L'arto descrive cioé ampie arcate in cui il moto anteroposteriore del femore ha parte capitale; il moto d'attrazione della tibia sul femore, caratteristico nell'animale normale, ha perduto d'importanza. Ove esso sussista (poi che talora é affatto obliterato) agisce in opposto senso, non con il risultato attrattivo, cioé, di un moto flessorio, ma con il risultato impellente di un moto estensorio. La tibia, mentre il femore compie la sua arcata d'avanti in addietro, si viene estendendo sul femore stesso, cosí da trovarsi estesa sulla sua linea allorché il femore abbia raggiunta l'estrema sua posizone. Si vede come in tal caso la linea d'azione dell'arto traversi diagonalmente il corsaletto dell'animale. Ma l'azione tipicamente impellente del primo arto destro si accompagna e si coordina ad un'azione attrattiva del primo sinistro, la quale rappresenta del pari un'alterazione dell'impiego normale.

In tale arto sono grandemente ridotti i moti ad arcata, moti im-
pellenti in senso anteroposteriore; talora essi sono affatto scomparsi; il femore non ha che piccole oscillazioni in un piano orizzon
tale e la maggiore attivitá viene esplicata dalla tibia che ha
esagerato il suo moto contrattivo sul femore.   La tibia, e con essa
la linea dei tarsi, puó essere paragonata ad un uncino che si ag-
grappa e si ritrae.   Il suo moto avviene pressoché completamente entro il piano di simmetria del segmento femorale; il femore puó essere piú o meno inclinato sul piano sagittale; spesso
la sua inclinazione é uguale in valore a quella dell'arto simmetrico
all'estremo della sua corsa.   Cosí che le due azioni, impulsiva a
destra, attrattiva a sinistra, si sommano lungo una medesima
direzione, raggiungendo un effetto massimo.   I secondi arti
rivelano poco piú attenuato uno squilibrio simile nel loro impiego;
l' arto destro continua od esagera nello stesso senso la sua attivitá normale: da grandi colpi di arcata dall'innanzi all'addietro,
cui prendono parte tutti i segmenti dell'arto, dal femore in giú.
In esso sono scomparsi o ridottissimi i fenomeni di flessione della
tibia sul femore.   Dal lato sinistro predominano invece tali
azioni flessive; i moti ad arcata del femore sono di molto ridotti
(raramente scompaiono del tutto).   I moti della tibia collaborano
con quelli del femore.   Il piano, cioé, in cui muovesi la tibia, é
obliquo rispet o al piano di simmetria del femore; entrambi i
piani della locomozione vengono trasportati in una direzione ad un
dipresso normale a quella del piano di simmetria del femore (moto
anteroposteriore del femore) il che fa credere che la tibia compia
moti ad arcata piú cospicui del reale.   Aggiungasi che le linee
tarsali degli arti destri descrivono quelle rotazioni che abbiamo
rilevate nell'individuo normale, mentre i tarsi di sinistra vengono
spostati parallelamente a se stessi.   Nei terzi arti infine, nei
quali lo squilibrio é meno visibile, data la loro funzione eminente-
mente impulsiva, si puó notare che gli angoli massimi e minimi
(di contrazione e di estensione) fra la tibia ed il femore hanno
valori superiori nell'arto destro, il quale veramente "impelle."
L'arto sinistro o da deboli impulsioni, oppure permane in uno
stato di flessione pressoché costante, spostandosi all'avanti a suo
turno, di quel tanto che é richiesto dallo spostamento generale
del corpo senza esercitare impulsione veruna.

Rapidamente abbiamo cosí tratteggiate le linee di impiego degli arti nel moto di maneggio, ponendone in luce due momenti fondamentali che per ora daremo come fatti di osservazione, senza occuparci della loro interpretazione:

1. Una differenza quantitativa di attività muscolare fra gli arti dei due antimeri.

2. un predominio dal lato illeso di atteggiamenti flessivi delle leve dell'arto.

Riteniamone la nozione di una dissimmetria dinamica fra gli antimeri, intesa come differenza della complessiva quantitá di lavoro fornita dagli arti di ciascuna banda e domandiamoci: é tale dissimmetria come puro fatto meccanico condizione sufficiente a determinare il maneggio? Non meglio si puó tentare di rispondere a tale problema, che studiandosi di provocare, in animali integri, i moti in circolo, per alterazione delle condizioni meccaniche della loro locomozione.

6. LA RIPRODUZIONE SPERIMENTALE DEI MOTI DI MANEGGIO

Esperienze di questo tipo meriterebbero di venire riprese ed analizzate dettagliatamente; se ne guadagnerebbero forse interessanti notizie sui processi di autoregolazione dell'organismo. Diró rapidamente delle poche che io ho allestite. Giá esse, nel loro sistematico fallire, ci provano quel ch'era di leggieri prevedibile; come l'organismo in locomozione non sia una semplice macchina in moto, ma bensí qualcosa di piú delicato e complesso. Il che—beninteso—non tange i procedimenti della ricerca, ma puó solamente distribuirli in una seriazione le cui gradazioni corrispondono ad una diversa approssimazione alla complessitá reale delle relazioni. Poi che il momento piú appariscente del maneggio sembra consistere nella differenza di attività degli arti dei due antimeri abbiamo tentato, ricordando anche le conclusioni del Bethe, in *Pimelia* ed in *Blaps* normali, di provocare un simile squilibrio, operando l'immobilizzazione degli arti da un lato, mediante legatura loro, in un solo assieme. Ma con questo grossolano procedimento vengono turbate le condizioni generali di equilibrio dell'animale nello spazio, in tal modo che esso, nonché locomoversi, neppure riesce a mantenere la stazione eretta.

Poiché non sembrava praticamente possibile, in un individuo normale, altrimenti sopprimere l'intera attivitá di un antimero, abbiamo provveduto ad abbassare il valore locomotorio, accorciando la lunghezza delle leve che, puntellandosi al terreno, tramettono il movimento al corpo. Abbiamo mozzati, cioé, i tarsi e parte delle tibie sino a due terzi della loro totale lunghezza agli arti da un lato, cosí che da quel lato essi incontrassero il terreno in regioni assai piú prossime al corpo che dall'opposto., tentando cosí di riprodurre all'incirca una delle condizioni occorrenti nel maneggio e di cui abbiamo detto. In seguito alla mozzatura degli arti sinistri era quindi preveduto un maneggio sinistrorso. Previsione non confermata dall'esperienza; l'animale ha tenuta marcia variamente orientata ed eccitata, senza tracce di giro in circolo, verosimilmente supplendo al disturbo arrecatogli con una piú intensa impulsione agli arti sinistri e con una maggior distensione di questi e retrazione degli opposti. In un *Blaps* che seguiva maneggio destrorso, una simile operazione non ha che allungato insensibilmente e fugacemente il circolo descritto.

Sempre al medesimo fine, in altro *Blaps* abbiamo immobilizzate tutte le articolazioni degli arti destri mediante un avvolgimento spirale attorno all'arto, di un sottile e robusto fil di ferro, cosí da irrigidire l'arto in una posizione di anormale estensione. Analogamente ed inversamente abbiamo agito con una simile legatura che costringesse l'articolazione tibiofemorale sinistra in un angolo fisso minore del normale. In entrambi i casi era preveduto un giro in tondo a sinistra ed in entrambi l'animale si é invece faticosamente mosso sia in linea retta, sia descrivendo svolte varie in diverso senso. E' notevole come animali siffattamente operati mostrino spesso nella deambulazione periodi di netta incoordinazione dei moti e—naturalmente— periodi frequenti di sosta e di immobilita'. La slegatura degli arti ripristina immediatamente la deambulazione normale. Per non inceppare immediatamente i moti delle leve dell' arto, abbiamo indi aggiustati alle tibie destre di una *Pimelia* tre leggeri fuscelli, di lunghezza appropriata, allo scopo di allontanare dal corpo il punto di applicazione dello sforzo motorio. Talora l'animale ha dato cenn di una certa tendenza ad obliquare sulla sinistra ed ha descritto pure qualche largo giro sinistrorso, ma senza costanza alcuna. I periodi di immobilitá abbondano ed é sufficiente una eccitazione per pressione sulle elitre, per provocare marcia in linea retta, talora non coincidente con l'asse sagittale, ma in direzione ad esso obliqua sulla destra.

Qualche risultato positivo é possibile ottenere legando fra di loro le estremitá delle allunghe assicurate alle tibie del secondo e terzo arto destro. In tal modo non viene gran che compromesso

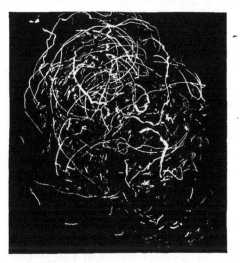

Fig. 34 Maneggio artificialmente provocato in *Pimelia*. Si noti la fisionomia molto differente da quella dei maneggi naturali.

l'equilibrio generale del corpo, inoltre vengono completamente paralizzate le attivitá dei due arti, medio e posteriore, di destra. Grazie all'attivitá, ora isolata, degli arti di sinistra, allorché essi si muovono, il corpo viene sul posto o con lievi spostamenti, ruotato sulla destra. Il primo arto destro, che é rimasto libero nei suoi movimenti, ora compie tentativi di raddrizzamento della marcia, arrancando nel modo che gli é consueto ed ora inverte il senso del proprio moto ed eseguisce una vera marcia indietro (fig. 34).

E' notevole la difficoltá che si incontra a porre in moto l'animale cosí obbligato, anche con violenti stimoli di percussione sulle

elitre. Sembra cioé, che la soppressione delle possibilitá mec
caniche della locomozione coordinata (legatura di due arti conti
gui e soppressione della loro indipendenza di moto) tragga seco
difficoltá generali nella stessa generica deambulazione.   (59)

Tali esperienze in genere negative, tranne che in un caso
banale, mostrano come nel maneggio uno squilibrio appunto
dinamico delle condizioni delle locomozione non esaurisca il
fenomeno e—se pure ne costituisca un momento necessario—non
sia peró sufficiente a determinarlo. D'altronde giá dalla pura
descrizione della meccanica locomotoria degli arti abbiamo
appreso come a quel momento "meccanico"—per cosí dire—si
aggiunga e mi si passi l'espressione—un momento "fisiologico,"
con che intendo accennare alle generali e fini alterazioni dell'impiego di ciascun arto, esplicantisi sopratutto nel predominare di
attivitá flessorie nell'antimero illeso. Ora, non v'é che un sistema dell'organismo la cui lesione sappia produrre alterazioni
tanto dettagliate e generali dell'attivitá muscolare di quello. Ed
é il sistema nervoso. In particolare, poi che la lesione é inferta
al capo, il sistema nervoso cefalico.

Riserveremo quindi il nome di maneggio ai fenomeni di alterata motilitá, implicanti una rotazione dell'organismo attorno ad
un asse qualsiasi, passante o non per il corpo, causati da una
alterazione patologica delle regioni nervose cefaliche, avvertendo
che l'alterazione del cervello puó non essere morfologicamente
evidente; é peró sufficiente che essa sia un'alterazione della sua
simmetria funzionale.

L'espressione di maneggio é stata imfatti diversamente usata
dai varii autori e ad opera di taluno ha subito notevoli e—a
parer nostro—non ben giustificate estensioni. Passiamo rapidamente in rivista quello che gli autori hanno appunto osservato in
proposito.

Tra le alterazioni della simmetria sensoria che non interessino
direttamente i centri nervosi, abbiamo giá elencate le esperienze
di opacamento delle cornee compiute dal Dolley ('16), dal Parker,
analoghe a quelle escogitate e praticate su varie specie di artropodi dallo Holmes ('01–'05) da Brundin e da McGraw ('13) suoi

allievi, da Carpenter ('18) sulla *Drosophila*, con risultati incostanti e similmente dal Radl sulla *Musca domestica* ('03). Nel 1901 il Rádl aveva giá pubblicate le osservazioni di moto in circolo relative all'idrofilo, per estirpazione di un occhio. Lo Hadley, nel 1908, distruggendo nei gamberi la cornea di un occhio, otteneva, oltre a movimenti in circolo, anche rapide rotazioni intorno all'asse longitudinale dell'animale. Un accoppiamento dei due moti: di traslazione circolare e di rotazione, aveva giá osservato il Demoor nel 1891 in un *Palaemon serratus*, per lesione, non di un organo ricettivo, ma di una porzione laterale del cervello. Rimanendo sempre nel campo delle alterazioni della simmetria sensoria, ricorderó, oltre a quelli giá citati, di Barrows e di Kellogg, i moti in circolo per amputazione di un' antenna, osservati dal Dubois ('86) nel *Pyrophorus*, previa asportazione bilaterale degli occhi.

Sembra che anche gli otocisti possano determinare, se asportati unilateralmente, fenomeni di moto in circolo e di rotazione introno all asse longitudinale, secondo riferiva Yves Delage nel 1887 a proposito della *Mysis*.

L'Herrera, nel 1893, presentava alla Société zoologique de France una breve nota, in cui, con molta parsimonia di dettagli e di documenti, riferiva di aver provocati moti in circolo in mosche ed in altri insetti (sic!) introducendo nel corpo dell'animale, per una ferita laterale, qualche cristallino di bromuro di potassio, la cui azione ipostenica era peró di breve durata. Il medesimo autore avrebbe ottenuti maneggi anche per alterazione delle condizioni puramente meccaniche della locomozione: ponendo a cavalcioni dell'insetto una sorta di bilanciere fatto con un ago incurvato e lateralmente caricato di un peso (di una pallina di cera) ed anche amputando completamente da un lato gli arti dell'animale. Quanto noi stessi abbiamo visto sui nostri coleotteri, ci fa alquanto dubbiosi circa la semplicitá dei mezzi e la facilitá dei risultati delle esperienze dello Herrera. Anche Demoor riferisce ('90) di un moto in circolo osservato in un *Carabus monilis var consitus* per deficiente impiego di un arto, l'anteriore destro. Tale deficienza non era sperimentale, ma dovuta a condizioni patologiche spontaneamente insorte nell'animale.

Ma i moti di rotazione osservati da piú antica data e verificati di maggiore costanza, sono quelli dovuti ad alterazioni sperimentalmente apportate al sistema nervoso centrale e particolarmente ai gangli sopraesofagei. I dati relativi al ganglio sottoesofageo sono infatti scarsi, l'aggressione di quest'ultimo ganglio essendo in realtá cosa assai piú complessa ed irta di difficoltá tecniche, che nol sia la lesione dei gangli dorsali all'esofago. Demoor riferisce in proposito ('91) una esperienza del Faivre in

Fig. 35   Maneggio di *Aromia*.

cui la lesione unilaterale del sottoesofageo nel ditisco ha prodotto un breve maneggio.

Si puó dubitare, in base ai fatti esposti in altra nota di chi scrive, se tale breve maneggio sia effettivamente legato ad una lesione nervosa.   La bibliografia circa i moti in circolo per lesioni unilaterali dei sopraesofagei é assai vasta; la si puó vedere nei lavori citati onde sarebbe inutile che io la riportassi.

Gli autori hanno descritti diversi tipi di rotazioni:

1. in un piano secondo curve chiuse o spirali o ad ansa (moti di maneggio propriamente detti) (fig. 35).

2. moti di rotazione dell'organismo intorno all'estremo cefalico od all'estremo aborale od attorno ad assi passanti per regioni

del corpo prossime a queste, moti cui il Demoor aveva assegnato ('91) il nome di moti a raggio di ruota (*en rayon de roue*), togliendo a prestito l'espressione al Beaunis, che precedentemente l'aveva

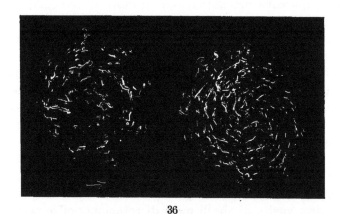

36

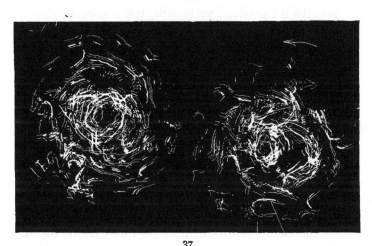

37

Fig. 36   Due maneggi della medesima *Blaps* mostranti il passaggio dal maneggio p.d. al maneggio in posto
Fig. 37   Maneggi in posto di *Carabus*.

usata per i vertebrati e che piú recentemente e pittorescamente furono dagli autori francesi chiamati moti di valzer (fig. 36–37).

3. moti di rotazione intorno all'asse longitudinale in entrambi i versi.

4. ed infine moti che il Demoor ('91) ha chiamati di *culbute*, effettuantisi secondo curve chiuse contenute in un piano verticale e da lui osservati nel *Palaemon serratus*.

Una complessiva denominazione di moti di maneggio assegna il Bohn ai moti rotatorii che egli provoca per disuguale illuminazione oculare in alcuni organismi marini e dei quali abbiamo giá fatto cenno.

Vanno ricordati infine, benché non siano propriamente moti di rotazione, i movimenti a rinculoni segnalati giá dal Dubois nel 1886 per l'emisezione sagittale dei cerebroidi e per lesioni in regioni prossime, sui quali é tornato il Comes, che crede di averli osservati per il primo negli artropodi decapitati.

Quest'ultima forma, come, noi stessi abbiamo osservato nei *Carabus* e nelle *Blaps* si puó combinare al moto in circolo, dando i maneggi a rovescio, o retrogradi.

Il quadro, vasto, di simili moti di rotazione, offre certi tratti comuni in tutto il tipo degli artropodi, tratti comuni che non mi sembrano ancora ben saldi rispetto al loro condizionamento anatomico e fisiologico.

Espressamente ho taciuto dei casi simili osservati, ad esempio, per lesioni dell'orecchio interno nei vertebrati, onde non ista bilire connessioni che possano essere illusorie.

7. ASPETTI DEL MANEGGIO: L'ORIENTAMENTO DEL CORPO. LA VARIAZIONE DEI MOTI

Ci é qui impossibile dettagliatamente esporre tutte le varianti e gli aspetti speciali che il moto di maneggio assume nei singoli casi. Tale studio analitico e minuzioso esige separata trattazione ed io la intraprenderó altrove, esaminando altresí il problema delle relazioni che intercedono fra l'ubicazione topografica e l'entitá della lesione ai sopraesofagei e la reazione motoria dell'organismo. I dati schematici che ho precedentemente esposti ci saranno sufficienti per un tentativo di interpretazione generale dei movimenti di maneggio nei coleotteri.

Nell'insetto che sta compiendo un movimento di maneggio, non solamente il portamento degli arti é soggetto ad un'anormale dissimmetria, bensí quello di tutte le parti dell'organismo suscettibili di reciproca mobilitá. Prescindiamo dalle antenne; l'afflosciamento dell'antenna dalla parte lesa é sovente in diretta connessione con la lesione, sia dello stesso nervo antennario,sia della sua inserzione al cerebron (l'antennarió si diparte poco avanti e poco sotto all'inserzione del lobo ottico), sia dei muscoli proprii dell'antenna. Tale immobilizzazione non é del resto, fenomeno costante. Pochi casi ho osservati in cui le antenne mostrassero un portamento diverso tra di loro, senza che alcuna lesione dei tessuti suindicati fosse palese. Il capo é invece, in tutte le specie cimentate, volto da quella parte da cui si effettua il maneggio e talora obliquamente piegato dalla medesima banda. Tale piegatura aumenta di valore allorché il giro di maneggio si faccia piú stretto e si stabilisce allorché l'insetto passa da una deambulazione rettilinea ad un moto in circolo. Ho mostrato peró come tale fenomeno non sia da annoverarsi tra le cause efficienti del maneggio. Esso é semplicemente un fatto concomitante.

Allorché il corsaletto sia alquanto mobile sull'addome, come nei carabi, esso pure si mostra incurvato nel medesimo senso del capo, rispetto all'asse sagittale dell'addome. Anche tale curvatura si stabilisce e si accentua durante il maneggio, analogamente a quella del capo. Ci sono ormai note le flessioni cui sono stati sottoposti gli arti dal lato illeso: maggiore accostamento dei femori alle pleure e delle tibie ai femori. Gli arti dal lato illeso sono relativamente piú flessi e piú retratti verso il corpo degli opposti. In conseguenza di cio', l'orientamento stesso di tutto il corpo nello spazio viene turbato; se la lesione sia stata praticata a destra, il corpo si mostra piú o meno sbieco sulla sinistra. Tale abbattimento laterale del corpo non é legato alla lesione di qualche organo di senso statico od a quella di qualche apposita funzione nervosa, bensí alla sola condizione meccanica della maggiore flessione degli arti da un lato.

Nei carabi normali l'orientamento generale del corpo sembra vincolato a quello del piano anteroposteriore che passa per gli equatori delle cornee. Ledendo infatti la muscolatura del collo ad un carabo, cosí da impedire al capo movimenti spontanei e da poterlo per qualche tempo obbligare in posizioni determinate, ho osservato che l'abbattimento del corpo sulla destra era dovuto ad una rotazione del capo sulla sinistra. Rotazione, cui l'animale, privato del controllo muscolare sul capo, non poteva altrimenti reagire. Per effetto di tale rotazione, il piano oculare veniva ad essere obliquo sulla sinistra; l'evidente conato dell'animale di riportarlo orizzontale causava l'abbattimento a destra del corpo. Abbattimento che infatti cessava, cosí che l'animale riprendeva la postura normale, allorche' io ruotavo il capo in senso inverso orientandolo come negli individui integri. Che peró tale reorientamento automatico, del quale giá numerosissimi e svariati esempi offre la bibliografia in argomento, fosse dovuto ad una autoregolazione in certo senso cenestetica, se non visiva,

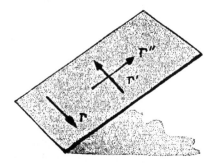

Fig. 38   Vedi testo.

ma non ad un ipotetico senso autonomo della spazialitá (quale sembrano indicare, ad esempio, le esperienze del Cornetz sulle formiche) dimostra il fatto che l'animale posto su di un piano inclinato non si muove secondo una parallela al lato d'appoggio e nel senso che meglio serva a correggere l'obliquitá del piano oculare, cioé secondo la direzione r della fig. 38, ma indifferentemente secondo le direzioni r' ed r'' in cui l'osservatore lo ponga. Il reorientamento sembra quindi legato qui a sole percezioni cenestetiche, interne all'animale, in discordanza con le conclusioni di Lyon ed Uexküll. Un caso piú complesso e piú difficile da analizzare mi é stato offerto da un ditisco leso al sopraesofageo destro. Ad una natazione sinistrorsa si accompagnava un distinto obliquamento del corpo all'avanti ed in basso ed uno sbandamento, pure nettissimo, del corpo sulla sinistra, il quale ritmicamente ed a non lunghi intervalli si mutava in una rotazione dell'animale intorno all'asse sagittale, sinistrorsa per chi guardasse l'animale a posteriori. Che tale rotazione non fosse dovuta che ad alterati moti degli arti, i quali non ho peró potuti individuare, é dimostrato dal fatto che una successiva lesione alla parte sinistra del cervello ha riequilibrato l'animale. Neppure qui, dunque, trattasi dell'álterazione di un senso specifico.

Tutte le anomalie che abbiamo cosí riassunte si lasciano ricondurre ad un'unica espressione; predominio nell'antimero illeso delle attivitá muscolari flessorie ed in genere dei gruppi muscolari adduttori e contrattori. La flessione del capo e del corsaletto, la flessione degli arti, sono diverso aspetto di una medesima condizione fisiologica; lo stesso addome, ove non fosse costretto entro una capsula chitinosa, probabilmente seguirebbe quelle medesima flessione dell'asse sagittale che fa il corpo curvo verso il lato illeso. E qui ci aiutano i risultati delle esperienze del Matula sulle larve di *Aeschna*, larve fornite di un tegumento debolmente chitinizzato ed articolato, sui segmenti dell'addome. Tali larve, per estirpazione di una metá laterale del cervello, mostrano l'addome piegato cosí che la parte concava guarda dal lato illeso del capo. Di piú, in tali organismi, il citato Autore ammette che lo scervellamento produca un aumento di tono dei gruppi muscolari flessori all'indietro ed una diminuzione dei flessori all'avanti, ossia un meccanismo analogo a quello che io suppongo nei coleotteri lesi ad una regione laterale del cervello. Né é molto diffcíle agire sulle condizioni generali del muscolo, anche prescindendo da lesioni del sistema nervoso, in modo da provocare simili esclusivismi di dati gruppi muscolari. Loeb, Garrey e Maxwell, ad esempio, studiando le reazioni galvanotropiche dell'*Amblystoma* e del *Palaemonetes* hanno trovato che l'animale, in un campo percorso da una corrente elettrica, tende a spostarsi verso il catodo o verso l'anodo, con alterazioni caratteristiche del portamento degli arti, cui sono impediti i movimenti in un senso determinato, per il predominare dell'azione di gruppi di muscoli antagonisti.

Ma vi é tutta una serie di fatti assodati dalla fisiologia della innervazione muscolare e che va sotto il nome di innervazione reciproca dei muscoli antagonisti, la quale ci può dare ancora qualche lume per l'interpretazione dei fenomeni di maneggio. Allorché una regione mobile dell'organismo sia sollecitata da muscoli antagonisti, tali cioé che la loro contrazione induca in essa movimenti reciproci, allorché l'uno dei muscoli é in contrazione, l'antagonista si mostra rilasciato.

Tale rilasciamento non é puramente passivo e dovuto sola mente alla distensione operata sul muscolo dall'azione dell'antago nista, ma la condizione di rilasciamento viene determinata in esso per via riflessa e sussiste anche allorché le connessioni meccaniche fra i due antagonisti vengano obliterate. Cosi' che entrambi i processi, nel gruppo antagonista, sono, in certo senso attivi. Ed il fatto é stato sperimentaménte provato dallo Sherrington sui muscoli retti del globo oculare di rana. Fenomeni simili sembrano largamente diffusi, benché il Dubois-Reymond ne abbia infirmata una facile generalizzazione. Il Fröhlich ha constatata anche nella muscolatura dei cefalopodi la contemporaneitá della contrazione e del rilasciamento nei gruppi antagonisti. Il fenomeno si verifica tanto eccitando il centro nervoso relativo, quanto eccitando il nervo lungo tutto il suo decorso. Lo Sherrington ha supposto che il rilasciamento riflesso sia dovuto ad un arco destato dalla contrazione dell'un muscolo negli organi muscolo tendinei del Golgi e deprimente il tono del muscolo antagonista. Il Verworn é riuscito a tradurre in tracciati i fenomeni descritti dallo Sherrington.

Ora, negli arti flessi di un insetto che descriva un moto di maneggio, ci troviamo indubbiamente in presenza delle medesime relazioni, poi che i segmenti dell'arto dell'insetto sono posti in movimento da una serie di coppie di muscoli antagonisti che regolano il loro spostamento nei due opposti sensi (fig. 4). In esso insetto i muscoli contrattori degli arti dal lato illeso saranno in condizione di esaltato tono e gli estensori loro antagonisti in condizioni opposte di tono depresso. Ossia, le condizioni complessive della muscolatura dell'antimero illeso sono alquanto complesse. Come si può quindi genericamente parlare di una esaltazione globale del tono muscolare, in un antimero, per lesione unilaterale di centri nervosi, come della principale causa del moto di maneggio—cosí come é nel pensiero, ad esempio, del Bethe?

Tutto questo insieme di alterazioni a carattere dissimmetrico, che accompagna l'effettuarsi del moto di maneggio, dipende anch'esso dal sistema nervoso, indubbiamente, ma non sembra vincolato *sempre ed esclusivamente* al maneggio, benché sia durante il maneggio appunto che esso trova la sua piú netta manifestazione. In taluni casi—i piú— in cui il maneggio non abbia carattere di continuitá, le vediamo definirsi allorché il maneggio interviene; in altri, in cui il maneggio manca, le vediamo abbozzarsi, accennarsi. Sono frequenti infatti gli individui lesi, che, pur non girando nettamente in tondo, manifestano leggere curvature del corpo e del capo da quel lato ove gli arti si mantengono un poco piú flessi, senza che questi lievi sintomi si acuiscano tanto da passare ad un vero moto di maneggio. In altri casi le vediamo persistere, affievolendosi, qualche tempo dopo che l'animale, trascorso un periodo di passeggero maneggio, ha ripresa la deambulazione normale. Sembra cioé che la loro determinazione non sia rigorosamente localizzata nella medesima regione nervosa, la cui lesione provoca il maneggio, ma che l'alterazione di zone circostanti sia anche sufficiente a destarle, in forme, peró, meno precise. Non si tratta, beninteso, che di una supposizione. Ma ció che é sopratutto da ritenere si é che esse appaiono con varia nettezza tutte le volte che un moto di maneggio interviene, piú spiccate nei maneggi tipici, meno nei transitorii ed atipici. Sono, queste, manifestazioni tali, crediamo, che una teoria dei maneggi non possa prescindere dalla loro considerazione. Gli autori che hanno descritti moti di maneggio li hanno un poco dipinti come qualcosa di fatale come un ananke rotatorio che sospinga incessantemente l'animale su di un binario circolare sino alla consumazione dei suoi giorni. Solamente la Drzewina, ch'io mi sappia, ha riferito di un ripristinamento della deambulazione normale, contraddicendo il Bethe, nel *Carcinus moenas* e nel *Pachygrapsus marmoratus*, in cui la guarigione avviene dopo piú che una decina di giorni. La *Lygia oceanica* ed una specie di

Palaemon le hanno dato risultati analoghi, benché meno precisi. Ma la Drzewina non da particolari suff cienti sulla tecnica dell'aggredimento del ganglio, né ha compiuti controlli anatomici sui suoi esemplari.

In realtá—ed il fatto risulta evidente dai dati relativi ai singoli casi di maneggio che altrove dettagliatamente esporró, il moto di maneggio é ricco di sfumature e di varianti. Raramente, anche prescindendo dai casi di ripristino temporaneo e totale della locomozione retta, esso maneggio é ininterrottamente continuativo; sempre offre una curvatura variabile che va dai giri in posto ad ampie circonferenze. In un esemplare leso e lasciato a se medesimo, non é raro osservare di giorno in giorno e talora di ora in ora qualche variazione, sia della forma che dell'ampiezza e della velocitá e talora persino, quando turbazioni secondarie si sovrappongano alla lesione originaria, del senso del maneggio. In casi atipici si possono osservare i moti di maneggio insorgere spontaneamente ad interrompere una deambulazione retta od un diverso comportamento anormale. In casi parimente atipici si puó pure osservare la cessazione definitiva di maneggi durati piú o meno a lungo, talora momentaneamente ripristinati da appositi stimoli, il che ha osservato anche la Drzewina nelle citate specie di crostacei. In casi di mancata lesione del cervello, ma di lesione di parti finitime—lesione mancata volontariamente od involontariamente da parte dello sperimentatore—l'organismo puó rispondere al trauma con pochissimi o con un sol giro di maneggio, talora con alcuni giri successivamente descritti con opposto verso.

Esiste, negli organismi lesi, una rigenerazione fisiologica? Il problema ha diverso senso a seconda del modo con cui lo si imposta, a seconda cioé che l'attenzione si porti esclusivamente sulla presenza di un tipico moto di maneggio o sull'insieme delle reazioni dell'organismo leso, nelle quali il maneggio puó rappresentare un momento piú o meno duraturo. Fra i numerosi casi tipici di durevole maneggio che ci si sono offerti nelle nostre esperienze, non ne possiamo annoverare che due, nei quali vi sia qualche cenno, non di rinormalizzazione del comportamento, ma di desistimento dal giro in circolo. Né i casi analoghi si presentano del tutto scevri da dubbio.

Nei quadri sintomatici piú complessi dei casi atipici, invece, per una lesione caduta fuor della regione frontale del cerebron, non é raro il trovare, a scadenza piú o meno lontana, ristabilita la deambulazione normale. Dall'insieme dei nostri dati ci sembra poter per ora concludere che nei coleotteri da noi presi in esame e per le durate medie dei periodi di osservazione, non segua rigenerazione fisiologica ai casi di lesioni bene localizzate e corrispondenti ad un quadro costante di alterazioni locomotorie (maneggio tipico). Ma quale valore puó avere allora il rinormalizzarsi del comportamento in casi di ferite anche piú gravi, ma non frontali? Quale meccanismo nervoso gli presiede? Il problema é aperto, né certamente é dei meno interessanti che offra la fisiologia del sistema nervoso in questi insetti.[11]

Riassumendo, per quanto ce lo permette, di fronte alla complessitá ed alla imponenza del problema, la frammentarietá di molti lati delle nostre osservazioni, ci sembra che ad una lesione unilaterale generica del cervello, l'organismo risponda con un vario e complesso assieme di fenomeni fra i quali prende posto anche quella serie di rotazioni nello spazio che abbiamo convenuto di chiamare maneggio. Serie, la quale offre di molte sfumature e di molte varietá, cosí che il maneggio tipico ne sembra a sua volta, un caso particolare, legato a certa gravitá della lesione ed al concorrere di certi particolari ed anormali atteggiamenti del corpo e degli arti.

La differenza fra il punto di vista abituale agli autori ed il mio sta appunto qui, nel considerare il maneggio, non come la risposta obbligata o la risposta in certo senso normale, regolamentare dell'organismo, ma come un particolare punto ed un particolare aspetto fra gli anormali sintomi che seguono alle lesione. E' lecito domandarci: cosí come ci sembra che il maneggio corrisponda, nei casi tipici, ad una ben delimitata lesione di una regione del cervello, non potrebbo essere che ciascuno di quegli aspetti anormali del comportamento che costitiscono il quadro morboso dei casi atipici, cioé della maggioranza dei casi, corrisponda ad una diversa e parimenti localizzata lesione di

---

[11] Questi accenni non posson essere chiariti senza uno stretto riferimento alla conoscenza di dati relativi a ciascuno dei casi che ho presi in esame.

una regione nervosa? La localizzazione della regione lesa varierá sempre e conformemente al variare del comportamento anormale? Lo potranno eventualmente dire ulteriori ricerche, ma é nostro concetto, genericamente confortato anche da recenti conclusioni della neuropatologia umano di guerra, che la considerazione delle localizzazioni non debba andare disgiunta da quella delle condizioni fisiologiche generali dell'organismo.

Ma, senza misconoscere al sistema nervoso le sue attribuzioni di unificazione funzionale dell'organismo, poniamo sopratutto attenzione, come ad un immanente condizionamento dell'estrinsecarsi della sua attivitá a quello che nell'organismo i francesi chiamano "état physiologique" e che, come assieme, é determinato dalle attivitá e determina le attivitá di ogni sua regione.

Come é possibile, su queste basi, una teoria del maneggio? Poi che essa non puó che limitatamente tener conto di questo complesso, singolare ed ancor troppo poco noto condizionamento, non puó essere una teoria del maneggio-tipo, che nella schematicitá costante delle manifestazioni che costituiscono quest'astrazione di fenomeno, tenti di fissare e di interpretare le ragioni del permanere di una data alterazione nell'organismo.

E' quello che ora vedremo, esaminando quanto possano contribuire alle ipotesi giá enunciate da altri autori, le nostre osservazioni.

8. IL DETERMINISMO DEI MOTI DI MANEGGIO

Ció che ha sopratutto colpiti gli osservatori che hanno indagati e teorizzati i moti di maneggio, si é l'esistenza di una dissimmetria quantitativa, se pure non misurata, nelle prestazioni degli arti dei due antimeri, piú che quella di una dissimmetria di impiego, d'una dissimmetria, vorrei poter dire, morfologica e qualitativa negli atti degli arti. Convien ricordare che le nostre conclusioni non vogliono che riguardare le specie di coleotteri che abbiamo studiate, né noi sapremmo generalizzarle, né eventualmente precisare l'estensione di una loro generalizzazione. Poi che se una dissimmetria quantitativa nel lavoro muscolare di due parti del corpo in quanto si riduca ad una differenza di attivitá nel tessuto muscolare nelle due regioni e sia indipendente dalla conformazione delle parti mobili cui il muscolo si inserisce é piú comprensibilmente generalizzabile, l'impiego dell'arto é cosa tanto strettamente connessa alla sua morfologia, da essere necessarie constatazioni singole e casuistiche per definirne le condizioni.

Il Dubois, nello studiare i moti in circolo descritti dal *Pyrophorus* per lesione laterale dei gangli cerebroidi, aveva nettamente affermato il carattere puramente "relativo" della dissimmetria locomotoria e propendeva anzi a riferirne il valore ad una diminuita attivitá degli arti dal lato sano, rispetto all'attivitá normale degli arti dal lato leso. Egli dice infatti "L'observation directe et l'examen des graphiques montrent clairement que les membres du côté opposé á celui de la lésion ne sont pas paralysés, ils sont seulement atteints de parésie: les mouvements ont moins d'amplitude et leur énérgie étant moins grande, l'action des membres du côté opposé devient prédominante; l'insecte est alors posé du côté le plus faible."

Il Dubois non accenna a diversitá nell'impiego stesso degli arti, ma la interpretazione che egli propone é corretta, anche da un punto di vista assai generale, poiché egli non ha alcuna difficoltá ad ammettere che la lesione di una regione laterale dei cerebroidi possa riflettersi sulla regione laterale opposta del corpo ed esplicitamente parla di un incrocio delle azioni fisiologiche, corrispondente ad un decorso chiasmatico delle fibre nervose, alludendovi con l'espressione di "paralysie croisée."

Poiché dalle ricerche dello Yung in poi ('78) era stato sostenuto il concetto che la trasmissione delle azioni nervose fosse strettamente laterale e che, rispetto alla loro innervazione, i due antimeri dell'artropodo si comportassero in modo affatto indipendente ed autonomo, non solamente rispetto alla trasmissione attraverso la catena subintestinale, m rispetto allo stesso cervello. "Aucun fait—diceva lo Yung—ne permet de supposer un entrecroise ment de fibres dans le cerveau."

Concetto sul quale insisteva anche il Demoor nella sua memoria del 1891. Entrambi questi autori hanno tratte le loro conclusioni da ricerche eseguite sui crostacei decapodi. Certamente negli insetti una tale affermazione non potrebbe venire sostenuta, benché riflessi di quella idea, di una azione omoantimerica, cioé delle lesioni cerebrali, si ritrovino anche nella teoria del Bethe.

L'Herrera, che, indipendentemente dal Dubois, aveva compiuto su insetti le varie esperienze di cui ho sommariamente riferito, proponeva pure ('93) una molto schematica teoria del maneggio, tutta meccanica, di cui faró parola per dovere di imparzialitá. Avendo constatato nell'insetto leso un abbassamento del corpo verso il lato della lesione, ed assumendo che a tale posizione sbieca sia dovuto l'avvicinamento al corpo degli arti dal lato sano, egli opina che la combinazione di questa forza attrattiva laterale degli arti sul corpo con la spinta in avanti di cui un corpo stesso è dotato, sia sufficente a chiarire la comparsa dei moti di maneggio. L'osservazione che la somma di due vettori ortogonali in un piano, rappresentanti due spostamenti rettilinei ed uniformi in moto, non puó essere una traiettoria circolare, basterá a far porre da banda l'interpretazione dell'Herrera.

Ometteró di ricordare qui le interpretazioni centriste e psicologiche anteriori, di Faivre, Burmeister, ecc., giudicandole interpretazioni non fisiologiche. Il Binet, nel 1894, si accontentava di una espressione piú generica, assumendo che i moti in circolo fossero dovuti ad una *ineguale eccitazione* degli arti dai due lati.

Né accenneró alla sommaria interpretazione del Matula, che per la sua particolaritá ed incompletezza non si presta ad una discussione generale. Occorre giungere sino al Bethe per trovare un'espressione piú precisa del determinismo dei moti di maneggio, espressione, la quale traduce in veste teoretica quella visione sopratutto quantitativa del moto di maneggio che abbiamo visto soffermarsi come sul momento fondamentale del moto in circolo, sulla dissimmetria delle attivitá propulsive degli arti dalle due metá del corpo. Non che il Bethe, accurato osservatore, non abbia notate differenze nell'impiego degli arti, poi che egli espressamente rileva nei suoi esemplari una positura anormale degli arti stessi, provocata da una disuguale tensione dei flessori e dei rotatori, ma egli non la pone in istrètta relazione con il maneggio, né con la lesione laterale del cervello (egli parla di *Ausschaltung des Gehirns*) e neppure accenna all'influenza che essa ha sul compimento dei moti locomotorii degli arti. Il Bethe ammette bensí che il cervello eserciti una generica azione inibitoria sugli automatismi e sui riflessi segmentali dei metameri dell'organismo posteriori a quelli cefalici. Esso cervello inoltre agisce come tonificatore dei muscoli; la sua azione si esercita—ed in questo il Bethe si collega alle vedute dello Yung e del Demoor— con uno spiccato lateralismo: ogni metá del cervello inibisce e tonifica la muscolatura del corrispodente antimero. Gli impulsi che partono dal cervello sono trasmessi all'indietro senza sottostare a smistamenti od a deviazioni chiasmatiche. Per lesione di una porzione laterale dell'organo inibitore viene tolta questa sorta di controllo ai muscoli degli arti dal lato corrispondente, i quali, movendosi senz'essere inibiti, dispiegano un'attivitá maggiore di quella degli arti dal lato opposto; questo disquilibrio di attivitá produrrebbe il moto in circolo dal lato sano.

Ecco le parole testuali del Bethe: *"Der Kreisgang, nach der gesunden Seite ist lediglich auf die Ungehemmtheit der operierten Seite zurückzuführen."*

Ne segue—ed il Bethe infatti lo afferma—che il moto in circolo non sia un moto coatto: compensando in qualche modo questa deficienza di inibizione dal lato leso, si deve poter raddrizzare la deambulazione. Ossia, il Bethe inverte quella che era la posizione del Dubois, il quale indeboliva il lato dell'animale opposto alla lesione. Mentre il Bethe, per assenza di tonificazione indebolisce il lato corrispondente.

Riferendoci sempre a quanto abbiamo veduto sulle nostre specie di coleotteri, confesseremo che l'interpretazine del Bethe ci sembra contenere bensi' elementi di veritá e rispondere in parte alle condizioni che effettivamente si verificano nel maneggio,ma sembraci pure che essa non descriva che un aspetto del fenomeno e non possa rappresentare una teoria completa del moto in circolo. Eccone le ragioni.

Anzitutto, constatazioni anatomiche compiute senz'alcuna preoccupazione fisiologica da morfologi puri, il Viallanes, il Cuccati, il Berlese, hanno assodato nel cervello degli insetti la presenza di chiasmi. Nelle fig. 39–40 riproduciamo taluni disegni del Berlese, togliendone il decorso di taluni fasci che s'incrociano prima di partirsi dal cerebron per incanalarsi nelle commissure della catena subintestinale.

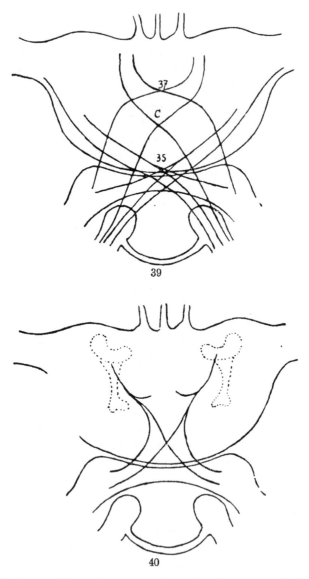

Fig. 39 Decorso di fasci chiasmatici nel cerebron degli insetti (da Berlese)
Fig. 40 Il chiasma ottico olfattivo nel cerebron degli insetti (da Berlese)

Il fascio indicato con il N° 37, si diparte dalla superficie del protocerebron da un lobulo di cellule gangliari, si incrocia con il simmetrico posteriormente alla sutura dei lobi protocerebrali e si reca ai lobi dorsali del deutocerebron. I morfologi lo hanno espressamente designato come *cordone chiasmatico*. Vi é ancora il *chiasma ottico olfattivo*, che abbiamo schematizzato nella fig. 40 e che dipartendosi dai calici si suddivide in tre rami di cui uno va al corpo centrale, l'altro al deutocerebron dalla propria parte ed il terzo, incrociandosi con il simmtrico, molto dietro il corpo centrale, al deutocerebron dalla parte opposta. Ma vi sono fasci che dal deutocerebron passano direttamente al tritocerebron e di qui alle commissure longitudinali della catena subintestinale.

Uno di essi fu descritto da Cuccatí, l'altro, denominato *fascio chiasmatico*, parte dal deutocerebrom, indi si biforca e mentre uno dei rami incrociandosi con il simmetrico, termina nel deutocerebron dalla parte opposta, l'altro, pure dopo essersi incrociato, va al tritocerebron e passa nelle commissure sottoesofagee.

Ve n' é ancora uno, proveniente dai lobi ottici, che, radendo il confine tra la sostanza punteggiata e la sostanza corticale passa al tritocerebron dalla banda opposta.

Tutte le regioni laterali del cerebron sono quindi in mutua relazione anatomica.

In realtá, l'effetto fisiologico della lesione laterale non é ristretto all'antimero corrispondente, come provano le informazioni che abbiamo precedentemente esposte e che dipingono un quadro generale di lesioni cui nessuna parte del corpo sfugge e che in ispecie per quel che riguarda l'impiego degli arti nella locomozione si accentua negli arti situati dalla parte opposta a quella della lesione. Né é chiaro, fondamentalmente, il concetto medesimo di inibizione. Il fatto capitale e piú volte verificato si é che l'assenza del cervello e la sua lesione provocano la comparsa di moti pendolari degli arti, di automatismi, di gesti ritmici e continuativi. Ma sappiamo noi in qual misura si dividano il determinismo di questi fenomeni, il sistema nervoso centrale e le disposizioni neuromuscolari del rimanente organismo? Ed il parlare di centri e di inibizione puó anche essere giustificato, allorché, con esse espressioni, per mancanza di una migliore conoscenza e di una piú adeguata interpretazione dei fatti, si intenda designare un complesso di reciproche influenze fra le parti inner- vate, le cui relazioni, traducentisi in un definito comportamento, vengono alterate in un deteminato senso dalla mancata fun- zionalitá di un anello della catena, del ganglio, cioé.

Altri fenomeni di ordine analogo a quelli che hanno suggerito il concetto di una inibizione del cervello sui gangli della catena, hanno condotto infatti al concetto simmetrico di una inibizione della catena sui gangli cerebrali, concetto giá accennato dal Bethe e che, nell'interpretazione del Comes ('12) ("inibizione reciproca dei gangli"), ha assunto importanza ed ampiezza pari a quella del primo.

E benché noi non mettiamo ın dubbio il valore dell'interpreta- zione per dimostrare come all'espressione di inibizione vada asse- gnato anzitutto un valore schematico e simbolico, ritorneremo un istante sulla nota esperienza del Normann, che tagliata trasver- salmente una *Allobophora*, osservava come la metá anteriore, "inibita" dal cervello tuttora presente, non alterasse il compor- tamento proprio all'individuo integro, mentre la metá posteriore non piú inibita nei suoi riflessi, si divincolava incessantemente. Ripetuta l'operazione sulla metá anteriore, il fenomeno si ripete identicamente.   Ma identicamente si ripete pure operando simil- mente sulla metá posteriore, per la quale non puó piú venire invocata l'azione inibitoria di un cervello.

Vi é forse infine una contraddizione nel modo con cui il Bethe descrive le funzioni del cervello negli artropodi.   Esso é ad un tempo organo tonificatore ed inibitore.   Quindi gli arti dal lato leso, benché apparentemente godano di una esaltata atti- vitá, dovranno avere minore efficacia dinamica, minore energia, almeno in quanto energia é attitudine alla produzione di lavoro meccanico, mentre gli arti dal lato illeso dovranno avere con- servato la loro primitiva efficienza.   Il corpo dell'insetto do- vrebbe quindi essere piú debolmente sorretto e guidato e spostato dal lato leso che da quello sano, e, se un disquilibrio vi é e se tale disquilibrio é sufficiente a turbare la dirittezza della deambula- zione, ció dovrebbe avvenire in senso inverso a quello che é indi- cato dai fatti ed é preveduto dalla teoria del Bethe.   Né ci si obbietti che basti il solo fatto esteriore della maggiore distensione delle arcate dal lato leso ed il maggiore distanziamento dei punti di appoggio, ivi, degli arti sul terreno, per provocare il moto in circolo. Poiché si é visto come l'insetto sappia autoregolarsi e correggere disquilibrii puramente meccanici analoghi e di valore di gran lunga superiore a quelli naturalmente provocati da un esaltato divaricamento degli arti.

Ma ci si puó chiedere se tale mancanza di inibizione realmente esista, se cioé esista una reale esaltazione della motilitá limitata agli arti dell'antimero che ha sofferta la lesione. Il Bethe pone l'affermazione senza suffragarla di dati numerici, frutto di esperienze quantitative. Ma a noi sembra, benché anche per noi l'esecuzione di simili misurazioni sia programma di ulteriori ricerche, che ben poco valore debbasi accordare all'osservazione immediata di un insetto che compia un moto di maneggio. Il nostro occhio é sopratutto giudice, in tale osservazione, non di velocitá angolari, bensí di velocitá lineari (moti degli estremi degli arti, rispetto al centro di curvatura del maneggio). Ed in realtá non si saprebbe decidere qual sia la piú adeguata fra le due questioni; se l'insetto si muova in circolo perché gli arti esterni sono dotati di maggiori velocitá lineari degli interni, o se detti arti esterni siano dotati delle suddette maggiori velocitá perché l'insetto si muove in circolo. Né la questione potrebbe essere detta oziosa. D'altro canto, ció che nella considerazione delle azioni dinamiche cui bilateralmente é sottoposto l'organismo, ha importanza, non é il valore assoluto di esse dall'uno dei lati, bensí il rapporto di quelle da un lato a quelle dall'altro. Moltiplicando per un uguale fattore i valori degli impulsi forniti al corpo dagli arti dei due antimeri, muterá la velocitá di trasporto di esso corpo, ma non la forma della traiettoria. Ora, osservando un insetto che compia il maneggio, siamo impossibilitati ad istituire un simile confronto fra gli arti dal lato leso e quelli dal lato illeso, poiche gli arti vengono, dai due lati, impiegati diversamente e poiché un possibile aumento della motilitá—che negli arti dal lato leso si traduca in una maggiore ampiezza delle arcate in senso anteroposteriore, piú facilmente apprezzabile ad un'osservazione dorsale—dal lato illeso puó manifestarsi, ed in realtá si manifesta, come ho detto a piú riprese, con una maggiore ampiezza delle arcate attrattive tibiofemorali, in un piano traversale al corpo dell'animale.

L'aumento della motilitá pur essendo esteso a tutto l'organismo, non verrebbe quindi apprezzato che come aumento della motilitá in senso anteroposteriore degli arti posti esternamente. Se poi tali variazioni realmente corrispondano, o meno, a variazioni dell'energia intrinseca, del tono e delle condizioni fisiolo giche del muscolo, non potrá essere deciso che da ricerche ergografiche istituite sui muscoli dei due antimeri.

Misurando infine l'ampiezza delle arcate sui reogrammi, cioé il valore numerico di una quantitá proporzionale alla dissimmetria propulsiva, ai due lati del corpo, abbiamo visto che in taluni casi—senza peraltro generalizzare—allo stabilirsi di essa dissim metria maggiormente contribuiscono le diminuite attivitá pro pulsive dal lato illeso, che un loro aumento dal lato oposto.

Ancora un'osservazione a proposito dell'inibizione cerebrale sui gangli della catena. Per toccamento elitrale le *Pimelia* presentano la nettissima reazione di una immobilizzazione riflessa che deve essere interpretata come una inibizione energica del cervello su ogni impulso motorio coordinato dei gangli della catena. In un animale normale, tale riflesso é simmetrico, cioé, si esercita contemporaneamente ed ugualmente su tutti gli arti del corpo. In un animale privato dei gangli sopraesofagei e quindi delle azioni inibitorie sulla corrispondente metá del corpo, tale immobilizzazione riflessa non potrá stabilirsi normalmente che nell'antimero illeso. Od almeno, fra i due antimeri dovrá potersi notare qualche divario nell'assunzione della posizione di immobilitá e nella sua durata. Tale divario, in realtá, non si nota, come se, cioé, non si fosse verificato alcun fatto di inibizione mancata.

D'altro canto l'insufficienza dell'interpretazione del Bethe, che uno squilibrio propulsivo sia bastevole ad originare e ad intratenere il moto di maneggio, é giá stata dimostrata nella mancanza di simili moti in quelle esperienze in cui tale disquilibrio era stato artificialmente provocato, prescindendo da lesioni nervose.

Vi é ancora una difficoltá e sta nel concepire le modalitá della distribuzione della "mancanza di inibizione" nell'organismo.

Essa dovrebbe infatti colpire tutti i gruppi muscolari ugualmente il che non si verifica, poiché abbiamo visto come, a prescindere dai muscoli degli arti, il capo, lo stesso asse sagittale del corpo siano flessi cosí da formare concavitá verso il lato illeso. E perché nell'antimero illeso, in cui, in grazia della lateralitá dell'-azione cerebrale ammessa dal Bethe, nessuna condizione nuova avrebbe dovuto insorgere, dovrebbero invece predominare i gruppi di muscoli flessori che originano quelle curvature?

E per quanto riguarda la muscolatura degli arti, giá abbiamo fatto osservare come essa sia costituita di muscoli antagonisti. Come si potrá concepire in questi una distribuzione della mancata inibizione? Entrambigliantagonisti dovrebbero essere ugualmente non-inibiti e la loro azione dovrebbe liberamente ed accentuatamente esplicarsi in ogni possibile senso.   Come si interpreteranno allora quelle particolari modificazioni d'impiego e quelle determinate ed obbligate direzioni assunte nello spazio dai moti degli arti, che abbiamo particolareggiatamente descritte?

Sono, queste, altrettante difficoltá che offre l'interpretazione del Bethe e che forse non si oppongono all' interpretazione che mi ha suggerita l'osservazione dei moti di maneggio nelle note specie di coelotteri

La esporró per sommi capi, senza dettagliarla, come concezione d'assieme.

1. La lesione di una regione laterale dei gangli sopraesofagei trae seco un'alterazione nel portamerto dell'animale, che, in armonia con le disposizioni chiasmatiche osservate nei fasci nervosi cerebrocatenali non é ristretta ad una metá laterale del corpo, ma interessa l'intero organismo e con particolare evidenza si rivela nell'antimero opposto a quello della lesione.

2. Detta lesione provoca molto probabilmente un accrescimento generale della motilitá, particolarmente visibile negli arti dal lato leso.

3. Provoca infine nell'organismo uno squilibrio nel normale uso della muscolatura, predominando in essa l'attivitá dei muscoli flessori nell'antimero illeso, attivitá resa manifesta dalla permanente maggiore flessione degli arti illesi e dall'incurvamento del corpo verso il lato sano.

4. Tali abnormi condizioni fisiologiche della muscolatura influenzano il normale svolgersi dei moti locomotorii, cosí da generare il movimento in circolo.

5. Il moto in circolo é dovuto prevalentemente—(ma non esclusivamente, poi che esso é un movimento cui attivamente partecipa tutto l'organismo)—alle azioni attrattive predominanti dei due primi arti del lato illeso, variamente orientate, rispetto alla direzione dell'asse sagittale. E'coadiuvato e facilitato dai moti propulsivi ad arcata, caratteristicamente proprii di tutti gli àrti dal lato leso e particolarmente dei primi due.

6. Che tali azioni flessive siano il movente principale del giro in circolo é provato

a) dall'impossibilitá di provocare moti in circolo per esagerazione artificiale delle attivitá propulsive da uno dei lati.

b) dal ristabilirsi di locomozione retta od oscillante, allorché, per l'amputazione dei segmenti articolati degli arti, venga resa impossibile l'esecuzione dei moti attrattivi.

c) dal permanere del moto di maneggio allorché siano impediti i soli moti propulsivi ad arcata degli arti dal lato leso.

7. Il moto di maneggio risulta quindi bensí da un'alterazione delle condizioni generali di simmetria dell'organismo, come avevamo assunto dapprincipio, ma da tale alterazione che interessi tutto l'organismo stesso, assumendo vario aspetto nelle diverse sue regioni.

Quest'interpretazione dei moti di maneggio, che crediamo piú adeguata alla complessa realtá dei fatti, non contiene alcuna ipotesi, né va al di lá dei fatti medesimi, opportunamente coordinati e collegati.

Essa quindi non ci dice nulla di preciso circa le modalitá dell'alterata influenza del sistema nervoso sui gruppi muscolari. Né peraltro attribuisce a gangli condensazioni verbalistiche di proprietá non altro che supposte. Essa, per quanto non sia che un'approssimazione a quella che dovrá poter essere la teoria dei moti di maneggio, offre peró ancora alcuni vantaggi da questo punto di vista, dai quali si potrá trar partito, indagando, come é in animo di chi scrive, le alterazioni istologiche del sistema nervoso che accompagnano le alterazioni fisiologiche del comportamento.

L'ammettere infatti la propagazione di una inibitivitá e di una azione tonificante dal complesso cerebrale alla muscolatura ed alla sua innervazione é concetto che, ove anche fosse esatto, sarebbe troppo generico per prestarsi a precisazioni anatomiche. Né il considerare solamente la soppressione dell'inibizione e del tono (di essi, infatti non si puó pensare, al piú, che una graduazione di intensitá) puó rendere conto degli svariati effetti che la diversitá della localizzazione lesiva induce come sperimentalmente abbiamo constatato accadere.

Converrá invece esaminare il decorso dei fasci e veder di porre in relazione la rottura di continuitá anatomiche e la soppressione di contiguitá funzionali oppure la deviazione di archi riflessi, che possano render ragione dele particolari condizioni di innervazione che rivela la fisiologia del maneggio. E basterebbe, a provare l'insufficienza di quel concetti, la constatazione di lesioni laterali del cerebron non accompagnate da moti di maneggio.

Siamo ben lungi dal poterne precisar le ragioni, ma noi non vediamo altra via—per quanto questa sia irta di difficoltá—per giungervi.

Riuscirá pure chiaro come nella nostra interpretazione, altrettanto bene come in quella del Bethe, possano rientrare i moti di maneggio provocati per sezione unilaterale delle commissure fra sopra e sottoesofageo cosí come eventuali maneggi provocati da lesioni del sottoesofageo stesso.

Ma conviene lasciare libertá completa di indirizzo alla futura ricerca sperimentale e non costringerla entro predeterminate linee suggerite dalla teoria.

La quale potrebbe anche venire contraddetta da nuovi fatti, fors'anche per avere troppo voluto rimanere aderente ai fatti medesimi.

SUMMARY OF CONCLUSIONS

1. Injury to one side of the supraoesophageal ganglion brings about a change in the behavior of the animal (beetle), which, in harmony with the chiasmatic arrangement of the fibers running from the brain to the lower ganglia, is not restricted to one lateral half of the body but involves the whole organism and

reveals itself with special clearness in the antimere opposite to that of the lesion.

2. This injury provokes very probably a general increase of movement which is manifest particularly in the appendages of the injured side.

3. The final effect is to produce a disturbance of equilibrium in the normal functioning of the musculature, in which the activity of the flexor muscles of the uninjured side predominates, an activity rendered manifest by the greater permanent flexion of the uninjured appendages and by the bending of the body toward the uninjured side.

4. These abnormal physiological conditions in the musculature so influence the normal course of the locomotor movements as to produce movement in a circle.

5. The movement in a circle is due for the most part (but not exclusively, since it is a movement in which the whole organism participates actively) to the traction predominatingly exercised by the first two appendages of the uninjured side, this traction being variously directed with respect to the sagittal plane. It is helped and facilitated by the propulsive arcuate movements, characteristic of all the appendages of the injured side particularly of the first two.

6. That the movements of flexion are the principal cause of circus movements is proved:

a) By the impossibility of eliciting such movements by the artificial exaggeration of the propulsive activity of one side.

b) By the re-establishment of straight or oscillatory locomotion, when, by the amputation of the segments of the leg, the execution of the movements of traction is rendered impossible.

c) By the persistence of circus movements where only the propulsive arcuate movements are prevented in the legs of the injured side.

7. Thus circus movement does indeed result from an alteration of the general conditions of symmetry of the organism, as we have assumed from the beginning, but from such an alteration as affects the entire organism, assuming various aspects in different regions.

GLI AUTORI CITATI

BAGLIONI Die Grundlagen der vergleichenden Physiologie des Nervensystems und der Sinnesorgane. Winterstein's Handb. d. vergl. Physiol., 4er Bd. Fischer, Jena, 1913.
Physiologie des Nervensystems. Ibidem.

BARROWS The reactions of Drosophila ampelophila to odorous substances. Jour. Exp. Zool., vol. 4, 1907.

BAUDRIMONT Note sur la marche des insectes. Procès Verbaux Société Linn. Bordeaux, T. 65, 1911.

BETHE Vergleichende Untersuchungen über die Funktionen des Centralnervensystems der Arthropoden. Pfluger's Archiv., 68. Bd., 1897.

BOHN Mouvements de manège en rapport avec les mouvements de la marée. C. R. Soc. de Biologie, Paris, 1904.
Mouvements rotatoires chez les larves des crustacés. Ibidem, T. 59, 1905.
Mouvements rotatoires d'origine oculaire. Ibidem, T. 58, 1905.
L'éclairement des yeux et les mouvements rotatoires. Ibidem, T. 59, 1905.

BRUNDIN Light reactions of terrestrial amphipods. Journal Animal Behavior, vol. 3, 1913.

CARPENTER Some reactions of Drosophila with especial references to convulsive reflexes.

COMES Sui movimenti di maneggio e sul loro significato nella teoria segmentale. Biologisches Centralblatt, Bd. 30, 1910.
Effetti della decapitazione in Calotermes flavicollis ed in altri artropodi. Ibidem, Bd. 32, 1912.

CORNETZ Les explorations et les voyages des fourmis. Flammarion, Paris, 1914.

DELAGE Sur une fonction nouvelle des otocystes. Arch. de Zool. exp., IX série, T. 7, 1887.

DEMOOR Recherches sur la marche des arachnides et des insectes. Arch. de Biol., T. 10, 1890.
Étude des manifestations motrices des crustacés au point de vue des fonctions nerveuses. Arch. de Zool. exp., IIe série, T. 9, 1891.

DOLLEY Reactions to light in Vanessa antiopa with special references to circus movements. Jour. Exp. Zool., vol. 20, 1916.
The rate of locomotion in Vanessa antiopa in intermittent and in continuous light of different illuminations and its bearing on the continuous theory of orientation. Ibidem, vol. 23, 1917.

DRZEWINA Mouvements de rotation et retour à la marche normale après section unilatérale du système nerveux. C. R. Soc. de Biol., T. 65, 1908.

DUBOIS Application de la méthode graphique à l'étude des modifications imprimées à la marche par les lésions nerveuses expérimentales chez les insectes. Bull. Soc. Biol. (8), I, 1885.
Les élatérides lumineux. Bull. Soc. Zool. de France, T. 11, 1886.
Leçons de physiologie générale et comparée. Carré et Naud. Paris, 1898.

DÜRKEN Experimental-Zoòlogie. Springer, Berlin, 1919

GARREY Light and the muscle-tonus of insects. The heliotropic mechanism. The Journal of General Physiology, 1918.

HADLEY Reactions of blinded lobsters to light. American Journ. Physiol., vol. 5, 1901.

HERRERA Sur le mouvement de manège chez les insectes. Bull. Soc. Zool. de France, 1893, vol. 18.

HOLMES Phototaxis in the amphipoda. American Journ. Physiol., vol. 5, 1901. The reactions of Ranatra to light. Jour. Comp. Neur., vol. 15, 1905.

KAFKA Einfuhrung in die Tierpsychologie auf experimenteller und ethologischer Grundlage. Barth Leipzig, 1914.

KELLOGG Some silkworm moths reflexes. Biol. Bull., vol. 12, 1907.

LOEB Die Tropismen. Wintersteins Handb. der vergl. Physiol., Bd. 4. Fischer, Jena, 1913.
Fisiologia comparata del cervello e psicologia comparata. Sandron, Palermo (1908).

LYON A contribution to the comparative physiology of compensatory motions. Americ. Journ. of Physiology, vol. 3, 1900.

MATULA Untersuchungen uber die Funktionen des Centralnervensystems bei Insekten. Pfluger's Archiv., Bd. 138, 1911.

McGRAW AND HOLMES Some experiments on the method of orientation to light. Journ. Anim. Behavior, vol. 3, 1913.

NORMANN Durfen wir aus den Reaktionen niederer Tiere auf das Vorhandensein von Schmerzempfindungen schliessen? Pfluger's Arch., 67. Bd., 1897.

PARKER AND PATTEN. The physiological effect of intermittent and continuous light of equal intensities. Americ. Journ. Physiol., vol. 31, 1912.

RADL Untersuchungen über die Lichtreaktionen der Arthropoden. Archiv. für die gesamte Physiologie, Bd. 87, 1901.
Ueber den Phototropismus einiger Arthropoden. Biol. Centralbl., 21 Bd., 1901.
Untersuchungen ueber den Phototropismus der Tiere. Leipzig, 1903.

YUNG Recherches sur la structure intime, etc. Arch. Zool. expér., T. 7, 1878.

Abstracted by Francis B. Sumner and Henry H. Collins, authors
Scripps Institution for Biological Research, La Jolla, California.

Further studies of color mutations in mice of the genus
Peromyscus.

Three recessive color mutations (already discussed in previous
papers) are more fully described and the behavior of these in
hybridization considered. The mutant races are characterized
as 'yellow,' 'pallid,' and true albino. The first and last of these
appeared in stock belonging to pure subspecies, while the other
appeared after a subspecific cross. The three depend upon
changes in distinct genetic factors. Any two give the wild type
in the F_1 generation, with the wild type and both mutants in
the F_2. In respect to relative numbers, these last follow dihybrid
ratios, at least in some cases. As regards the yellows, however,
irregularities are found, both in respect to the proportions which
emerge from dihybrid crosses and the tendency of certain yellows
to produce offspring which intergrade with the wild type. The
causes of these irregularities have not yet been cleared up. Also,
at least two distinct strains of yellows have been encountered,
which differ from one another in respect to their mean color
values. The differences between these are hereditary and crosses
between the two give an intermediate condition. The pelage
of these various mutant races, as well as individuals of the wild
type, has been subjected to color analysis by means of the Hess-
Ives tint-photometer. This has rendered possible a fairly exact
quantitative expression of the various color differences concerned.
Three colored plates of skins are included.

FURTHER STUDIES OF COLOR MUTATIONS IN MICE OF THE GENUS PEROMYSCUS

F. B. SUMNER AND H. H. COLLINS

TWO PLATES (NINE FIGURES)

INTRODUCTION

The several sports or mutations to be considered in the present paper have already been mentioned or discussed in previous papers by the senior author (Sumner, '17, '18, '20, '22), and two of them—the 'yellow' and 'pallid' races—have been rather fully described. We shall here report the results of later observations and breeding experiments upon these mice. Crosses have been made between the chief mutant races, and their 'genetic behavior' tested according to customary mendelian methods. More accurate color determinations have been rendered possible through the purchase by the Scripps Institution of an efficient colorimeter.[1] In addition to this, we have thought it desirable to publish for the first time colored illustrations of several of the mutant strains.

[1] Reference is made to the Hess-Ives tint-photometer. The use of this instrument for determining the color values of mammalian pelages has already been briefly discussed by Sumner ('21). In using this apparatus, light reflected from the object to be examined is viewed in juxtaposition to light from a pure white block of magnesium carbonate, the two being seen through the same color screen. Three of these color screens are employed in succession, these being of such wave lengths as would give pure white light (or neutral gray) if the transmitted rays were combined. In making the reading with each of these screens, the light reflected from the 'magnesia' block is cut down by a diaphragm to a point at which its intensity is exactly equal to that of the object to be examined. At this point the two halves of the visual field are of equal illumination, and likewise (owing to the color screen) are of the same color. The illumination of the entire field is rendered homogeneous by a special series of rapidly rotating lenses, in consequence of which the area of pelage under examination appears of an absolutely uniform tint. Specially prepared flat skins are used, these being first thoroughly cleaned in benzine to remove grease.

289

THE JOURNAL OF EXPERIMENTAL ZOOLOGY, VOL. 36, NO. 3

The situation as regards one of the color varieties—the 'yellows'—has proved to be less simple than was at first supposed. This variety has appeared in several independent descent lines, and two of these lines seem to differ from one another characteristically in respect to the exact shade of 'yellow' which is manifested. Furthermore, neither of these last-named races appears to behave strictly like a monohybrid recessive. It seems likely either that there are independent 'modifying' factors concerned, which segregate according to principles not yet ascertained by us or that the 'yellow' factors themselves are unstable and undergo changes of some sort.

We are quite aware that the observations here presented are decidedly fragmentary in comparison with the more exhaustive investigations of many recent mendelian students. This has been due to several circumstances. In the first place, our studies of color mutations have throughout been regarded as incidental to our main program of work, which has concerned itself with the characters of 'natural' subspecies. In the second place, Peromyscus is not well adapted to experiments in which rapid breeding is an important consideration. Not only is it commonly impossible to obtain more than two or three generations in a year, but the cage-born mice show a considerable percentage of sterility, which may at any time bring some valuable descent line to a close. Questions which one could promptly settle by appropriate matings with more favorable material must await many months for an answer or must even remain unanswered. Finally, we must mention that these studies were interrupted when far from complete, and that a large part of the stock was killed at that time, although certain lines of experimentation were resumed later. This interruption was due to the absence of the senior author from La Jolla during the greater part of one year and to the junior author's permanently severing his connections with the Scripps Institution in the fall of 1919.

These explanations, we trust, will temper the criticisms of those who may be disposed to wonder why we have failed to settle certain obvious problems or even to attempt certain obvious experiments. The authors may fairly ask to be credited with having thought of some, at least, of the numerous lines of possible experimentation which will occur so promptly to the reader.

In the ensuing pages the several color mutations will first be discussed separately. Later, the results of crossing these mutant stocks will be considered.

THE 'YELLOW' COLOR VARIETY

The earlier history of one of the 'yellow' stocks here considered has been recorded in a previous paper (Sumner, '17), and a description of the hair has likewise been given (Sumner, '18). It is of some interest that this and five other independent outcroppings of 'yellow' in our stock have all been derived from the La Jolla strain of Peromyscus maniculatus gambeli. This may be due in part to the fact that more of these mice have been reared than those of any other local race—more, perhaps, than all of the others combined—and that the chances of encountering such infrequent sports have thus been increased. On the other hand, it should be remarked that the ancestors of the 'yellows' were all trapped within an area of a few square miles, so that the outcroppings of this character may not have been wholly independent of one another.

With the exception of a single aberrant individual,[2] which was probably a juvenile yellow, none of this color variety have been found by us among our wild stock, although we have trapped at least a thousand mice of this subspecies at La Jolla. In each case the yellows have appeared for the first time either in the first or the second cage-born generation. We do not, however, infer from this that the 'mutation' has been due to the artificial conditions of captivity. The probable occurrence of at least one wild yellow renders this unlikely, as also the fact that several wild specimens proved to be heterozygous (according to accepted standards) when mated with cage-bred yellows.

[2] Also trapped at La Jolla (Sumner, '17).

The 'a' strain of yellows (fig. 2)

The three chief independent strains of yellows will be designated as 'a,' 'b,' and 'c,' respectively, according to the order in which they appeared in our stock.   It is the 'a' strain which has been referred to in earlier publications.

As already stated, this strain appeared among the offspring cf two males and three females, these five being derived from a single pair of wild individuals (P ♀ 46 and P ♂ 16). The parents and grandparents were all of normal appearance. Unless the 'mutation' appeared for the first time in the germ-cells of the parent generation, these five parents of the original yellows must have all been heterozygous, since each gave rise to at least one yellow.   The total number of their offspring (excluding those dying very young) was 24, of which 8 (1 ♂, 5 ♀ and 2 of unknown sex) were yellows, while 16 (8 ♂, 5 ♀, and 3?) were of the wild color.   Furthermore, three of these parents, when mated later to yellows, gave rise to 5 yellows and 5 of the wild color.   If they were not actually heterozygous in origin, they must, at least, have been producing 'yellow' and normal gametes in about equal numbers.   Two matings of yellows of this strain resulted in 10 offspring, all yellow.   For the most part, however, the matings of these mice were made with non-yellows or with yellows of another strain, as will be described below.

As previously stated, these mice "are of a peculiar yellow-brown hue, probably lying between the 'cinnamon buff' and the 'clay color' of Ridgway, and not unlike the most highly colored parts of the hair in P. m. sonoriensis."[3]   The peculiar hue was attributed to two causes: 1) the larger number of banded ('agouti') hairs, in proportion to the all-black, and, 2) the greater proportional extent of the yellow region on these banded hairs.[4] As regards the first point, it should be added that strictly 'all-black' hairs (i.e., those entirely devoid of a paler cross-band)

[3] Sumner ('17).—Such a comparison with any set of color standards is admittedly extremely crude, since the pelage is very far from being a uniformly tinted surface.

[4] Sumner ('18).

COLOR MUTATIONS IN MICE OF PEROMYSCUS   293

are nearly or quite lacking in some yellows, and that, when present, they are probably confined to the darker, middorsal region of the body. It is not improbable, also, that the yellow pigment of the agouti hairs is more abundant or more highly concentrated in this color variety. This is yet more likely in the case of the '*b*' strain to be described next.

As also stated in earlier papers, the ventral pelage of the 'yellows' is more intensely white than that of the 'wild type.' This is due to the greater length of the terminal pigmentless zone of the hairs on the under surface of the body. Indeed, in the midventral line, the basal 'plumbeous' zone of the individual hairs is commonly quite lacking, the pelage being entirely white. There appears to be no reduction, however, in the depth of pigmentation of this basal zone throughout the body, nor do the tail, feet, ears, eyes, etc., appear to show any diminution in the amount of pigment normally present.

Mice of this strain differ considerably from one another in shade. In some specimens the entire pelage, or certain areas, exhibits a decidedly 'dusky' hue, owing, apparently, to the presence of a larger proportion of the all-black hairs or of hairs heavily tipped with black. Likewise, the richness of the color varies very much, the brighter specimens displaying considerable orange-yellow, the duller ones having a 'washed-out' appearance, suggesting that of jute or some similar vegetable fiber. These differences are probably due to the amount of the quality of the yellow pigment in the individual hairs, though no careful microscopic examination has thus far been made.[5] On the whole, the mice of the '*a*' strain are of a decidedly less rich hue than are those of the '*b*' strain next to be described.

Only eight adult skins of the former strain are available for color tests with the tint-photometer. These give the following mean and extreme readings, in terms of the three 'primary' colors of the color screens. The computed proportions of black,

[5] Studies of the quality and distribution of hair pigment, in relation to subspecies, color mutation, and behavior in hybridization, are being undertaken in this laboratory by Mr. R. R. Huestis. The matter has already been dealt with, to some extent, by Sumner ('18).

white, and 'color' (in this case a yellow-orange) are likewise shown in table 1.[6]

These figures may be profitably compared with similar ones giving results for a series of the normal wild gambeli race. Table 2 is based upon ten skins of adult individuals, five of which were trapped in December and January, five in June. Care was taken to select a representative series including lighter and darker specimens.[7] The conspicuously darker shade of even the paler mice of the wild type is indicated by the higher percentages of black, while the comparative lack of color is also obvious. The ratio of red to green on the other hand (see below), is very close to that in the 'a' yellows, being 3.09, as compared with 3.17. Thus the difference between these 'yellow' mice and the normals appears to depend chiefly upon the relative proportions of black and of yellow pigment, not upon the character of the latter.

The 'b' strain of yellows (figs. 3, 4, and 8)

A single pair of normally colored wild mice (P ♀ 15 and P ♂ 59) in the cultures of the junior author became the parents

---

[6] These last figures represent the proportional magnitudes of three sectors on a color wheel which would combine to produce the shade in question, assuming that the black disk was of zero luminosity, and the white equal to that of the standard, and that the colored sector was of maximum saturation and intensity. As a matter of fact, the commercial cardboard disks are very far from fulfilling these conditions, so that large corrections (over 25 per cent in the case of 'white') must be made, in order that color wheel and photometer determinations shall agree.

For reasons which cannot here be discussed, the proportion of black is regarded as equal to the difference between the highest color-screen reading and 100 per cent, the proportion of white being equal to the lowest color-screen reading, and the 'color' constituting the balance. In an earlier paper (Sumner, '21) the proportion of black was computed differently, following the instructions contained in a pamphlet issued by the manufacturers. The values thus obtained were somewhat too high. The procedure here adopted has the authority of Mr. F. E. Ives, the inventor of the instrument, and, furthermore, the figures thus arrived at correspond fairly well with those obtained by the color-wheel method (due corrections being made).

[7] It is true that this series includes none as pale and as highly colored as certain extreme variants of the 'normal' stock. At least one of these last (resulting from selective mating for two generations) resembles an average 'yellow' in appearance, though undoubtedly quite different genetically.

of 22 cage-born offspring, of which 14 (5 ♂ and 9 ♀) were of
the wild type, while 8 (5 ♂ and 3 ♀) were 'yellows.' As in
the case of the 'a' strain, the proportion of recessives among the ·
offspring of these presumably heterozygous parents was somewhat too high, though, as before, the total number of individuals

TABLE 1

'A' yellows

	HIGHEST	LOWEST	MEAN
Color-screen readings			
Red..................................	32.0	25.5	28.44
Green.................................	18.5	14.0	16.75
Blue-violet............................	13.0	9.0	11.37
Equivalents in black, white, and color			
Black.................................	74.5	68.0	71.56
White.................................	13.0	9.0	11.37
Color.................................	19.5	15.0	17.07

TABLE 2

Normal gambeli

	HIGHEST	LOWEST	MEAN
Color-screen readings			
Red..................................	19.0	14.5	16.15
Green.................................	13.0	9.5	11.25
Blue-violet............................	10.0	7.5	8.90
Equivalents in black, white, and color			
Black.................................	85.5	81.0	83.85
White.................................	10.0	7.5	8.90
Color.................................	9.0	5.0	7.25

was not sufficient to justify any conclusions from this fact.
It will be noted that in this case the proportion of males among
the yellows was much higher than among the wild type, the reverse being true of the first strain. Such differences are probably accidental.

The female parent of these yellows (P ♀ 15) when mated to one of her yellow offspring bore one normal and three yellow young. The male parent (P ♂ 59) was mated to nine other females[8] of the wild type, these being either wild mice or cage-bred ones, unrelated to the yellow stock. Eight of these matings resulted in the birth of twenty-four young, all of normal color. The other female (a wild specimen) gave birth to four normals and one yellow. She was evidently another heterozygous (or 'mutating'?) individual, unrelated so far as we know, to the male.

Matings between yellows of this strain yielded over fifty recorded offspring, the sexes being represented in about equal proportions. All of these mice have been entered as 'yellows,' though this characterization is subject to the qualification to be discussed presently.

Matings of yellows with those known (on the basis of parentage) to be heterozygous[9] gave 39 offspring, of which 21 are recorded as 'yellows,' 11 as 'normal,' and 7 as 'doubtful.' If those of the last class were all to be included among the 'normals,' we should have a reasonably close approach to the 50:50 ratio. Otherwise, there are too many yellows.

Only two matings are recorded between individuals, both known from their parentage to be heterozygous. These gave 7 normal, 1 yellow, and 1 of uncertain type.

It was early noted that the 'a' and 'b' yellows differed quite perceptibly in respect to their mean color tone. The second strain is, on the average, of a richer color, there being fewer black hairs in the pelage, and the pigment of the 'ticking' being redder. Indeed, the term 'yellow,' as applied to the 'b' strain, is in most cases a decided misnomer. The brighter specimens

[8] These figures relate only to fertile matings. Throughout this work many matings were made which yielded no results.

[9] Excluding the large number of related individuals which we know to have been heterozygous only from the fact that they produced one or more yellow offspring. The inclusion of these would, of course, be unwarranted unless it were possible also to include such heterozygous individuals as gave rise to no recessive offspring. There is no way of identifying these.

are not far from the 'ochraceous tawny' of Ridgway's "Color
Standards," while those of the 'a' strain approach more nearly
the 'clay color,' though the latter comparison is quite misleading.

It soon developed also that many of the 'yellows,' both of the
'a' and 'b' strains, were of a much less intense color and contained more black hairs than those which had been first examined
(fig. 4). These latter have, for the sake of convenience, been termed
'atypical yellows.' This expression is quite arbitrary, however,
since one may arrange among the offspring of 'yellow' parents
a graded series between the most 'typical' yellows and specimens
closely resembling the paler and more buff-tinted individuals of
the wild type. Indeed, where we are dealing with the offspring
of heterozygotes, it is not in every case possible to distinguish
the pure recessive 'yellows' from the others.

Regarding the genetic status of these 'atypical yellows' we
are at present far from clear. That they are not heterozygous
individuals, resulting from blended inheritance (imperfect dominance), seems certain. This we conclude both from the fact
that such individuals have never resulted from the mating of
yellows with pure dominants, and from the fact that two of these
'atypical yellows' have never produced offspring of the wild type.[10]
Unfortunately, the distinction between 'typical' and 'atypical'
individuals was not always recorded in our earlier entries, and
these cover perhaps the major part of our yellow stock. Our
records seem to show, however, that whereas very clear ('typical')
yellows tend to produce offspring like themselves, 'atypical'
offspring have occasionally been born to two perfectly 'typical'
parents. They have likewise resulted from the mating of a
'typical' yellow with a heterozygous animal whose yellow parent
was also 'typical.' In a number of instances, too, it is known
that 'typical' and 'atypical' mice have occurred among the offspring of the same parents, and even within the same brood.
That two 'atypical' individuals have ever produced 'typical'
young we have no clear evidence. On the contrary, abundant

[10] Unless we regard as atypical yellows certain of the 'doubtful' specimens
which appeared in the F_2 generation of the yellow-pallid cross (see below).

records show that the offspring of two 'atypical' animals generally resemble their parents in this respect.[11]

It is not impossible that we are concerned merely with differences of the type which are commonly called 'phenotypic' or
'somatic' (i.e., non-hereditary). But such an explanation is
hardly consistent with the facts just cited. It seems more likely
that we have to do with the presence or absence of 'modifying
factors,' or possibly even with the existence of unstable factors,
or departures from the simple 'factorial' scheme of heredity.

Table 3 gives the average and extreme values for the color
determinations of twenty-four adult pelages of the 'b' yellows.
About two-thirds of these were listed as 'typical' or 'nearly
typical,' the others as 'atypical.'

From tables 1 and 3 it appears that the 'a' strain shows a
slightly higher percentage of white, while the 'b' strain shows a
slightly higher percentage of black, the values for 'color' being
nearly the same for the two. The significance of these differences
is doubtful. Of far more importance is the fact that the ratio
of red to green[12] is distinctly higher for the 'b' strain than
for the 'a,' the mean figure being 3.59 for the former and 3.17
for the latter. A study of the frequency distributions shows
that this difference is probably a real one, and indeed a casual
comparison of the skins reveals it to the eye. Further evidence
of such a difference was derived from the examination of living
specimens, many of which were not skinned.

Hybrids between the 'a' and 'b' yellows

Matings of these two strains resulted in all cases in offspring
which were listed as 'yellows.' Three 'a' females were mated
to two different males of the 'b' strain. The resulting twelve
F₁ hybrids appeared, on the whole, as intermediate between

[11] The interruption of these studies above referred to is largely responsible
for these uncertainties.

[12] The values for red and green employed for this purpose are the excess of
each that remains after deduction of the amount which combines with the other
colors to constitute the 'white.' The lowest color-screen reading (in this case
the blue-violet) also serves to indicate the amount of white, there being no 'free'
blue-violet.

the parents in their mean color tone, though they presented a considerable range of variability. A single mating between a 'b' male and a heterozygous 'a' female led to the birth of one offspring of the wild color and two decidedly 'atypical' yellows.

It is thus evident that these two shades of yellow do not result from the modification of independent genetic factors, as is the case with the other color mutations to be described presently. Assuming that the differences between the 'a' and 'b' strains are hereditary at all—which seems fairly certain—we may, on the one hand, have to do with a case of 'multiple allelomorphs,' the two 'yellows' representing slightly differing modifications

TABLE 3

'B' yellows

	HIGHEST	LOWEST	MEAN
Color-screen determinations			
Red....................................	30.5	22.5	27.15
Green.................................	18.0	11.5	15.00
Blue-violet...........................	7.5	12.5	10.27
Equivalents in black, white, and color			
Black.................................	77.5	69.5	72.85
White.................................	12.5	7.5	10.27
Color.................................	19.5	11.5	16.88

of the same color factor. Or the primary factor concerned may be the same in the two cases, the difference being due to the presence in one variety of a secondary 'modifying' factor.

The mature pelages of twelve of these hybrid yellows were subjected to the color analysis. The mean values for black, white, and total 'color' are close to those for the two 'pure' strains, which, as stated before, agree closely with one another in these respects. Furthermore, these hybrids are strictly intermediate as regards the spectral position of their yellow pigment, as is indicated by the ratio of red to green. This is true not only of the mean value of this ratio (3.33), but of its range. As already stated, this intermediate condition is apparent to the eye.

Further evidence for the same interpretation of these color varieties is derived from back-crosses. Matings between F_1 hybrids and 'a' yellows resulted in ten offspring. Five prepared skins are available from this lot. These give us 3.2 as the mean ratio of red to green—a figure lower than all but four of the twenty-four 'b' yellows and almost identical with the average of the 'a' race. Notes were made, furthermore, upon the other specimens, either when living or freshly killed. These indicate that they were, for the most part at least, of a dull yellow or buff appearance. In some cases they were expressly likened to the 'a' yellows.

Matings of some of the foregoing back-cross individuals inter se—((a-b) - a) - ((a-b) - a)[13]—resulted in the birth of thirteen young, all of which were listed as 'yellows,' though they are described in much the same terms as their parents. The ground-color was a very dull yellow or buff, darkened by a considerable admixture of black hairs. They resembled the duller specimens of the 'a' yellows, and in no case approached the more ruddy hue of the 'b' strain.

The other back-cross (i.e., between the F_1 hybrids and the 'b' yellows) gave a quite different result. Of the two mature skins which were preserved, both are closely similar to the brightest 'b' yellows in appearance, giving red : green ratios of 3.5 and 4.0, respectively. These lie altogether outside the range of the 'a' series, and the larger figure almost reaches the extreme for the 'b' series. Furthermore, notes made upon the entire lot (thirteen in all), when living or freshly killed, show that at least six resembled the 'b' strain more nearly than the 'a,' while several others are listed as 'intermediate.' There is no record of a specimen's having the predominant appearance of the 'a' strain.

---

[13] We have employed hyphens instead of multiplication signs in designating these various crosses, since reciprocal crosses were commonly made, and we wish to be non-committal as to which parent belonged to which race. When the multiplication sign is used, it is commonly understood that the female parent is named first.

Independent lines of 'yellows'

As already stated, the presumably heterozygous male parent of the '*b*' strain produced one yellow and four dark offspring when mated with one of the nine females (P ♀ 91) which were used in addition to the mother of the '*b*' yellows. There is no record of any member of this brood having left any descendants, the skin of the yellow was not preserved.

Three yellows and three normals were produced by the mating of a '*b*' yellow (C_1 ♂ 78) with an unrelated wild-type mouse (C_1 ♀ 96), which was considerably paler than the average but was not regarded as a 'yellow,' even an 'atypical' one. These mice likewise left no descendants.

Another independent outcropping of the 'yellow' mutation consisted of a single individual, which appeared in a brood of three, whose parents and grandparents were known to be of the 'wild color.' This chanced to occur in the course of an experiment in which paler and darker strains of the 'normal' mice were crossed. We are not, however, disposed to attribute any significance to the latter fact. As in the preceding cases, the parents of the yellow individual were doubtless heterozygous for the yellow factor, unless they were themselves producing 'mutant' germ-cells de novo. This strain, likewise, was not continued further.

Yet another independent appearance of the 'yellow' color variety, which we may call the '*c*' strain, occurred in a lot of La Jolla gambeli which are believed to have been trapped at least a year later than any of the preceding ones. A pair of wild mice ('selection series' P ♀ 118 and P ♂ 22) gave rise to eight young, of which three were yellows (one of these being doubtful), the remainder being of the normal color type. The female was likewise mated to a yellow male of the '*a-b*' lot, and the male was mated to three yellow females of the '*a*,' '*b*,' and mixed strains. Among the thirteen young thus produced, only three yellows appeared, instead of six or seven as would be the expectation from such a mating. As in the case of all of the yellows, subsequent to the '*a*' and '*b*' strains, these lines were brought to a close with the generations just referred to. No careful comparisons were made between any of these mice and the 'a' and 'b' yellows, and only one of the skins was saved, so that it is now impossible to make such a comparison. It is our recollection that the 'c' mice resembled the 'b' rather than the 'a' strain, though the single preserved skin is probably intermediate.

It is worth noting, though perhaps not significant, that the heterozygous (?) parents of five out of six of our yellow strains gave numbers of yellows in excess of mendelian expectation. Of the 68 mice thus produced, 44 were normal, 24 yellow, giving a ratio of 1.8:1, instead of 3:1. Such a departure from the

TABLE 4

	DARKER	LIGHTER	MEAN
Juvenile 'yellows'			
Black.....................................	84.5	80.0	82.25
White.....................................	12.0	16.0	14.00
Color.....................................	3.5	4.0	3.75
'Normal' juvenile gambeli			
Black.....................................	91.0	85.0	88.00
White.....................................	8.0	12.0	10.00
Color.....................................	1.0	3.0	2.00

'expected' condition may well be accidental, however, particularly in view of the fact that the 'heterozygous' parents of the 'a,' 'b,' and 'c' strains gave an excess of wild-color offspring when mated with yellows (viz., 16:11).[14] Whether these parents were in reality heterozygous, rather than original producers of mutant germ-cells, is not definitely shown by our records. The data given seem compatible with either interpretation.

Juvenile yellows

As stated in earlier papers, the yellow variety is nearly or quite as distinguishable in the juvenile pelage as in the mature (figs. 7 and 8). Table 4 gives the proportions of black, white, and color in

[14] This apparent excess of yellows may be due to another cause (see Summary and Conclusions).

two juvenile 'yellows,' a darker and a lighter specimen. In comparison with these are shown the corresponding values for darker and lighter specimens of the wild type, the latter extreme being taken from a selected strain of pale or 'buff' animals. The wide differences between both the mean and the extreme values of the two series are sufficiently obvious. On the other hand, there are but trifling differences between the darker 'yellow' and the paler 'normal' individual.

THE 'PALLID' COLOR VARIETY

These mice were first referred to as 'partial albinos' (Sumner, '17), but this designation was plainly at variance with customary usage, so that the non-committal term 'pallid' was later adopted.

This 'mutation'—if it did arise de novo during these experiments—appeared among a lot of F_2 hybrids between Peromyscus maniculatus rubidus and P. m. sonoriensis. Four pallids and seven of the wild type resulted from the mating of an F_1 male and his two sisters, each of these last producing two pallids. It is certain that this was no simple segregation phenomenon, due to the recombination of factors regularly present in the two subspecies which were crossed. Up to the present time, more than 300 F_2 and F_3 hybrids between these two races have been reared and no other case of the pallid mutation has come to light.

Since the pallid-color variety has been described in some detail in earlier papers (Sumner, '17' '18), a brief account will suffice here. It is characterized primarily by the lack of most of the black (or sepia) pigment found in normal mice (figs. 5, 9). This lack appears in the absence of all-black (i.e., non-banded) hairs from the pelage, and the extreme reduction of pigment in the basal zone of the others. The latter is of a pale ashy hue instead of slate-colored. Furthermore, the eyes are dark red instead of black, the ears are not appreciably pigmented, and the dorsal tail stripe (normally due to dark hairs) is scarcely perceptible. A further peculiarity of this strain is the fact that the eyes are smaller, or at least less protruding, than in the wild type. The pallid mice are pale gray when young, developing a considerable admixture of yellow or orange when adult.

In a previous paper several peculiarities were pointed out in the microscopic appearance of the individual hairs: 1) A considerable proportion of these are practically devoid of pigment in the zone which is ordinarily yellow, the rest being normal in this respect; 2) the normally dark surface pigment of the terminal portion of the hairs is nearly or quite invisible; 3) in the basal zone, the normal black pigment bodies are represented by groups of small irregular granules or flocculent dark masses.

While the range of variation in the pallids is not as great as that of wild mice belonging to some of the natural subspecies, well-marked individual differences are none the less present. Some specimens have a considerably greater amount of yellow in their pelage than others, presenting a richer color on this account. These are perhaps intermediate between the 'cinnamon' and 'cinnamon buff' of Ridgway. On the other hand, specimens have appeared in which the pelage is noticeably darker than the average. Between these two extremes all gradations may be found.

Even wider variations have been met with in respect to the eye color of the pallid mice. In some individuals this is scarcely darker than that of the eyes of true albinos. In at least two specimens, on the other hand, it is nearly as dark as in normal animals of the wild type, though even here careful comparison reveals an undoubted difference. In the great majority the eyes may be characterized as dark red.

Whether or not these individual differences in coat color or eye color are hereditary has not been determined.[15] It is of possible significance that both of the dark-eyed variants arose as 'extracted recessives,' after a cross with other mutants (albino and yellow). One of them possesses, in addition to dark eyes, the darkest pelage of any pallid yet noted.

Aside from the occurrence of these variations of a possibly genetic nature, the pallid mutation has behaved, in every respect, as a simple monohybrid, recessive character. Despite the existence of plainly darker specimens, no true 'intermediates' or

---

[15] One of the two dark-eyed individuals referred to is apparently sterile; the other is not yet old enough for breeding purposes.

doubtful cases have been encountered. It is not worth while to detail the numerous matings which have been made within the pallid stock. It should be pointed out, however, that the same excess of recessives (here four out of eleven) was found among the offspring of the original heterozygous (?) parents, as in the case

TABLE 5

Adult pallids

	HIGHEST	LOWEST	MEAN
Color-screen readings			
Red..................................	40.5	31.5	35.54
Green................................	26.0	19.5	22.04
Blue-violet...........................	19.0	14.0	16.17
Equivalents in black, white, and color			
Black.................................	68.5	59.5	64.46
White................................	19.0	14.0	16.17
Color................................	26.0	15.0	19.37

TABLE 6

Juvenile pallids

Color-screen readings (mean)

Red..	31.75
Green...	25.00
Blue-violet...	22.00
Equivalents in black, white, and color (mean)	
Black...	68.25
White...	22.00
Color...	9.75

of all but one of the strains of yellows. At best, however, these combined figures merely suggest a possibility.

The following figures (tables 5 and 6) express the results of color determinations of twelve adult and two juvenile skins of the pallid mice (see also pl. 2). It is evident that the juvenile animals surpass the adult in black and white (i.e., are more gray), being relatively deficient in color. The ratio of red to green is 3.30 for the adult pelages, 3.25 for the juvenile. Thus the quality of the yellow pigment appears to be closely alike in the early and later pelages of the pallid mutant. Likewise, there are no very wide differences in this respect between the pallids, yellows and normal gambeli.

TRUE ALBINOS

True albinism has appeared but once in our cultures, although many thousand mice have been reared since the commencement of these studies.[16] Two broods from a single pair belonging to the first cage-bred generaton of P. m. gambeli (La Jolla race) consisted of two albinos and six of the wild type. There was one albino of each sex, both being fertile. At least one of the normal color proved to be heterozygous; the others were either sterile or their condition was not ascertained.

These albinos have, as was to be expected, behaved as simple mendelian recessives. A few matings between individuals known to be heterozygous (other than the parents of this strain) have given 7 normal and 6 albino young, an obvious excess of the latter. Matings between albinos and heterozygotes, on the other hand, gave 17 normal and 18 albino, which is as close an approach as possible to the 'expected' number.

The mice of this strain (fig. 6) are plainly complete albinos, comparable with ordinary white mice, which are albinic house-mice (Mus musculus). Indeed, the white Peromyscus are very similar to the latter in appearance, differing chiefly in having larger and more protruding eyes and somewhat longer ears. They likewise differ in being almost completely odorless, as are also normally colored specimens of Peromyscus.[17] As already stated, the eyes of the albinos average much paler than those of the pallids, the former being pink, the latter commonly dark red. In both of these varieties, they are distinctly smaller (or at least protrude less) than in normal individuals.

---

[16] We have had as many as 1500 mice at one time in our 'murarium,' most of these being cage-born. The experiments have lasted seven years.

[17] A single cage of ordinary white mice will impart a pronounced 'mousy' scent to an entire room even when well ventilated. This is not true of a thousand specimens of Peromyscus.

Two skins of the white race of Peromyscus have been tested with the tint-photometer. Although these pelages are probably as 'white' as those of any albino mammal, they are obviously not of so intense a white as the standard magnesium carbonate. Few persons, however, would probably expect to find such a considerable reduction in luminosity as actually occurs. The percentage of white proved to be only 75.7, as compared with the standard, leaving 16.7 per cent of black and 7.5 of 'color,' in this case a yellow (R:G = 1.58). Neither black nor yellow, however, are due to the presence of the ordinary hair pigments, since these appear to be entirely lacking. The black results chiefly from the loss of light which passes through and between the nearly colorless hairs, and is not reflected back to the eye. The proportions of the primary colors are not quite the same, however, as in the standard, owing doubtless to a faintly yellowish tint in the keratin of the hair.

YELLOW-PALLID CROSSES

Matings were made between eight yellows and three pallids, resulting in the birth of twenty-two young. These were all of the wild type, of a medium shade, and fairly uniform in color, presenting about the same appearance and range of variation as a similar number of local gambeli. The yellows employed were partly of the 'a' strain, partly crosses between the 'a' and 'b' strains.

Of the F_1 generation, 5 males and 11 females were successfully mated, yielding 64 young. There thus chanced to be exactly four times the minimum number required for the proportional representation of all classes in a dihybrid cross. Of the 64 F_2 individuals, 36 were males, 27 females, and one of unknown sex. Since there were no significant differences in the distributions of genetic classes, according to sex, males and females will be combined in the treatment below.

On the assumption that we had to do here with a typical dihybrid cross, involving complete dominance, the most probable distribution of classes for this number of individuals would be:

36 normals (wild type)

12 yellows

12 pallids

4 double recessives.

Since the peculiarities of both the yellows and the pallids depend upon the absence of pigment, and since the former do not lack any pigment which is not likewise lacking in the latter, it would seem probable that the double recessives would be indistinguishable from ordinary pallids. The proportions above stated would accordingly become 36:12:16.

The actual numbers[18] proved to be:

36 normals

6 yellows

7 doubtful (normals or yellows?)

15 pallids.

The exact agreement of the first of these figures and the almost exact agreement of the last are sufficiently striking. If it were permissible to regard the 'doubtful' specimens as being genetically 'yellows,' a single further transposition would bring these figures into exact conformity with 'expectation.' Unfortunately, the case is not as simple as this.

The thirty-six 'normal' mice, while differing among themselves, presented no greater range of variability than would a similar random collection of wild gambeli, which race, indeed, they resembled pretty closely in appearance. In this respect, they agreed with their F_1 parents. Despite evident differences of color, likewise, the pallid specimens probably showed no greater range of variation than was to be found in the original pallid stock. The 'yellow' class in the list includes only those concerning which no doubt was felt. While none of the pelages were as rich in color as the brightest of the 'b' strain of yellows, they were probably an average lot.

The difficulty here relates to the status of the 'doubtful' specimens, intermediate in appearance between yellows and normals. One of these died early, so that its status could not

[18] The classification of individuals was made without any reference to these totals, which were not computed until later.

be tested. Of the remaining six, two were entered as 'probably yellows,' though 'atypical' (p. 297), while four were entered as 'probably not yellows.' They were, however, recorded as being much paler and more buff than the average gambeli.

These six specimens were mated in various combinations, and some twenty-five descendants were born.[19] The latter were a rather nondescript lot. Five of them were pallids—a result which was not unexpected, since even some of the 'normals' were doubtless heterozygous for the pallid factor. It is of interest that these F_3 'pallids' included the dark-eyed specimen, with darker pelage, referred to above (p. 304). The others, for the most part, differed rather widely from the average normal gambeli. A few might have passed for paler (more buff) representatives of the latter subspecies, and had nothing in their appearance to suggest the 'yellow' variety. One or two looked as if they might be 'atypical yellows.' The majority, while not at all uniform, were of a curious, somewhat ruddy, appearance, having little resemblance to the typical yellows of either of the chief strains discussed above, but also probably unlike any wild mice which we have trapped. Even some of the descendants of the F_2 pair which had been listed as 'probably yellows' showed no nearer approach to the yellow type than the others. One of these, indeed, is entered as 'medium gray.'

This case is, of course, complicated by the fact that we are dealing with a mongrel combination of three subspecies, as well as with two different color mutations. The 'pallids,' as already stated, sprang from a sonoriensis-rubidus cross, while the 'yellows' were of pure gambeli stock. This intermixture of subspecies does not, however, affect the clear segregation of the pallid and the albino factors. It is only the 'yellow' factor (or factors) concerning which there is any question.

These difficulties cannot be cleared up until, 1) the genetic behavior of the 'yellow' color variety has been far more thoroughly

[19] These births occurred during the protracted absence of the senior author. Upon his return, it was not in many cases possible to distinguish the children of a given pair from their grandchildren, or from broods resulting from the mating of parents and offspring. If the parent mice had been pure recessives, however, this fact would, of course, have made no difference in the result.

tested and, 2) the behavior of the color differences entering into subspecific crosses has been determined with more precision. Investigations are under way which may settle some of these questions.

ALBINO-YELLOW CROSSES

Matings of three yellows and four albinos resulted in the birth of thirty F_1 offspring, of which twelve males and eleven females have lived to maturity. All are of the wild color, and would pass for normal gambeli.

Thus far, the F_2 generation consists of 83 individuals, of which 52 are normal (wild color), 13 yellow, and 18 albino. On the assumption that these factors are not linked, the 'expected' numbers are 47, 16, and 21, respectively. The departure from expectation is doubtless accidental. In any case, there is no evidence of linkage, the occurrence of which would have reduced the proportionate number of dark individuals, instead of increasing it.

The number of F_2 albinos and yellows which have thus far been tested for linkage is very small, but it is of interest that the proportion of recombinations is even greater than would be expected from random assortment (Sumner, '22). Thus the meager data at hand make it plain that no considerable degree of linkage, if any, exists between these factors.

ALBINO-PALLID CROSSES

Matings of five pallids and two albinos resulted in the birth of nineteen young (11 ♂, 8 ♀), all fully pigmented mice, having the appearance of ordinary normal gambeli.

It is of interest that conclusive evidence was found for a high degree of linkage between the pallid and the albino factors (Sumner, '22).

Only sixteen F_2 young have thus far been reared, derived from simple $F_1 \times F_1$ matings, though a considerable number of a somewhat more complex pedigree were obtained. Of these sixteen, 9 were dark, 6 pallid and 1 albino. The abnormal proportions of the two recessive classes is doubtless a chance result due

to inadequate numbers.   Other types of matings give no grounds
for expecting the number of albinos to be deficient here.

The really important tests, as stated in another paper, have
been made, 1) by mating 'extracted' albinos of the F_2 generation
with 'pure' pallids (i.e., those known to be free from the factor
for albinism); 2) by mating extracted pallids with pure albinos,
and, 3) by mating extracted albinos with extracted pallids.
There were likewise a number of matings in which the pedigrees
were somewhat less simple than here indicated.

Eighteen F_2 mice were involved in these tests.   The total
number of their offspring was 135, the number per pair ranging from three to twenty-six.   Not all of these parents, taken
singly, have thus far given birth to a sufficient number of young
to prove their genetic composition with any certainty.   But the
cumulative testimony of all of these matings is overwhelming.
Not a single pallid mouse and only two albinos have appeared
among the 135 young which have thus far been born.   Had there
been a normal proportion of 'carriers' among the parents, these
matings should have yielded fifty-five of the recessive types.
That all of the offspring with two exceptions (these being sibs)
were of the wild type is evidence of a high degree of linkage (in
this case 'repulsion') between the albino and the pallid factors.

SUMMARY AND CONCLUSIONS

1. Three distinct color 'mutations' have been described, which
first appeared in captive stock of the commonest species of
California deer-mouse, Peromyscus maniculatus.

2. Two of these, the 'yellow' and albino varieties appeared in
cultures of P. m. gambeli, originally trapped in the vicinity of
La Jolla.   The third, or 'pallid' variety, first appeared in the F_2
generation of a cross between the subspecies rubidus and
sonoriensis.

3. The albino and pallid varieties arose but a single time each
in our cultures.   Of the 'yellows' there were six independent
outcroppings.

4. At least two of these independent yellow strains differed
from one another in the mean color tone displayed, and this

difference proved to be, hereditary. Since the hybrid offspring of these two strains were all 'yellows,' and displayed, on the whole, an intermediate tint, we probably have to do either with a case of 'multiple allelomorphs' or a case in which one or more 'modifying factors' condition the difference.

5. In none of these cases is the evidence sufficient to show whether the actual mutation, or modification of a genetic factor, occurred in our own cultures, or whether the mutant factor had been present for many generations in a simplex condition.

6. The number of 'mutants' originally produced was considerably higher than would be expected, on the assumption that the parents were both heterozygous. This excess was found in five out of six of the independent outcroppings of yellow. It was also found in the pallid strain, but not in the very small number of original albinos. Combining all the offspring of these original heterozygous parents, we have 87 individuals, of which 57 were of the wild-type and 30 were mutants (recessives), giving a ratio of 1.9:1, instead of 3.:1. The departure from the normal is not, however, of very probable significance.[20] Furthermore, there is another possible interpretation of this excess of recessives. We are necessarily dealing only with the offspring of parents known to have produced some recessives. It may well be that there have been other pairs of heterozygous parents in the stock, which have not been recognized as such owing to their having produced only normal offspring. Inclusion of these last would increase the ratio of dominants to recessives.

7. All of these mutations, like the vast majority of those described by previous writers, plainly involve the loss of something normally present. In the case of the albinos, all pigment has been lost, both from the hair, the skin, and the retina. The pallid mice have lost most of their dark pigment, and probably some of their yellow, and here also the loss has been general, affecting all pigmented parts of the body. In the yellows there has been an almost complete suppression of the all-black (un-banded) hairs and a shortening of the basal, deeply pigmented zones of the others. In the 'agouti' hairs of these mice on the

---

[20] The probability is only about nine out of ten.

other hand, the amount of yellow pigment has certainly been increased, so that there has been a partial compensation for the loss of dark pigment. The eyes, ears, and feet of the yellows are as dark as those of the normal.

8. These three mutant types are all recessive to the wild type. The albinos and pallids breed true and exhibit comparatively little variability. Likewise they segregate clearly in crosses with the wild type or with one another. The yellows, on the other hand, exhibit a wide range of variability, intermediates being found between the typical yellow and the normal condition. They also display other irregularities which will be discussed in another section.

9. These three mutations relate to quite distinct genetic factors. Any two, when crossed, give rise to the wild type in the F_1 generation. In the F_2 the wild type and two mutant types segregate clearly, except for certain irregularities with respect to the yellows.

10. Albino-pallid crosses reveal the existence of a high degree of linkage between these two factors. On the contrary, no linkage appears to exist between the albino and the yellow factors. The yellow-pallid cross was not tested in this respect.

11. In the F_2 generation of the yellow-pallid cross, the proportion of yellows proved to be considerably too small, though these discrepancies may perhaps be accidental. There were, in addition to the true 'yellows,' about an equal number of 'doubtful' individuals, approaching the normal in appearance, which appeared to be genetically neither true yellows nor true normals. Likewise, in the original yellow cultures, there occurred, as stated above, many somewhat intermediate individuals, these being sometimes found in the same brood with typical ones. The genetic status of these 'atypical yellows' and other 'doubtful' individuals of the same stock has not yet been determined by us. We are certain, however, that they are not merely mice which are heterozygous for the yellow factor. Heterozygotes are commonly as dark as the 'wild' type of gambeli.

12. The 'pallid' mice, though far more regular in their genetic behavior than the yellows, nevertheless show a quite evident

variability in their coat color and an even wider variability in their eye color. In respect to the latter, they range from a pink only slightly darker than that of the albinos to a shade only slightly paler than the full black of the wild type. Indeed, both of these extremes have been found among the derivatives of an albino-pallid cross. Whether these differences are due to 'modifying factors' or to the 'contamination of factors,' or whether they are 'purely somatic' (whatever that may mean!) we are unable to conjecture at present.

13. There are included in the preceding pages the results of numerous color analyses made by the senior author with the aid of a Hess-Ives tint-photometer. The principal (macroscopic) differences between the 'yellow' and 'pallid' mutants and 'wild-type' mice of the subspecies gambeli were found to be due to different proportions of black, white and a 'color' of tolerably constant quality. The spectral position of this last, as judged by the red: green ratio, was found not to differ very widely in any of these forms. This may be taken as evidence of a considerable degree of uniformity in the 'yellow' pigment of the hair, though the latter is doubtless not the only factor concerned in the gross results. Small, though well-marked, differences were found, on the other hand, even in this red : green ratio, the most noteworthy case being that of the 'a' and 'b' strains of yellows.

14. To what degree the color mutations here discussed correspond with those which have been described for house-mice or other domesticated rodents, we cannot state with certainty. The authors have not had the opportunity to make the necessary comparisons either with living specimens or skins of such races.

On first thought, it might seem that our 'yellows' are of the same type as the familiar recessive, black-eyed yellows, whose condition is attributed to the replacement of the 'extension' factor, 'E', by its recessive allelomorph, 'e' (Castle, '20, p. 124). It must be recalled, however, that the black pigment in our races is far from being restricted to the eyes, but is present in full intensity in the bases of the body hairs, in the ears, feet, tail, and some other parts.

Castle ('16, '20) records that "the occurrence of yellow sports among wild meadow mice (Microtus) has been observed by Cole, Barrows, F. Smith and others," and Dunn (1921) lists several wild rodents in which 'restricted yellow' is said to occur.[21] Unfortunately, no further particulars are available relative to the pigmentation or genetic behavior of these.

It is hardly necessary to point out that our 'yellows' have no relation to the race of dominant yellow house-mice, which has been so much discussed in genetic literature.

Regarding the possible relationship of our 'pallids' with other described color varieties of rats and mice we can speak somewhat more definitely. The almost (though not quite) complete lack of black pigment, together with the presence of abundant yellow pigment, in the hair and the dark red color of the eyes would suggest that our 'pallid' race corresponds, in some sense, to the 'red-eyed yellows,' discussed by Castle ('14 and later papers) and others for rats.[22] This surmise is greatly strengthened by the fact, referred to above, that in both species there is a high degree of linkage between the factor for the red-eyed mutation and that for albinism (Castle, '14, '16 a, '19; Castle and Wright, '15; Dunn, '20).

The occurrence of true albinos among Peromyscus has already been recorded by Castle ('12). Castle's specimens belonged, however, to a different species from ours (P. leucopus noveboracensis), the progenitor of the strain having been sent him from Michigan. Doctor Castle kindly furnished us with a pair of these mice, but the latter, like all the surviving members of his strain, proved to be sterile.

The 'identity,' in terms of the factorial hypothesis, between albinism in Peromyscus and that in the house-mouse cannot, of course, be taken for granted. Since the crossing of these genera seems to be impossible, this question can never perhaps be conclusively answered.

[21] 'Peromyscus maniculatus gambeli' is included in this list. If Dunn here refers to the 'yellows' discussed in the present paper, and reported earlier by Sumner, he is probably not justified in assigning them to this class.

[22] First referred to as 'black-eyed yellows.'

LITERATURE CITED

CASTLE, W. E.   1912   On the origin of an albino race of deer-mouse.   Science, vol. 35, pp. 346–348.

1914   Some new varieties of rats and guinea-pigs and their relation to problems of color inheritance.   American Naturalist, vol. 48, pp. 65–73.

1916   Genetics and eugenics.   Harvard University Press.

1916 a   Further studies of piebald rats and selection, with observations on gametic coupling.   Carnegie Institution Publication no. 241, pt. III, pp. 163–190.

1919   Studies of heredity in rabbits, rats and mice.   Carnegie Institution Publication no. 288, pp. 1–56.

1920   Second edition of Genetics and eugenics.

CASTLE, W. E., AND WRIGHT, S.   1915   Two color mutations of rats which show partial coupling.   Science, vol. 42, pp. 193–195.

DUNN, L. C.   1920   Linkage in mice and rats.   Genetics, vol. 5, pp. 325–343.

1921   Unit character variation in rodents.   Journal of Mammalogy, vol. 2, pp. 125–140.

RIDGWAY, R.   1912   Color standards and color nomenclature.   Washington: published by the author.

SUMNER, F. B.   1917   Several color mutations in mice of the genus Peromyscus.   Genetics, vol. 2, pp. 291–300.

1918   Continuous and discontinuous variations and their inheritance in Peromyscus.   American Naturalist, vol. 52, pp. 177–208, 290–301, 439–454.

1920   Geographic variation and mendelian inheritance.   Jour. Exp. Zoöl., vol. 30, pp. 369–402.

1921   Desert and lava-dwelling mice and the problem of protective coloration in mammals.   Journal of Mammalogy, vol. 2, pp. 75–86.

1922   Linkage in Peromyscus.   American Naturalist.

WRIGHT, S.   1917   Color inheritance in mammals. II. The mouse.   Journal of Heredity, vol. 8, pp. 373–378. III. The rat.   ibid., pp. 426–430.

PLATES

317

PLATE 1

EXPLANATION OF FIGURES

1 Normal Peromyscus maniculatus gambeli of about medium shade.
2 Typical specimen of '*a*' yellow.
3 Typical specimen of '*b*' yellow.
4 'Atypical' *b* yellow.

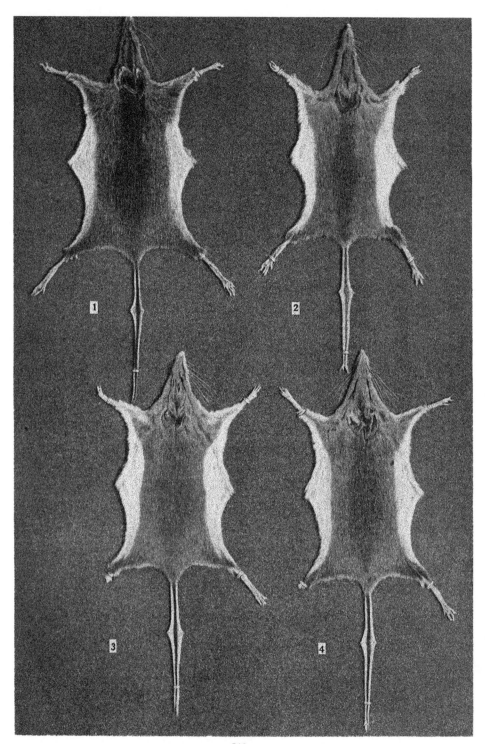

319

PLATE 2

EXPLANATION OF FIGURES

5 Adult pallid, of about medium shade.
6 Adult albino.
7 Juvenile gambeli ('wild type').
8 Juvenile yellow (from same brood as preceding).
9 Juvenile pallid.

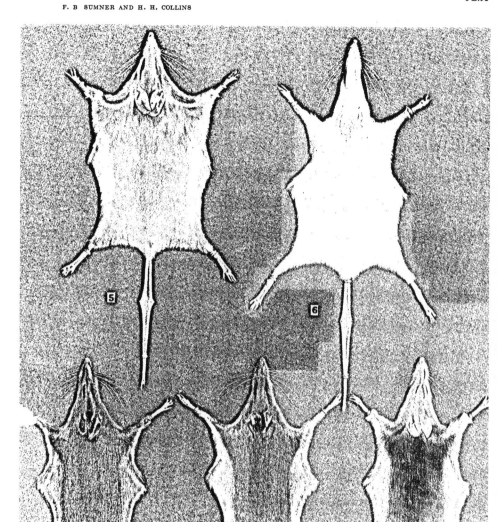

Abstracted by George Howard Parker, author
Harvard University.

The crawling of young loggerhead turtles toward the sea.

Newly hatched loggerhead turtles find their way from their nests to the sea in consequence of at least three factors: first, positive geotropism, as shown in their tendencies to move down slopes; second, their response to their retinal images, in that they move toward regions in which the horizon is open and clear and away from those in which it is interrupted by complicated masses, and, third, their probable response to color, in that they move toward blue areas rather than toward those of other colors (Hooker). These animals are not appropriately described as phototropic, for they do not move either toward a source of light or away from it, but they are to be regarded as exhibiting a more complex condition in that they respond to the details of their retinal images rather than to these images as wholes.

AUTHOR'S ABSTRACT OF THIS PAPER ISSUED
BY THE BIBLIOGRAPHIC SERVICE, JULY 24

THE CRAWLING OF YOUNG LOGGERHEAD TURTLES
TOWARD THE SEA

G. H. PARKER

Zoölogical Laboratory, Harvard University

It is difficult to imagine a more striking and invariable reaction among higher animals than the crawling of newly hatched logger-head turtles (Caretta caretta Linn.) toward the sea. For a long time this response has excited the interest of field naturalists, and within recent years it has been studied with especial care by Hooker ('08 a, '08 b, '09, '11) and commented upon by Mayer ('09, p. 121). To see a dozen of these newly hatched creatures, that have had no previous experience with the ocean, scramble toward it, notwithstanding that it may not be within the range of their vision, is a sight never to be forgotten. Any attempt on the part of an observer to check them in their course seems only to excite them to further effort which does not cease till they have reached the water. To the observer they seem to be drawn toward the sea by an influence as mystical as it is impelling.

On July 5th a considerable number of these turtles were hatched at the Miami Aquarium, Miami Beach, Florida, and it was here that I had an opportunity of studying their reactions not only on the day of hatching, but also over a subsequent period of a week or so. My thanks are due to the officers of the Miami Aquarium Association for the privilege of working at the aquarium labora-tory and for their generosity in providing me with all that was necessary for my investigations.

After my attention had been called to the remarkable regularity with which the young turtles on the aquarium wharf went toward the water, I began some more or less systematic observations and experiments. A piece of smooth paper-board, about a meter square, was laid down horizontally on the wharf and surrounded on all four sides by a low wooden fence some 15 cm. high. At the

323

middle of the pen thus constructed turtles were liberated one at a time and with their heads pointed in sequence north, east, south, and west. The water and the late afternoon sun were both to the west and the turtles almost' without an exception took that course.

To ascertain whether the sun or the water was the effective element in the situation, I carried the paper-board and about a dozen turtles across the narrow key on which the town of Miami Beach is situated to the opposite shore. Here the water was to the east while the sun of course remained in the west. On re-setting the pen on the beach and liberating the turtles as before at its center, they were found to go as regularly to the east, i.e., toward the water, as they had previously gone to the west. It was therefore plain that the sun had no significant influence over their movements, but that these were related to the body of water.

On my return from the ocean on the east side of the key to the bay on the west side I stopped in a field about midway between the two bodies of water, and having set up the pen here with its floor horizontal, I proceeded to a third set of tests. I was greatly interested to see that under these circumstances the turtles were not disposed to move much from the center of the board and that when they did move they were as likely to go in one direction as in another. During these tests the sun was still high in the western sky and the results showed again that this source of light was not a significant factor in determining the direction of motion. Not only did these observations demonstrate that the sun was ineffective, but they showed also that the water, at least at the distance of about a quarter of a mile, was also ineffective, for the turtles went as often to the north or the south as they did to the east or the west. Had they been under the influence of the water, they should have turned even in this intermediate position to the east or the west, but not to the north or the south. Apparently the middle of the key was a region in which the turtles were as in-different to the influence of the water as they were in all places to that of the sun. So far as the sun is concerned, my observa-tions agree entirely with those of Hooker, who declared both in his preliminary ('09, p. 124) and in his final report ('11, p. 70) that the sun has nothing whatever to do with the direction in which the turtles creep.

During these preliminary trials it became perfectly evident that the young turtles were very responsive to the slope of the surface upon which they moved; in other words, that they were geotropic to a marked degree. Care was therefore taken that the paper-board on which the tests were carried out was always horizontal and attention was given to the geotropism of the turtles as such. In all my experience the young loggerhead turtle was always positively geotropic. It regularly goes down slopes, notwithstanding the fact that it has ample energy and strength to go up them, and it is responsive to even so slight an inclination to the horizontal as 10°. On more considerable inclinations the animals go down with a rush. Hooker ('11, p. 72) states that "under ordinary circumstances the young turtles are negatively geotropic, but if the possible descents have been exhausted, they become positively geotropic." As a descent is evidence of positive and not negative geotropism, Hooker has apparently confused terms and, if this is so, his observations agree with mine except that I have never seen any evidence whatever of negative geotropism (movement against gravity). In all my tests of turtles from the time of their hatching till the end of their first week of life, they have been consistently positive in their geotropism. When placed on the natural slope of a beach, even though they cannot see the water, they travel downward at a considerable rate.

How they escape from the shallow nest in which they are hatched I do not know, for I have never had the opportunity of studying them under these circumstances. Newly hatched turtles in the laboratory were from the beginning always positively geotropic, so that I have no reason to believe that there is any change in their geotropism. Such nests as I have seen were always shallow, and it is quite possible that turtles on hatching pass up their slopes and over their rims under other influences than those having to do with geotropism and thus escape. But concerning the details of this question I have no facts to offer.

In all the tests I carried out, as already stated, the turtles were uniformly positively geotropic.

Having determined that young turtles are positively geotropic, I next turned my attention to their responses to water. That they are not directed in their movements by water vapor in the air, by the smell of water, or of materials associated with it, or by the sound of waves, etc., has been abundantly shown by Hooker ('11), and on these points I can confirm practically all his statements. If next a course over which young turtles are making their way toward the sea a shallow vessel of water is placed so that they might enter it, they pass it by without showing even a deflection in their line of march. Had the water in itself possessed a quality attractive to the turtles, some change in their course would certainly have been evident. That water does not thus attract them is also evident from the fact that turtles liberated in the pen on the laboratory wharf in the darkness of night did not move toward the water as they did in the daylight, but crept about indiscriminately.

If water in itself does not influence the direction of movement of the young turtles and if sources of light, such as the sun, are equally ineffective in this respect, what is the factor that determines the course that they take on a horizontal surface? When turtles were liberated at the center of the horizontal pen already described as set out on the laboratory wharf and their courses toward the water were closely observed, it was found that these courses were not directly toward the water, which was almost exactly west, but they were a little to the north of west. When the surroundings were inspected to ascertain what might be present to cause this peculiarity, it was seen that the main mass of the aquarium building loomed up a little to the south of east from the pen and that if a straight line was drawn from this mass through the center of the pen westward it would point to a very open unobstructed horizon. Such a line included the course taken by the turtles. It therefore seemed possible that the young turtles took a course which led away from the most considerable mass upon the horizon and toward the most unobstructed and open parts of that line. To test this view I tried an experiment almost exactly like that described by Hooker ('11, p. 72) in that I set up the horizontal pen in an open field to the east of a low hedge west of which was a sea-wall and the bay. On liberating the turtles here they all took easterly courses away from the hedge, and incidentally away from the water. Their courses were toward the open field. Evidently, the mass of the hedge and the open field influenced the direction taken much as the laboratory buildings and the open water had done, but in this instance it was perfectly evident that the water had no necessary part in determining this direction. These results agree exactly with those obtained by Hooker.

I next set up the pen in a square open field, three sides of which were bounded by well-grown trees and the fourth fairly open. The turtles went with reasonable regularity toward the open side. I then transferred the pen to a very open sandy plane and piled up next one side of it a number of boxes and tin cans till a wall nearly a meter high was formed. From this the turtles regularly moved away. These various tests led me to conclude that any large mass interrupting the horizon forms a center from which young loggerhead turtles retreat.

But these animals not only retreat from conspicuous masses on the horizon; they also move toward the open horizon with great certainty. I was persuaded of this by an accidental observation made while I was testing a very different matter. I had placed the horizontal pen between and in front of two well-grown bushes to ascertain whether the turtles would orient to one or other of these or give a combined reaction to both. Much to my surprise the turtles did not retreat from the bushes, but took a course in the opposite direction as though they were going directly through the opening between the bushes. On looking through this opening I observed what had escaped my attention before, a considerable stretch of free horizon. I repeated this test several times and always with the same result, the turtles took a course between the two bushes. Consequently, I concluded that though young loggerhead turtles move away from large interrupting masses on the horizon, they also move with great certainty toward a section of open horizon even though this may be relatively small.

In a final set of experiments designed to test the effects of the overhead and the horizontal fields of light, I had an excellent opportunity to observe the exact method of response of the turtles to their illuminated surroundings. Two large collecting tubs were placed upon a sea-wall, one right side up and the other bottom up. The tubs had a diameter of about 50 cm. and a height of a little over 30 cm. Five turtles were tested by being placed alternately on top of the inverted tub at its center and inside the upright tub likewise at its center. From the top of the inverted tub the turtle could see the whole landscape; from inside the upright tub it could see only the overhead sky. From both positions the turtles were free to move in any horizontal direction. In each set of tests a given turtle was headed successively north, east, south, and west. The sea-wall on which the tubs were set ran approximately north and south; to the west was open water; to the east a field with trees and shrubbery. In the twenty tests inside the tub the turtles remained stationary for five minutes in fourteen instances and in the remaining six they took various courses which may be roughly described as twice to the northwest, twice to the east, once to the northeast, and once to the southwest, showing that the overhead sky had no effect on their orientation. In the twenty tests on the outside of the tub the turtles went invariably to the west, away from a horizon interrupted by trees and shrubbery and toward one of open water. But the interesting part of these tests was not so much the direction taken by the turtles as the way in which this direction was apparently discovered. When the young turtle was set at the middle of the inverted tub, it rested there quietly about half a minute, raised its head high in air, made a complete circle or more in a very restricted area, and then moved off immediately to the west. The preliminary circular movement, almost always a complete circle or more, was made irrespective of the position in which the turtle was set. To all appearances it seemed as though the animal tested first the whole horizon and then moved in a direction away from large masses and toward greatest openness.

In describing the photic responses of the young loggerhead, Hooker uses the terms photophilism and phototropism ('11, p. 71) without, however, making very clear what is meant by these, and finally concludes that, though the animals are not influenced in their movement by the sun, they are nevertheless positively phototropic ('11, p. 75). He compares them with certain positively phototropic animals studied by Cole ('07) in which responses to the area of illumination rather than to the intensity of the light was the determining factor in their locomotion. In my own tests on the turtles I have never seen any evidence that they are, strictly speaking, phototropic. Thus I have never been able to get a turtle in a dark room to creep toward a light such as happened with all the animals tested by Cole. Hence I think it improbable that the young loggerheads are correctly described as positively phototropic. To me they seem to be an example of a much more complicated set of relations than those seen in phototropism. Their retinal images are immensely complex as compared with those in many of the lower forms, and they respond more to the details of these images than to the images as wholes. That part of the image which represents the region of the horizon is much more important in determining the direction of locomotion in the young turtle than any other part. If a portion of the horizon is interrupted by many masses rich in detail, such as trees, shrubbery, houses, etc., it may form a center from which the turtle will move away. If a portion of the horizon is uninterrupted and very uniform, as where sea and sky meet, that part may form a center toward which the turtle will go. These conditions indicate that the details of the retinal image in the turtle are the significant features in determining the direction of its creeping rather than the image as a whole and that consequently the turtle possesses a kind of vision more like that in the human eye than like that in the eye of purely phototropic animals, even if we include among these the peculiar instances pointed out by me ('03) and by Cole ('07) in which the size of the illuminated area is as significant as the intensity of the light.

Hooker's contention ('09; '11, p. 74), that turtles move toward blue rather than toward other colors and thus reach the sea, may perfectly well be correct. When I made my tests I unfortunately had no adequate means of experimenting with colors, and consequently I am not in a position to add anything to this aspect of the subject. I am nevertheless convinced that beside color the photic complexity and photic simplicity of the region of the horizon, as has already been detailed, are factors of first importance in determining the direction of locomotion, for in a number of my tests, as for instance those in the open field, natural blues even in the sky were often absent. Hence I conclude that, quite aside from the effect of color, the direction of locomotion of young loggerhead turtles is significantly influenced by the interruptedness or openness of the region of the horizon.

On beaches locomotion down a slope, away from an interrupted horizon and toward an open one, as well as toward masses of blue, would almost invariably lead to the sea. These doubtless are the chief factors that influence the course of the newly hatched loggerhead turtle whereby it reaches the ocean. Although water is the environment to be attained, water in itself plays no part in directing the movements of this animal which are indirectly influenced by those features of the environment just enumerated. It would be interesting to ascertain whether any of these factors affect Fundulus in its escape over the beach to the sea from small pools as described by Mast ('15).

CONCLUSIONS

Newly hatched loggerhead turtles find their way from their nests to the sea in consequence of at least three factors: first, their positive geotropism as shown in their tendencies to move down slopes (Hooker); second, their response to their retinal images in that they move toward regions in which the horizon is open and clear, and away from those in which it is interrupted, and, third, their probable response to color in that they move toward blue areas rather than toward those of other colors (Hooker).

These animals are not appropriately described as either negatively or positively phototropic, but are to be regarded as exhibiting a more complex condition, in that they respond to the details of their retinal images rather than to these images as wholes.

LITERATURE CITED

Cole, L. J. 1907 An experimental study of the image-forming powers of various types of eyes. Proc. Amer. Acad. Arts Sci., vol. 42, pp. 335–417.

Hooker, D. 1908 a The breeding habits of the loggerhead turtle and some early instincts of the young. Science, vol. 27, pp. 490–491.

1908 b Preliminary observations on the behavior of some newly hatched loggerhead turtles (Thalassochelys caretta). Yearbook Carnegie Inst., Washington, no. 6, pp. 111–112.

1909 Report on the instincts and habits of newly hatched loggerhead turtles. Yearbook Carnegie Inst., Washington, no. 7, p. 124.

1911 Certain reactions to color in the young loggerhead turtle. Papers Tortugas Lab., Carnegie Inst., vol. 3, pp. 69–76.

Mast, S. O. 1915 The behavior of fundulus, with special reference to overland escape from tide-pools and locomotion on land. Jour. Anim. Beh., vol. 5, pp. 341–350.

Mayer, A. G. 1909 Ann. Rep. Director Dept. Marine Biol. Yearbook Carnegie Inst., Washington, no. 7, p. 121.

Parker, G. H. 1903 The phototropism of the mourning-cloak butterfly, Vanessa antiopa Linn. Mark Ann. Vol., pp. 453–469.

Abstracted by William M. Goldsmith, author
Southwestern College, Winfield, Kansas.

The process of ingestion in the ciliate, frontonia.

The food of the ciliate, frontonia, is primarily diatoms, desmids, euglenas, filaments of oscillatoria, and various other microscopic plants. The mouth is normally very small, but may be expanded to approximately two-thirds the length of the body without injuring the organism. Five factors are involved in the process of ingestion of material longer than the expanded width of the body of the frontonia. A. Action of oral cilia: The cilia about the mouth of frontonia exert a direct pull upon the incoming food. B. Action of the locomotor cilia: The cilia of the body drive the organism forward and thus force the stationary food into the mouth. C. The rotation of the body axis: The end of the fiber usually enters the mouth and passes anterodorsally until it comes in contact with, and exerts a pressure upon, the aboral wall, after which the frontonia swings around and releases the tension. Points of contact between the ingested particle and the inner side of the body membrane are called tension points. D. Body contractions: A series of sharp contractions of the body wall assists in relieving certain other tension points. E. Cyclosis: Cyclosis probably aids by moving the end of the fiber around the wall. Unusual and fantastic figures are produced through the contortion of the organism by the ingested food.

THE PROCESS OF INGESTION IN THE CILIATE, FRONTONIA[1]

WILLIAM M. GOLDSMITH

Southwestern College, Winfield, Kansas

TWENTY-FIVE FIGURES (THREE PLATES)

INTRODUCTION

The present paper is primarily a record of a series of observations on the ingestion of various kinds of food by the ciliate, Frontonia leucas, and on the relation which this unusual method of ingestion bears to the variation in shape and habits of the organism. The points of chief importance and interest in connection with the present problem are as follows:

1. Frontonia frequently takes food consisting of filaments many times longer than itself. The manner of ingestion of such food is explained in detail.

2. The food of frontonia, is primarily diatoms, desmids, euglenas, filaments of oscillatoria, and various other microscopic plants. Various indigestible particles may also be ingested.

3. The mouth of frontonia is normally very small as compared with that of Paramecium and other common ciliates. However, it may be expanded to approximately two-thirds the length of the body without injuring the organism.

4. The normal shape of the body may be altered by certain characteristic contractions, by simple twisting and bending, and particularly by the presence of ingested materials.

[1] These investigations were carried on in the Zoölogical Laboratory of the Johns Hopkins University during the year 1919–20, in connection with the regular laboratory course in animal behavior. The writer is indebted to Prof. S. O. Mast for many valuable suggestions during the progress of the work. He is also under obligation to Prof. Asa A. Schaeffer, of the University of Tennessee, for reading the manuscript.

5. Food is not taken into the mouth by the usual ciliary action of the cytostome. A number of related factors are involved in this process. These are considered in the text and are listed in the summary.

GENERAL BEHAVIOR

Under normal conditions, in a quiet culture, frontonias may be seen swimming slowly here and there near the substratum apparently in quest of food. Any slight disturbance, such as the jarring of the container or the addition of weak chemicals to the culture, causes the organisms to rise from the bottom and to swim around more rapidly. However, after the removal of the stimulus they soon settle down to the bottom and continue the slowly swimming movements. Schaeffer ("Ameboid Movement," '20) says: "Frontonia feeds mostly by 'browsing,' that is by eating particles lying on or against some solid support." If an individual chances to come in contact with any object approximating the size of the food it is accustomed to eating, such as filaments of oscillatorias, diatoms, desmids, and various other microscopic plants, it usually pauses, places the oral opening near the object and proceeds to brush it with the oral cilia as though attempting to ascertain its nature (fig. 1). While the mouth is in close proximity with the object, the posterior part of the body frequently swings about this point as upon a pivot, sometimes turning through an arc of 90° or even entirely around. It is not uncommon for the organism to leave the object, move off for a short distance, then turn slowly about and swim here and there over the same area, eventually coming in contact with the same object and repeating the process.

THE MECHANICS OF INGESTION

Although frontonias ingest food particles of various shapes, the process of feeding can be more successfully observed when the food is in the shape of a filament. Before actual ingestion is begun the organism usually moves over the food and slowly swings around until the longitudinal axis is parallel with the long axis of the object to be engulfed. It then moves slowly forward with the oral cilia in contact with the object (fig. 1) until the end of the linear food particle is reached, when the mouth is slowly pushed over the end of the object and thus ingestion begins (fig. 2).

Just as the oscillatoria filament is about to enter the mouth, the frontonia bends the anterior end downward as shown in figure 2. This brings the plane of the mouth perpendicular to the long axis of the fiber, and thus permits the fiber to enter with the least resistance. As the fiber slowly enters the body, the granules and food particles suspended in the endoplasm are pushed aside, leaving a clear space on either side of the entering food. This space usually presents the general appearance of the ectoplasm. From all indications it seems quite certain that a small amount of water is taken in with the fiber which adds to the transparency of the surrounding space. This clear area is not definitely set off from the endoplasm material as is the case during the last two or three hours of digestion. The protoplasmic granules and smaller food particles may pass from one area to the other, and often crowd in and at times, and at certain places, obliterate the transparent space. The movement of these granules in front of and to the side of (fig. 3) the incoming food fiber is quite characteristic of the movements accompanying the entrance of any solid into a viscous medium containing particles in suspension. As will be suggested later, there is little streaming of the granular endoplasm unless there is first a movement of the incoming fiber or possibly a contraction or other movement of the body wall.

A. Ciliary action—first and second factors of ingestion

The mechanics by which the oscillatoria filament (or any other material) is caused to enter the body of the frontonia is of vital interest in connection with the present investigations. At the outset it should be emphasized that the customary method employed by the ciliates, namely, the sweeping of food particles into the mouth in a current of water created by the cilia, is obviously out of the question, since the food is oftentimes much longer than the organism. First, the oral cilia may be in actual contact with the food and exert a direct pull thereon and thus actually pull the object into the mouth. Secondly, the action of the cilia of the body may push the frontonia in the direction of the food and thus force the end of the object through the mouth opening. Prolonged study revealed the fact that both factors of ingestion are employed. The most conclusive evidence in support of the first possibility was found in the fact that the organism was oftimes seen to lie quietly while the food slowly entered the mouth; while, on the other hand, it was not uncommon for the oscillatoria fiber to remain comparatively still, while the frontonia slowly moved forward as the end of the fiber entered the mouth. This forward movement suggests that the frontonia either pushes the fiber into its body by swimming toward and around it or that the oral cilia, pulling upon one end of the stationary fiber, move the frontonia in the direction of the food. The fact that the ciliate at times moves forward when there is a concavity at the oral region (fig. 4) suggests that the push comes entirely from the locomotor organs. However, other situations are noted wherein the oral region moves along the fiber while the ciliate as a whole and the fiber itself are both stationary (figs. 10 and 11, h to l). Such observations would seem to establish the fact that both the oral and locomotor cilia play a part in the mechanics of ingestion.

B. Body movement—third and fourth factors involved in the mechanics of ingestion

In case the food particle is no longer than the expanded width of the body of the ciliate, the two factors heretofore considered suffice to explain ingestion. In the case of a diatom, for example, the body cilia force the frontonia forward while the oral cilia pull the food into the mouth. However, when one end of the food body is forced against the aboral wall, as at a, figure 3, and the other end still protrudes from the mouth, continued ingestion, if no other factors entered, would cause a rigid fiber to be thrust firmly against the aboral wall. Further ingestion is impossible without the play of other factors, and these appear to result from the stimulation due to the pressure of the end of the fiber against the body wall. It will be convenient to designate the points where this occurs as *tension points*. Specifically, however, the writer would define tension points as those points of contact between the ingested material and the body wall in which a sufficient pressure is exerted to be a stimulus. Such stimuli result in, (*a*) the changing of the angle between the body axis and the fiber, or in, (*b*) certain characteristic body contractions. It will be shown later that cyclosis is also an important factor in relieving the stimulation at the tension points, especially when flexible fibers are being ingested.

When the food fiber comes in contact with and protrudes the aboral wall, special effort seems to be exerted in an attempt to continue ingestion without altering the process. This outward pressure at the tension point seems to serve as a stimulus and to cause one or more things to happen. Sometimes partial or complete ejection of the food takes place, either suddenly by jerky movements, or slowly by reversing the ingestion process. Usually, however, the body of the frontonia goes through certain squirming movements and straightens out more in line with the axis of the fiber, thus aiding in the continuation of the ingestion process, (fig. 4). It will be noted from figures 1 to 8 that the frontonia is now turned through an angle of 180° from the position at the beginning of ingestion. When in this position the incoming food meets the least resistance, as the end of the fiber must now travel posteriorly along the aboral wall (*b*, *c* and *d*) rather than anterodorsally, as it did when it first entered the mouth. Thus, at the completion of this stage of ingestion the organism is turned completely around (compare fig. 1 and 8). It will be noted that during the early stages of ingestion the ciliate is usually directly over the fiber with the anterior end bent downward so that the mouth will come in contact with the end of the food body. This being the case, the end of the fiber which is being ingested is raised from the substratum.

The turning of the body, as shown in figures 3 to 6, causes the point of contact (fig. 3, *a*) of the end of the oscillatoria filament and the body wall to shift posteriorly (figs. 4, 5, and 6; *b*, *c*, and *d*). After the body reaches approximately the position shown in figure 5, the usual method of ingestion (the pull of the cilia of the mouth and the push of the locomotor cilia) carries the end of the fiber along the aboral wall to the posterior end of the frontonia (fig. 7, *e*). The continued pressure exerted from within not only makes more pointed the posterior end, but also causes an elongation of the entire organism (fig. 8). The anterior end now moves along the fiber, thus causing the mouth to be drawn well toward the anterior end of the organism (fig. 8, *f*). As the mouth is pulled forward the pressure at the posterior end becomes greater and greater. This is the second tension point in the process. The stimulus causes the organism to again undergo sharp body contractions. As in the former case, if the pressure is not relieved the oral cilia are relaxed, causing the body again to shorten. With a whirling backward movement the frontonia ejects the food particle (fig. 12). However, if the fiber bends or breaks, normal ingestion continues.

In the specific case under consideration, the oscillatoria filament was bent as indicated (fig. 9) and the mouth continued to move along the fiber (fig. 10, *h*, *i*, and *j*), causing the posterior end to be drawn toward the mouth. The whole animal was bent upon itself like a hinge (fig. 11). Since under the given pull the ciliate had now reached its limit of expansibility and, furthermore, since the fiber did not bend again, further ingestion was impossible. Accordingly, the frontonia suddenly contracted, whirled about the oscillatoria filament, causing the mouth (*m*) to be pried wide open, and flung itself from the food (fig. 12) With reference to this particular method of ejection, Schaeffer says: "If there are several coils of a filament whose other end is fast, rolled up inside of a frontonia, the mouth sometimes stretches antero-posteriorly until the coil as a whole without unwinding is thrown out of the body."

Rigid fibers were used extensively as food for experimental purposes, as the organism would continue to draw in one of them as long as possible and then eject it, only to repeat the process time after time. Since the fiber was longer than the expanded length of the organism and not sufficiently flexible to be wound up inside of the ciliate, the repeated attempts at complete ingestion were, of course, futile, but the process was nevertheless normal. This repetition made it possible to observe in detail again and again all of the movements involved in the ingestion of the same fiber by the same frontonia. These observations were furthermore especially valuable, as the fiber bent at a weak place at a distance from the ingested end equal to the expanded length of the ciliate. This bending permitted each process to continue more than twice as long as it otherwise would have done, since it was possible for more of the fiber to be ingested.

C. Cyclosis, the fifth factor of ingestion

Schaeffer emphasizes the fact that "in Frontonia leucas, rotational streaming is under the control of the organism, and special use is made of it in feeding." Although it will be shown later that cyclosis is effective, and in some cases essential, during the ingestion of certain flexible fibers, observations show that with some food material complete ingestion takes place without this fifth factor. A reconsideration of figures 1 to 9 will illustrate the point under consideration. In this case the rigid alga filament is forced into the body by the pull of the oral cilia and the push of the locomotor cilia of the body wall. The rotation of the body relieves the tension point (fig. 3, a) and permits the end of the fiber to pass down the aboral wall. Although there are slight indications of cyclosis during the ingestion of material of this type, careful observation makes it evident that rotational streaming is not essential. It might be concluded, then, that rigid material whose length is no greater than the expanded length of the frontonia can be, and is, ingested without the effective play of cyclosis.

VARIATION IN SIZE OF MOUTH

Such observations on ejection of partly ingested material as those considered above revealed some interesting facts regarding the nature of the mouth of frontonia. In many cases specimens which had ingested a sufficiently long fiber to produce a coil inside of the body suddenly whirled about and caused the mouth to be expanded sufficiently wide for the spiral to pass out without being first unwound. In many instances the mouth was stretched almost the length of the body. However, the author would conclude from his observations that the stretching of the mouth of frontonia is brought about by mechanical means through the twisting movements considered above rather than through being under the control of the organism itself. At first this unusual process was thought to be simply the breaking of the body wall, but by segregating individuals which had thus ejected rolls of oscillatoria filaments it was found that they were in no wise injured and that the stretching of the mouth was a normal process. Some of the unpublished notes of Schaeffer bear directly upon this point. He says in part: "In case the thread is too long, the coil does not always unwind but in nearly all cases, if the coil consists of many turns, the coil in its entirety comes out of the animal, the mouth apparently stretching for nearly the whole length of the animal. This is a normal process and does not hurt the frontonia." The mouth of the frontonia is not only expanded to an unusual width during ejection, but also many instances of ingestion, or attempted ingestion, have been noted in which the mouth was pushed open almost far enough to ingest objects as large as the animal itself. Figure 13 shows a frontonia attempting the ingestion of a mass of débris larger than its own body. Observations indicate that this enlarging of the mouth during ingestion is brought about by the play of the first two factors involved in ingestion, namely, by the pull of the oral cilia and by the forcing of the organism against and around the material being ingested by the action of the locomotor cilia of the body wall.

INGESTION OF LARGE DESMIDS

At times desmids which seemed to be ciliated (fig. 16) were seen to swim here and there through the frontonia cultures. Since the high power revealed a thin layer of protoplasm between the ciliated wall and the body of the desmid (Closterium), it was evident that these unusual organisms were frontonias which had engulfed desmids of more than twice their normal length.

Since Closterium was the largest rigid body known to be ingested by any frontonia, a study was made of the methods employed.

In order to expediate observation, rich cultures of frontonia were deprived of food for a number of hours (from twelve to twenty-four), and were then removed to depression slides containing numerous specimens of Closterium. Under these conditions, the ciliates readily attacked the desmids until many frontonias attempting ingestion could be observed at the same time. In practically all cases ingestion was indeed only an attempt, as complete ingestion was of very unusual occurrence as compared with the number of trials. For example, on December 10, 1919, at 8:00 A.M., numerous specimens of Closterium were added to a rich culture of hungry frontonias in a Syracuse watchglass, and the culture observed at brief intervals throughout the day. Although the ciliates spent the day in almost continuous attempt at ingestion, only three could be found at 5:30 P.M. which contained closteria.

The method of taking in these unusually large food particles was found to be almost identical with that involved in the eating of oscillatoria filaments as recorded in the earlier part of this paper, except that, of course, the mouth was more expanded. The average limit of linear expansion is shown in figure 15. At this point the organism either suddenly jerked back, whirled about, and left the desmid, or allowed the mouth to recede slowly down the desmid and completely ejected same, or relaxed as shown in A, figure 14, after which other attempts might be made before the food was completely ejected.

INGESTION OF SMALLER BLUE-GREEN ALGA FIBERS, OSCILLATORIA PROLIFICA, ETC.

The five factors considered in the earlier part of this paper are all noticeably effective during the ingestion of small flexible fibers. Figures 17 to 21 show the fibers being formed into a coil. The particular significance of this set of observations was in the further demonstration that cyclosis, regardless of the cause, is effective, if not essential, in some cases of ingestion. Without assuming that this is actually the case, it would be

very difficult to explain how the tension point at c, figure 19, could be relieved. Since the contractions, mentioned as the fourth factor, are direct compressions of the body wall, they would seem to force the end of the fiber through the body membrane. Therefore the movement produced by cyclosis would probably be the only factor which would alleviate this tension. When long fibers, as illustrated in figure 20, were completely ingested, the ciliate became very sluggish and discontinued the usual movements until digestion was nearly completed.

EFFECT OF THE INGESTED MATERIAL UPON THE SIZE AND GENERAL APPEARANCE OF THE BODY

Ingested food causes an unusual variety of shapes of the frontonia's body. Cultures taken directly from the brook have been found to contain individuals of almost every imaginable shape. The various shapes, of course, depend upon the variety of food available. After a few weeks' work the experimenter could cause to be produced many desired fantastic figures. For example, it was a very simple matter to produce the characteristic 'half-moon' frontonia shown in figure 23. This was done simply by cutting oscillatoria fibers into pieces slightly longer than the linear expanded length of the average frontonia. The imperfect 'half-moon' shown in figure 24 resulted from the ingestion of a longer fiber than was used in the case of the typical 'half-moon,' while the bow-and-arrow-like ciliate (fig. 22) is an unusual case in which a shorter piece of blue-green alga lodged perpendicular to the fiber which produced the 'half-moon.' The interesting case shown in figure 25 is simply a 'half-moon' frontonia in which the action of the digestive fluids caused the ends of the fiber to curl. Had the fiber given way in the center, the shape would have been markedly different. Since the food material includes not only hundreds of the smaller and more or less common fresh-water algae and the slowly moving protozoa, but also the limitless variety of foreign matter and débris which one finds in the sediment of a brook or which may be added to a culture, one is not surprised to find almost any imaginable shape. Moreover, the general appearance of the frontonia not only varies with the shape, size, color, density, and flexibility of the ingested material, but also with the arrangement as well.

Although material of almost any color might appear in the body of the frontonia, the more common colors, especially during the progress of digestion, are the various shades of brown, green, and blue.

DIGESTION

Although the problem of digestion need not necessarily be considered with the mechanics of ingestion, a number of simple observations were made along this line. As was suggested in the earlier part of this paper, the smaller solid particles of the cell are not usually found in contact with the long food fibers. This clear space forms the beginning of the future food vacuole. The digestive fluid attacks certain parts of the fibers, especially the ends, more readily than others, and this causes the replacing of the graceful curves by sharp bends, breaks, and general distortion. The walls of the oscillatoria filament give way and after two or three hours of digestion the free ends usually begin to roll up. Later the fibers break at various points and the pieces roll up until only small spherical food vacuoles containing irregular masses remain. The entire process consumes approximately six hours.

SUMMARY

1. Observations and experiments were made upon frontonias while these organisms were ingesting euglenas, diatoms, desmids, and oscillatoria filaments.

2. The ingestion of blue-green algae, especially oscillatoria filaments, furnished the most conclusive demonstrations of the method involved, as the process continued a greater length of time and involved more factors than did the ingestion of smaller organisms.

3. Five factors are involved in the process of ingestion of material longer than the expanded width of the body of the frontonia. In case of smaller particles, the third, fourth, and fifth factors mentioned below are not essential to ingestion.

A. *Action of oral cilia.* The cilia about the mouth of the frontonia exert a direct pull upon the incoming food.

B. *Action of the locomotor cilia.* The cilia of the body in general drive the organism forward and thus force the stationary food into the mouth.

C. *The rotation of the body axis.* The end of the fiber usually enters the mouth and passes anterodorsally until it comes in contact with and exerts a pressure upon the aboral wall (fig. 3), after which the frontonia swings around through an angle of almost 180°, using the mouth as a pivot. This change of position permits the fiber to pass dorsally along the aboral side of the ciliate. Through the play of either factor A or B, or both, ingestion continues until the fiber exerts such a pressure on the body wall at the extreme posterior end that the organism is extremely elongated and pointed (fig. 8). Such pressure on the body wall acts as a stimulus, causing movements that relieve the stimulation. The rotation of the body axis assists in relieving the stimulation at certain of these tension points.

D. *Body contractions.* A series of sharp contractions of the body wall assists in relieving certain other tension points.

E. *Cyclosis.* Cyclosis aids by moving the end of the fiber around the wall, thus making further ingestion possible (fig. 19).

4. Unusual and fantastic figures are produced through the contortion of the organism by the ingested food which varies in size, shape, density, elasticity, and color (figs. 21 to 25).

PLATES

345

PLATE 1

1   A normal frontonia approaching the end of an oscillatoria fiber prior to ingestion.

2   Oscillatoria fiber entering the body of the frontonia.   Characteristic shape of the organism during the early stage of ingestion of linear objects.

3 and 4   Turning of the body of the frontonia in order to relieve the stimulation at the first tension point (fig. 3, a).

4 to 6   The body cilia is pushing the frontonia in the direction of the food and thus forcing the end of the object posteriorly along the aboral wall.

7 and 8   The fiber reaches the posterior end of the organism and causes an elongation of the body of the frontonia.

9   The fiber breaks and ingestion continues by the oral cilia pulling the mouth along the object.

346

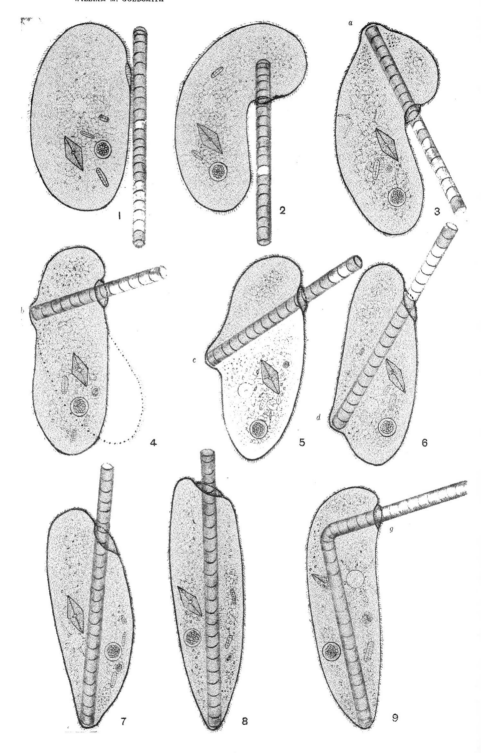

10 and 11   The mouth of the frontonia continues to be pulled along the fiber (*h, i, j, k, and l*) until the body is stretched to its maximum.

12   Further ingestion being impossible, the mouth stretches anteroposteriorly while the entire organism whirls about and flings the fiber from the body.

13   A frontonia attempting the ingestion of an object larger than its own body.

14 to 16   Ingestion of a large desmid (Closterium).

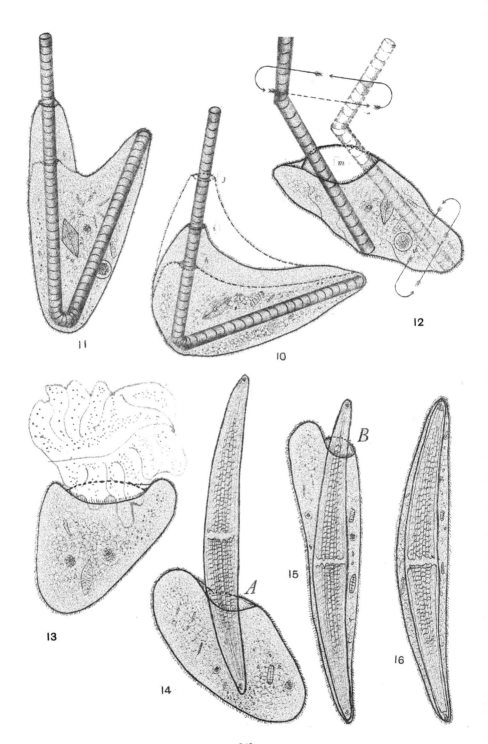

PLATE 3

EXPLANATION OF FIGURES

The ingestion of flexible fibers (Oscillatoria prolifera)

17 to 20   Method by which a number of coils of an alga filament are rolled up inside of a ciliate.

21   Two frontonias attempting to ingest the same filament.   In this particular instance the mouths met and the organism on the left slowly ejected the fiber as it passed into the mouth of the one on the right.

22 to 25   The ingested material alters the size, shape, and general appearance of the body of the frontonia.

350

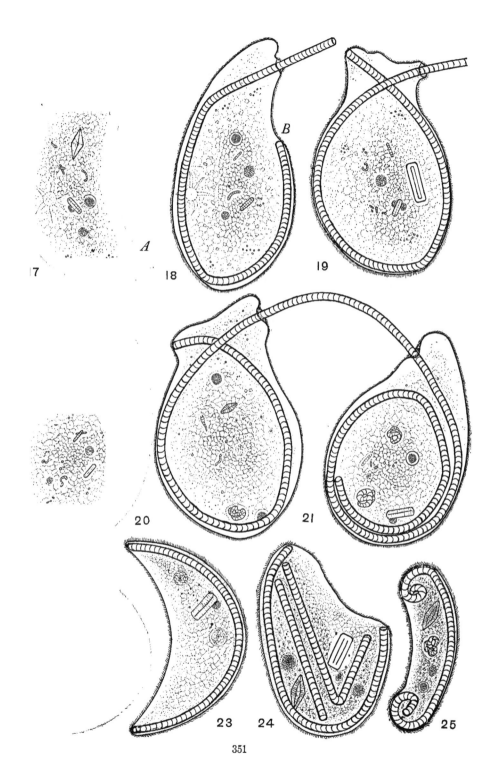

7

A

B

18

19

20

21

23   24   25

351

Abstracted by Donnell Brooks Young, author
Carleton College, Northfield, Minnesota.

A contribution to the morphology and physiology of the genus
Uronychia.

This paper records the results of regeneration experiments on
the hypotrich Uronychia. It shows that in this form the power
to regenerate parts lost by cutting or other injury is not dependent
upon the presence of a micronucleus, although parts without a
micronucleus do not divide. The ability to regenerate increases
with age, being least just after division and greatest just before
the process starts. The large cirri are so highly differentiated
that injury to them alone does not result in their regeneration,
although they do regenerate if the body of the animal is injured.
Injuries to the micronucleus frequently result in the formation of
monsters.

BY THE BIBLIOGRAPHIC SERVICE, JULY 24

A CONTRIBUTION TO THE MORPHOLOGY AND PHYSIOLOGY OF THE GENUS URONYCHIA

DONNELL BROOKS YOUNG

Carleton College

THREE TEXT FIGURES AND THREE PLATES (TWENTY-EIGHT FIGUR ES)

CONTENTS

INTRODUCTION

In 1910, while working at the biological station at Roscoff, Dr. G. N. Calkins demonstrated that the power of regeneration in the European species of Uronychia, Uronychia transfuga, varied considerably according to the time which had elapsed since the last division. It was at his suggestion that these studies were repeated on the American species. The following work was done at the Marine Biological Laboratory, Woods

353

Hole, during the summers of 1919 and 1920. I wish to take this opportunity to thank Doctor Calkins for the interest which he has shown and for the help and advice which he has given. I am indebted to my wife, Helen Daniels Young, for her painstaking preparation of the figures.

MATERIAL AND METHODS

Uronychia is a common marine hypotrich found at Woods Hole. Practically every salt-water culture examined during the summer was found to contain it, although all the forms which were used for experimentation were collected from two limited regions. It soon became evident that more than one species was present in these cultures, and the first problem was to make an accurate study of them so that they could be distinguished from each other. This led to a study of the genus as a whole with the results given in the next section.

No satisfactory culture medium was found, so that it was not possible to keep isolation cultures going for more than a few generations. However, for the purposes of this work an infusion made of eel-grass, flour, and a very little malted milk in sea-water was found to serve the purpose. Mass cultures were kept on hand, and the forms to be used for the various experiments were selected from these.

The method of procedure was as follows: the mass cultures were examined under the low power (10 × oc. and 55 mm. obj.) of a binocular microscope, and those individuals which were seen to be in the process of division were picked out by means of a small pipette and isolated in culture dishes in a few drops of the culture medium mentioned above. Such dividing forms can be identified quite easily because of the cirri which form precociously on the daughter cells. Usually there was but little difficulty in finding as many such individuals as could be used at one time. These were kept under observation and the time of division noted. Then at the desired time after division one of the daughter cells would be isolated in another culture dish and cut. . It was found that after a little practice the cutting could be done free-hand, in any plane desired, with a small sharp scalpel. Careful study of the fragments made the exact plane of cutting certain. In all cases the normal sister cell was kept as a control, and if for any reason it did not live and divide, the experiment was not listed as successful. Individuals experimented on were kept until they died or else divided. Drawings were made from life of some of the more interesting cases. Stained preparations were made for the study of nuclear conditions.

Specimens to be stained were put upon a cover-slip which had been smeared with a small amount of egg albumen and then a drop of the killing fluid added. The best killing fluid was found to be a saturated solution of bichloride of mercury in absolute alcohol. As soon as the specimen was stuck to the cover-slip it was put into a dish of the killing fluid and left for five minutes. It was then transferred to 95 per cent alcohol and then to water. Most of the individuals were stained with Heidenhain's haematoxylin, the short method. This gave most excellent results, for it not only stained the nucleus clearly, but it demonstrated the cirri as well. No other stain showed the new cirri in dividing individuals as sharply. Acid fuchsin, methylene blue, and Mallory's triple stain were used with indifferent success.

For the study of sections the following method gave satisfactory results. Uronychia transfuga was found in abundance in the zoogloea scum of old culture dishes. The only other hypotrich of the same size found here was a Euplotes. The scum from one dish was removed by rolling it into a ball, which was then fixed. Flemming's, Bouin's, Zenker's, and Schaudinn's fixing fluids were used. In this way a number of Uronychia could be handled together. This zoogloea was then embedded in the usual manner and sectioned ($5\,\mu$). Iron haematoxylin and Mallory's triple stain were used. The best results were obtained with Zenker's fixation and iron haematoxylin.

DESCRIPTION OF THE GENUS URONYCHIA (STEIN)

Diagnostic characters: medium-sized, colorless hypotrich with a constant body form; body oval, rigid, truncate anteriorly, somewhat rounded posteriorly; peristomal fossa broad and deep,

extending more than half the length of the body. The adoral
zone is represented by cirrus-like membranelles which originate
from pockets or pits on the anterior border. The mouth is situated in the left posterior part of the peristome cavity. Three to
six membranelles are present in the gullet. The peristome is
bordered by preoral and endoral undulating membranes. Three
bow-shaped cavities are hollowed out of the posterior end of the
carapace, two ventral and one dorsal. Originating from the
dorsal pit, which is situated on the right side, are three large
cirri which curve toward the median line. Inserted in the right
ventral pit are four or five straight cirri, while two sickle-shaped
cirri originate from the left ventral pocket.

Two distinct types of movement are observed, one a steady,
forward swimming or crawling and the other a backward or
somewhat sidewise jumping, darting, or spinning. In the first
type only the cirri of the adoral zone and the undulating membrane are at first seen to be in motion. Careful observation
shows, however, that there are some smaller cirri on the ventral
side which are used for this ordinary swimming. At the slightest
irritation the animal will vigorously contract one or more of the
large posterior cirri and dart backward with surprising rapidity.
Frequently these jumps are so rapid that the eye cannot follow
and the animal seems to have vanished as if by magic. After
a series of such jumps, the number and violence depending on
the strength of the stimulus, the animal will come to rest again
and resume its normal swimming.

The nuclear complex varies. The macronucleus is usually
broken up into two or more fragments. One or two micronuclei may be present.

A contracting vacuole has been described by Claparède and
Lachman as follows, "La vésicule contractile n'est point placé
du côté droit mais du côté gauche, immediatement en avant
des deux pieds dorsaux gauches." Calkins states that in U.
setigera the contractile vacuole lies between the two sets of
posterior cirri. These observations have not been verified,
and it is certain that in some species, U. binucleata and U. setigera, no contractile vacuole is present.

Kent ('81), following Stein ('59), defines the genus in these words: "Body oval, encuirassed, turgid, the sides rounded, truncate in front with a prominent membranous upper lip. The hinder extremity having developed on the ventral side two converging bow shaped fissures into which the short claw shaped anal and marginal uncini or styles are inserted; ordinary ventral or frontal styles entirely absent; the peristomial evacuation pocket shaped, closing sphincter-wise at will, its inner or right hand border bearing a band shaped undulating membrane."

Although this description holds in general, it is not correct in some points. The membranous upper lip is not found in all species and Kent's own figures (figs. 9, 10) fail to show it. At times the undulating membranes of the peristome protrude in front, and it may have been this which was seen. From the study of the living animal it is difficult to determine whether the posterior cirri are inserted in two ventral pockets or whether some are in a third pocket which is on the dorsal side. Stained preparations (fig. 14) confirm Maupas when he writes, "Stein place encore dans cette serie (i.e., with the ventral posterior cirri) les trois gros appendices de la région postérieure de bord droit de ces mêmes Infusoires. Mais c'est lá une erreur; car ces appendices sont séparés des cirres transversaux par une mince lamelle prolongement de l'extrémité postérieure ou caudale du corps par conséquent appartiennent á la face dorsale ainsi que Claparède et Lachman l'avaient déjà bien reconnus." The cirrus-like membranelles are somewhat differentiated into a central and two lateral groups. Ordinarily the central group is the one most easily seen. The lateral ones bend in towards the center and partially close the anterior end of the peristome. Occasionally they spread out, and it is their movement as they open and close over the end of the peristome that accounts for the appearance of the sphincter-wise closing of the peristome.

Only two species have been described for the genus, Uronychia transfuga (O. F. M.) and U. setigera (C.). This first is defined by Kent ('81), following Stein ('59), thus, "Body ovate, truncate in front, slightly narrower posteriorly, more usually obliquely truncate, angular and bent toward the left, but somewhat evenly rounded, the lateral margins symmetrical; the surface of the dorsal region sometimes smooth, sometimes longitudinally ribbed; anal uncini and styles variable in character and number, usually from three to seven or eight recurved and occasionally fimbriated uncini inserted in the right posterior cleft but not more than two or three in the opposite one; each of the fascicles occasionally supplemented by one or two fine simple setae."

This description is a very inclusive one, and it seems probable that some of the variations mentioned are in reality separate species. Sufficient data are not given, however, to be sure of this. Unfortunately, no mention is made of the nuclear complex or of the exact arrangement of the cirri.

An examination of the drawings which Wallengren ('02) (fig. 8) made in his studies of the regeneration of the cirri during division shows an animal which differs in shape and number and arrangement of cirri from that which Calkins used for his work at Roscoff (fig. 12). A detailed study of living material would probably show them to be different species.

Three forms of Uronychia are to be recognized at Woods Hole. These differ in size, nuclear complex, and arrangement of cirri. The largest of the three corresponds most closely to Stein's description of U. transfuga, and is probably the same one which Calkins used at Roscoff. This species was seen only once during 1919, but was found in abundance in 1920. It appeared too late, however, for experimental use. It may be described as follows (fig. 7):

Description of Uronychia transfuga

Carapace 130 to 170 μ long by 100 to 120μ wide and 25 to 40μ thick. Anterior and posterior ends rounded. Carapace arched and smooth. On the posterior ventral border is a cavity in which are inserted two heavy curved cirri and a cluster of from five to seven small straight ones. From the right ventral posterior cavity originate four or five straight cirri. From the right dorsal posterior cavity project three heavy curved cirri. The anterior cirrus-like membranelles, seven to nine in number, are set in slight depressions on the anterior edge of the carapace.

On the right and left sides of the carapace are other cirrus-like membranelles which are not as easily seen, but which are directed forward and which can be made to close over the opening of the peristome. The peristome which extends back about half the length of the body is bordered by two undulating membranes of about equal size. The mouth is situated in the left posterior part of the peristome cavity and has membranelles in it. The nucleus is made up of a macronucleus which is broken up into from eight to fifteen fragments arranged in the form of a horseshoe with the open side toward the right. A small micronucleus is located between two of the posterior pieces of the macronucleus.

Uronychia setigera (*Calkins, '02*) (figs. 1, 2, 3)

This species may be distinguished from the other species of the genus by the following characters: Carapace 40 to 50μ long by 25 to 35μ wide and 10 to 15μ in thickness. Anterior truncate, somewhat rounded posteriorly, with three or four ridges on the dorsal surface, which is but slightly arched. The arrangement of cirri is practically the same as that described for U. transfuga. The following points should be noted, however. In addition to the three heavy curved cirri which project from the right dorsal posterior cavity, there is also found a single slender straight one. Another very slender cirrus originates from the right ventral posterior cavity at the right of the four or five straight ones, and being as a rule at an angle to them, is difficult to see. On the left margin of the peristome are two very delicate, sickle-shaped membranelles. The peristome which is deep and extends backward for more than half the length of the body, is bordered by an undulating membrane on either side, the one on the left being the larger. Just anterior to the pocket-like mouth, which is situated in the left posterior portion of the peristome cavity, are two cirri which originate within the peristome. A third cirrus arises from behind the mouth and extends forward. Within the mouth are four membranelles (text fig. A). The nuclear complex is more or less band-like, with the macronucleus broken into two large pieces and the micronucleus situated between them (fig. 15). It is situated on the left side of the body.

Uronychia binucleata (*new species*) (figs. 4, 5, 6)

This species is distinguished from the others by the following characters: Carapace 60 to 80μ long, 50 to 55μ wide and 20 to 35μ thick. Dorsal surface arched and marked with small pits. The anterior end more rounded than truncate.

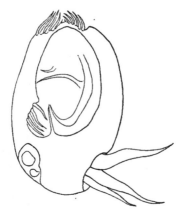

Text fig. A   Outline drawing of U. setigera, showing the position and arrangement of the peristome cirri.   From a stained preparation.

The posterior cirri differ from those of setigera in that never more than four are present in the right ventral pocket and that they are proportionately much shorter. No cirri originate in the peristome which, as in setigera, extends back for more than half the length of the body. The undulating membranes of the peristome are large and are seen frequently, as they are protruded balloon-like while the animal is swimming. Three of the delicate sickle-shaped membranelles are found on the left peristome border. The macronucleus is broken up into from three to five fragments and is in the characteristic position on the left side. Two micronuclei are present, one being located between the two posterior parts of the macronucleus, and the other between the two anterior parts.

In his study of U. setigera, Calkins ('02) did not recognize more than one species, and therefore his description is a composite one. As he states that no stained preparations were made, in all probability the nuclear complex which he mentions, a spheri cal macronucleus with a micronucleus beside it, is not nucleus at all but a food vacuole.

NORMAL CELL DIVISION IN U. SETIGERA
(Figs. 16 to 21)

Normal cell division was worked out in detail in U. setigera, both with living individuals and by means of stained specimens. Unusually fine material can be obtained, for not only do the nuclear parts stain clearly, but the cirri, especially those just form-ing, hold the stain and can therefore be studied in detail. The results of this study confirm Wallengren's observations ('02) in every respect. As is the case with all hypotrichs, all of the old cirri are absorbed and new ones, precociously formed, take their places.

The following observations made in the laboratory show the processes. 8.15 A.M. Individual isolated in a hanging-drop. 11.15. One of the right dorsal cirri of the anterior cell shows. 12.45. Several of the other precocious cirri of the anterior cell seen. 1.10 P.M. Some of the posterior precocious cirri show. 1.20. First sign of constriction noted. 2.10. Precocious pos-terior cirri larger than the old. 2.30. Complete separation of the cells. From this it will be seen that the whole process of division after the first appearance of the precocious cirri lasts about three hours. From eighteen to thirty-six hours elapse normally between successive divisions.

A study of stained individuals shows the process in more detail. Figure 17 shows a stained individual in the first stage of division. Precocious cirri can be seen just starting to develop a little above the middle of the body. The micronucleus has enlarged a little and is seen to be moving toward the outer edge of the cell. This individual shows that the cytoplasmic changes, such as the formation of the precocious cirri, and the nuclear changes proceed together. Figures 16 and 18 show ventral and dorsal views of a somewhat later stage. The micronucleus has moved out from the macronuclear band and has begun its mitotic division. The macronuclear fragments unite to form a band, but from these and many other preparations it is seen that there is considerable variation in this respect. Figures 19 and 20 show the precocious cirri still more advanced and the micronucleus separated into two parts which are, however, still connected by spindle fibers. Figure 21 shows the two cells almost ready to separate. From these drawings it will be seen that there is some variation as to time of absorption of the old cirri and as to the changes of the macronucleus.

Calkins ('11), in his figures of the division of U. transfuga, shows that the axes of the daughter cells change before the cells separate so that they are joined by a subterminal instead of a terminal protoplasmic connection. This causes all of the posterior cirri of the anterior cell to lie to the right of the uniting protoplasm. This shifting of axes was not observed in any of the forms studied at Woods Hole.

EXPERIMENTAL WORK

The following tables and summaries give the results of the various merotomy experiments. No mention is made of the controls, for unless these were normal the experiments were not recorded. The results are listed in tables according to the time after division at which the cells were cut and to the proportionate sizes of the pieces. The various planes of cutting are shown in text figure B and in the tables the cuts are recorded as indicated there. Thus 'transverse A' means that the cut was in a transverse plane and anterior to the center of the cell.

A complete record was not kept of the time which elapsed before the division of the operated cell. However, it was noted that in many instances the control cell divided two and occasionally three times before the operated one, and in two cases the control cells divided in eighteen and twenty-three hours, while the cut cells required forty-two and fifty-seven hours.

Thus the process of regeneration delayed the division rate.   The only exceptions to this were those cells which were cut during division.   In these cells division was completed before the regeneration of the missing parts and resulted in the formation of one normal cell and one which was abnormal in structure because of the cut.   The normal cell divided again at the usual rate, while in the operated one division was delayed by the proc-ess of regeneration.

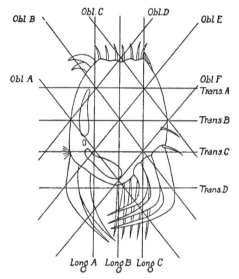

Text fig. B   Diagram showing the planes of cutting as listed in the tables following.

In order to describe the experiments it is necessary first to describe the way in which cells regenerate.   The first evidence of recovery after cutting is the closing over of the cut surface. This is accomplished to some extent by the drawing in of the edges of the body much as Holmes describes for Loxophyllum ('07).   However, in Uronychia the cuirass or pellicle is so rigid that this process is not very extensive.   Apparently a membrane is formed to cover the rest of the surface.   During this process there seems to be a tendency for the body to assume normal pro-portions, but probably the drawing in of the cut surface is enough to account for this appearance. Whether regeneration proper ever begins or not, this closing in of the cut surface takes place if the cell lives for more than a very few hours. The regeneration of the posterior end can be followed very easily because the cirri are so definite. At first they appear as swellings on the cut surface, but these elongate and finally show the characteristic shape. Frequently the normal cirrus apparatus does not develop, but lacks some of its parts or the parts are abnormally arranged. This is especially true of amicronucleate fragments. The regeneration of the anterior end is not as easily studied, for the anterior membranelles are not as clearly seen as the larger posterior cirri. At times it is difficult to be sure that the undulating membrane, which so often is protruded in front of the body, is not mistaken for new membranelles. However, the first evidence of regeneration is the appearance of slight pits from which the membranelles later develop, and by looking for them it is possible to be sure that new membranelles are forming.

When an animal was cut into two parts the power of coördinating the movements of the motile organs was destroyed in both pieces. For some this lack of control was very apparent, and each of the cirri moved independently, frequently in opposition to each other. This resulted in a very irregular and erratic movement of the fragment. Usually in the course of an hour the fragment became quiet and in a few hours the motor organs worked in harmony again.

In Euplotes, one of the closely related hypotrichs, Yocom ('18) has found a well-organized 'neuromotor' apparatus, which he believes acts as a coördinating center for the complicated sensory and motor organelles. If such a system should be found to exist in Uronychia, the reactions of the fragments could be easily explained. Although some time has been spent in a search for such a system, none has been seen as yet. As is discussed elsewhere, all that can be seen are fibers originating from the basal bodies stretching anteriorly. No center body or motorium is to be found (text fig. C, a and b).

The loss of coördination mentioned above could be accounted for by the severing of neuromotor fibers when an animal was

cut into two pieces, if such a neuromotor system exists in Uronychia. As the immediate effects of the injury wore off, the cirri
would become quiet and coördinated movements would again
be possible with the reëstablishment of a neuromotor center.

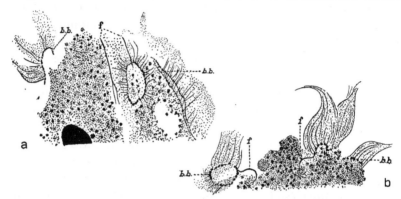

Text fig. C Drawings of two sections (a and b), showing the granular protoplasm, the basal bodies (*b.b.*), and the fibers which join with them (*f*). The
protoplasm is so granular that it is impossible to trace the fibers to any central
body or motorium.

Cells cut in first fifteen minutes after division

Summary of table 1. Thirteen individuals of the species
U. setigera were cut into practically equal pieces during the first
fifteen minutes after division. In only one case (no. 26) did
both pieces die without dividing. In six cases parts which were
supposed to be without micronucleus died without any trace of
regeneration. The pieces which were probably provided with a
micronucleus regenerated and later divided. In five cases the
fragment without a micronucleus showed slight regeneration.
That is, the cut surface closed over and new cirri apparently
began to form, but it is doubtful whether they would have completed regeneration, for some lived three to five days without
doing so. In only one case (no. 50) did the piece without a micronucleus grow to be at all normal. This individual was cut a
little to the left of the center in a longitudinal plane (long. B).
The amicronucleate fragment was somewhat larger and must
have had in it most of the posterior part of the macronucleus.

It lived for four days and regenerated one of the two left posterior cirri. It may be that the cut left the base of this cirrus attached to the right piece, and if so this accounts for the subsequent regeneration. The cell never became normal. The length of time which pieces lived after cutting seemed to depend on the amount of injury done by the cut and on the amount of food present in the cell. Several instances were noted in which the knife crushed one part of the cell more than the other. This crushing sometimes resulted in immediate death and sometimes in merely delaying the process of regeneration. The amount of food present in a cell could sometimes be told by the number of food vacuoles visible. In one instance (no. 47) there was a large food vacuole, and the piece having it lived for five days as against two to four for the others. Even though in some cases a new mouth may have developed in the amicronucleate piece, apparently no new food was taken into the cell. Evidently the pieces died of starvation.

Summary of table 2. Thirteen individuals of U. setigera cut into pieces of markedly unequal size during the first fifteen minutes after division are listed in table 2. In only two cases was any regeneration noted in the piece without a micronucleus. Both of these pieces had the posterior cirri attached and the anterior end not only closed over, but new anterior membranelles formed. These pieces lived more than twenty-four hours and shortly before death had the appearance of a very much truncated but otherwise normal individual. In ten cases no regeneration was seen in the small fragment and in most of them death was immediate. Experiment no. 122 is especially interesting, for in this case, although the large piece lived for thirty hours, no regeneration took place. The fragment was then killed and stained. No micronucleus was found (fig. 22). The small piece died very soon after cutting, without any signs of regenerating. The cut was recorded as transverse C. If the plane was a little farther forward than C, it seems probable that the micronucleus was injured or destroyed. From experiments which were performed and which will be described later, this is the probable explanation of this case.

TABLE 1

Uronychia setigera cut into practically equal parts during the first fifteen minutes after division

EX-PERI-MENT NUM-BER	TIME AFTER DIVI-SION	PLANE OF CUT	ANTERIOR OR RIGHT PIECE		POSTERIOR OR LEFT PIECE	
			Regeneration	Fate	Regeneration	Fate
	min.					
37	2	Trans. B.	None	Died at once	Complete	Divided
121	5	Obl. B.	Slight	Died	Complete	Divided
26	7	Trans. B.	None	Died	Complete	Died
39	7	Long. B.	Slight	Died	Complete	Divided
40	7	Trans. B.	Slight	Died	Complete	Divided
47	7	Trans. B.	Slight	Died	Complete	Divided
43	8	Trans. B.	None	Died at once	Complete	Divided
4	10	Long. B.	Slight	Died	Complete	Divided
41	10	Trans. B.	None	Died	Complete	Divided
120	12	Trans. B.	None	Died at once	Complete	Divided
109	13	Obl. B.	None	Died at once	Complete	Divided
115	13	Trans. B.	Slight	Died	Complete	Divided
50	15	Long. B.	Partial	Died	Complete	Divided

TABLE 2

Uronychia setigera cut into unequal parts during the first fifteen minutes after division

EXPERI-MENT NUM-BER	TIME AFTER DIVI-SION	PLANE OF CUT	ANTERIOR OR RIGHT PIECE		POSTERIOR OR LEFT PIECE	
			Regeneration	Fate	Regeneration	Fate
	min.					
44	5	Trans. A.	None	Died	Complete	Died
123	5	Trans. A.	None	Died at once	Complete	Divided
5	6	Obl. F.	Slight	Died	Complete	Divided
70	7	Trans. C.	None	Died	None	Died
64	10	Trans. A.	None	Died at once	Complete	Divided
125	10	Obl. D.	None	Died	Complete	Divided
119	10	Obl. A.	Slight	Died	None	Died
25	12	Obl. C.	None	Died at once	Complete	Divided
45	13	Obl. D.	None	Died at once	Complete	Divided
117	14	Obl. E.	Complete	Divided	None	Died
63	15	Trans. C.	Complete	Divided	None	Died
122	15	Obl. A.	None	Stained	None	Died
124	15	Trans. C.	Complete	Divided	None	Died

TABLE 3

Uronychia setigera cut into practically equal parts fifteen minutes to one hour after division

EXPERI- MENT NUMBER	TIME AFTER DIVISION	PLANE OF CUT	ANTERIOR OR RIGHT PIECE		POSTERIOR OR LEFT PIECE	
			Regeneration	Fate	Regeneration	Fate
	min					
111	16	Trans. B.	None	Died	Complete	Divided
118	16	Trans. B.	None	Died	Complete	Divided
112	19	Trans. B.	None	Died	Complete	Divided
36	20	Long. B.	None	Died	Complete	Divided
133	20	Trans. B.	Slight	Died	Complete,	Divided
42	22	Trans. B.	None	Died	Complete	Divided
135	22	Trans. B.	Slight	Died	Complete	Divided
134	23	Trans. B.	None	Died	Slight	Died
116	25	Trans. B.	None	Died	Complete	Divided
130	28	Trans. B.	Slight	Died	Complete	Divided
28	55	Obl. B.	Partial	Died	Complete	Divided
110	60	Trans. B.	Partial	Died	Complete	Divided

TABLE 4

Uronychia setigera cut into unequal parts fifteen minutes to one hour after division

EXPERI- MENT NUM- BER	TIME AFTER DIVI- SION	PLANE OF CUT	ANTERIOR OR RIGHT PIECE		POSTERIOR OR LEFT PIECE	
			Regeneration	Fate	Regeneration	Fate
	min.					
126	18	Trans. A.	None	Died at once	Complete	Divided
127	20	Trans. A.	None	Died	Complete	Divided
46	22	Trans. A-D.	None	Died	Partial Cirri reg. after div.	Divided
75	22	Obl. F.	Complete	Divided	None	Died
128	23	Trans. C.	Complete	Died	None	Died
136	24	Trans. C.	Complete	Divided	None	Died at once
62	25	Obl. F.	Complete	Divided	None	Died
129	25	Trans C.	Complete	Divided	None	Died
12	35	Obl. A.	None	Died	None	Died
1	40	Trans. A.	None	Died	Complete	Divided
27	50	Trans. A.	Slight	Died	Complete	Divided

Cells cut from fifteen to sixty minutes after division

Summary of table 3. Twelve individuals of U. setigera were cut into practically equal pieces from fifteen minutes to one hour after division. In one case (no. 134) both pieces died without either completing regeneration. Although neither piece was stained, from the position of the cut probably the micronucleus was injured. No regeneration of the amicronucleate piece was noted in seven instances (nos. 111, 118, 112, 36, 42, 134, 116). Death of the amicronucleate fragment usually came within a few hours of merotomy. In three individuals (nos. 133, 135, 130) slight regeneration was seen. All of these were cut in a median transverse plane (transverse B). New posterior cirri appeared as small swellings, but never went further than this in their development. The anterior amicronucleate fragment of one animal (no. 110) cut an hour after division regenerated almost completely. The posterior cirri began to form, but were irregular in their arrangement and never grew to normal size. This fragment died at the end of two days, before the other part had divided.

Summary of table 4. Eleven individuals were cut into quite unequal pieces from fifteen minutes to one-half hour after division. Both parts of no. 12 died almost at once. This was due to mechanical injury, for the larger piece was crushed and died before the smaller. One other experiment (no. 128) resulted in the death of both pieces, but before death the larger regenerated completely. The cause of death is not known. Regeneration in the small piece occurred in but one other instance (no. 27), but regeneration was incomplete, the posterior cirri appearing as tiny swellings; the small piece was from the anterior end. One very interesting experiment (no. 46) showed an unusual condition. The first cut failed to injure the body and removed only the last two-thirds of the posterior cirri. A second cut removed the anterior end of the cell. Regeneration was complete as far as the anterior membranelles were concerned, but the injured posterior cirri did not develop. At cell division the old, injured cirri were replaced by new ones as usual. This case will be discussed more fully in a later section.

Cells cut later than sixty minutes after division

Summary of table 5. Twenty-three individuals of U. setigera were cut into practically equal parts from one hour after division up to the time of the following division. In one case (no. 106) both pieces died without dividing, but in this instance the posterior fragment completed the missing parts before its death and the anterior one grew to be nearly normal. It may be that the micronucleus was destroyed, for the plane of cutting was within the limits of the position of this organelle. The amicronucleate fragments regenerated only partially in four instances.

TABLE 5

Uronychia setigera cut into practically equal parts later than one hour after division

EXPERIMENT NUMBER	TIME AFTER DIVISION		PLANE OF CUT	ANTERIOR OR RIGHT PIECE		POSTERIOR OR LEFT PIECE	
				Regeneration	Fate	Regeneration	Fate
	hrs.	*min.*					
30	1	15	Trans. B.	Partial	Died	Complete	Divided
33	1	15	Long. B.	Complete	Died	Complete	Divided
49	1	45	Trans. B.	Partial	Died	Complete	Divided
54	2	00	Trans. B.	Complete	Died	Complete	Divided
55	2	00	Long. B.	Complete	Died	Complete	Divided
56	2	00	Obl. E.-D.	Complete	Died	Complete	Divided
53	2	15	Trans. B.	Complete	Died	Complete	Divided
52	2	40	Trans. B.	Complete	Died	Complete	Divided
16	3	45	Trans. B.	Slight	Died	Complete	Divided
17	3	45	Trans. B.	Complete	Died	Complete	Divided
103	3	50	Long. B.	Complete	Died	Complete	Divided
106	4	00	Trans. B.	Partial	Died	Complete	Died
108	4	00	Trans. B.	Complete	Died	Complete	Divided
21	4	30	Obl. E.	Complete	Divided	Partial	Died
99	4	55	Obl. E.	Complete	Divided	Complete	Died
97	5	00	Trans. B.	Complete	Died	Complete	Divided
101	5	00	Trans. B.	Complete	Died	Complete	Divided
102	5	15	Trans. B.	Complete	Died	Complete	Div. unequal
85	10	30	Trans. B.	Complete	Died	Complete	Divided
86	10	40	Trans. B.	Complete	Died	Complete	Div. unequal
89	10	45	Trans. B.	Complete	Died	Complete	Div. unequal
90	10	45	Trans. B.	Complete	Died	Complete	Divided
91	10	45	Trans. B.	Complete	Died	Complete	Div. unequal

Three of these fragments were from the anterior end of the animal, and it was easily seen that the posterior cirri which they formed were not normal. In at least one the full number of cirri did not form, and the arrangement was therefore unusual. The one posterior amicronucleate piece which did not complete regeneration died at the end of about twenty hours. This may have been the result of mechanical injury. Eighteen experiments showed complete regeneration in both fragments, although the ones without micronuclei invariably died without dividing. These pieces lived sometimes five days. The parts with micronuclei completed regeneration and were, to all appearances, normal at the beginning of division. However, in four instances the division resulted in the formation of cells of unequal size.

Summary of table 6. Nineteen U. setigera were cut into very unequal parts from one hour after division up to the time of the following division. In four of these experiments both parts died, but not until after regeneration had begun in all. The plane of cutting in each instance was such that it probably injured or destroyed the micronucleus. In four other cases the smaller parts died very soon after the cutting, evidently because of the mechanical injury. In all of these instances the larger part regenerated completely and divided. In eight other experiments the smaller piece did not regenerate. The larger part invariably did regenerate and divide. In only two cases did the smaller parts regenerate completely, but these parts did not divide. In both of these the individuals formed from the small pieces were minute (in length only 10 to 15 μ), but normal in every respect as far as could be seen. In three of the cases mentioned above, the division of the part with micronucleus resulted in the formation of two individuals of unequal size, just as was observed in those instances listed in table 5. This unequal division indicates that a 'division zone' is formed during the growth of the cell long before the actual process of division begins.

Cells cut in early division

Summary of table 7. Nine individuals of U. setigera were cut in various planes during the early stages of division. At this period the micronucleus has not begun to elongate in a spindle, and any plane passing just above or just below the center leaves it uninjured. Two such cuttings were made. In experiment no. 3 the anterior end of the dividing cell was removed. Figure 23 was drawn from the stained preparation of the regenerated amicronucleate anterior fragment. Aside from the lack of a micronucleus the cell is complete. The posterior fragment completed its division before the missing parts were regenerated. This process took about ten hours and the complete sister cell divided twice before the regenerated one did once. In experiment no. 74 the animal was cut so that, as in experiment no. 3, the posterior part contained the micronucleus. However, in this case regeneration was completed before division. At the time of cutting the process of division had not progressed as far as in the first case. Evidently up to a certain stage in division the process can be stopped for the time being, but beyond that stage the momentum of division is sufficient to carry through that process regardless of other influences. Two individuals, nos. 34 and 65, were cut exactly in the center, and although the pieces regenerated in but little longer than it would have taken for the cells to have completed the division which they had begun, they never divided. It is quite certain that the micronuclei were destroyed. Figure 24 shows one of these cells, the anterior from experiment no. 65, which was killed and stained as soon as its sister cell died. No micronucleus was present. In the case of three individuals only small fragments were removed. In these division progressed normally and the lost parts were replaced only after the completion of the division. The time taken for regeneration was about the same as that which was needed for the regeneration of a similar part when the cut was made on a cell within the first few minutes after division. One (no. 83) was cut in a longitudinal plane. Both parts completed division and the two cells with micronuclei regenerated.

TABLE 6

Uronychia setigera cut into unequal parts later than one hour after division

EXPERIMENT NUMBER	TIME AFTER DIVISION	PLANE OF CUT	ANTERIOR OR RIGHT PIECE		POSTERIOR OR LEFT PIECE	
			Regeneration	Fate	Regeneration	Fate
	hrs. min.					
31	1 20	Obl. D.	None	Died	Complete	Divided
6	1 40	Obl. C.	None	Died	Complete	Divided
51	2 15	Trans. A.	None	Died at once	Complete	Divided
141	2 15	Trans. D.	Complete	Divided	None	Died
59	2 30	Trans. A.	Slight	Died	Complete	Divided
18	3 45	Obl. F.	Complete	Divided	Slight	Died
104	3 50	Obl. A.	Complete	Died	Complete	Died
105	3 50	Trans. A.	Slight	Died	Complete	Divided
107	4 00	Long. C.	Partial	Died	Complete	Divided
100	5 00	Obl. A.	Complete	Died	Partial	Died
98	5 00	Trans. C.	Complete	Died	Complete	Died
24	5 30	Obl. D.	Complete	Died	Complete	Divided
94	7 00	Obl. A-B.	Partial	Died	Complete	Div. unequal
96	7 00	Trans. A.	Slight	Died	Complete	Div. unequal
87	10 45	Trans. C.	Complete	Died	Complete	Died
88	10 45	Trans. A.	Partial	Died	Complete	Divided
92	10 45	Trans. A.	Complete	Died	Complete	Div. unequal
7	19 00	Trans. C.	Complete	Divided	Slight	Died
8	119 00	Trans. C.	Complete	Divided	None	Died at once

TABLE 7

Uronychia setigera cut in early division

EXPERIMENT NUMBER	PLANE OF CUT	ANTERIOR OR RIGHT PIECE		POSTERIOR OR LEFT PIECE	
		Regeneration	Fate	Regeneration	Fate
3	Trans. A-B.	Complete	Stained	After div.	Div. ant. cell small
34	Trans. B.	Complete	Died	Complete	Died
65	Trans. B.	Complete	Stained	Complete	Died
74	Obl. B.	Complete	Died	Complete	Div. after 24 hours
83	Long. B.	None	Div. and died	After div.	Divided
68	Trans. A.	None.	Died at once	After div.	Div. ant. cell small
81	Obl. F.	After div.	Divided	None	Died
82	Obl. D.	Slight	Died	After div.	Div. ant. cell small
78	Trans. A and C.	Small pieces both died no regeneration		Complete	Div. after 12 hours both were normal

The two amicronucleate pieces died without regenerating. One individual (no. 78) was cut into three pieces. The two end parts which were small died very soon. The center piece, however, completed its division and the halves regenerated in about twelve hours.

TABLE 8

Uronychia cut in mid-division

EXPERIMENT NUMBER	PLANE OF CUT	ANTERIOR OF RIGHT PIECE		POSTERIOR OR LEFT PIECE	
		Regeneration	Fate	Regeneration	Fate
19	Trans. A.	Slight	Died	After div.	Div. ant. cell small
38 71	Obl. C.	None	Died.	After div.	Div. ant. cell small
66 67	Trans. A.	Complete	Divided	Complete	Divided
69	Long. B.	Slight after div.	Divided	After div.	Divided
73 79 80	Trans. C.	After div.	Divided	Slight	Died
84	Obl. E.	After div. the post. cell died without reg. while the ant. cell reg. complete.		After div. the post. cell reg. completely, the ant. cell died without reg.	

TABLE 9

Uronychia setigera cut in late division

EXPERIMENT NUMBER	PLANE OF CUT	ANTERIOR OR RIGHT PIECE		POSTERIOR OR LEFT PIECE	
		Regeneration	Fate	Regeneration	Fate
20	Trans. C.	After division	Divided	None	Died
35	Trans. C.	None	Div. and post. died	Slight	Died
113	Long. B.	Slight after division	Div. and died	Complete after division	Divided
131	Long. B.	Abnormal	Died	Complete after division	Divided
132	Obl. B.	None	Div. and died	Complete after division	Divided
77	Obl. E.	Div. ant. cell died without reg., post cell reg. and died		Div. ant. cell reg. and div., post cell died without reg.	

Cells cut in mid-division

Summary of table 8. Ten individuals of U. setigera were cut at about the middle of the division process. At this time the micronuclei were somewhat separated and a single cut could not destroy both unless it was made in a longitudinal plane. Two individuals were cut in the plane of division. The resulting cells developed to be normal animals just as if the usual division process had been completed. In six cases where either the posterior or anterior end of the dividing animal was removed, division was not disturbed, but was completed before any regeneration took place. The small amicronucleate parts were probably injured by the cutting in one or two cases, for no regeneration was recorded for two of them and in the other four only slight regeneration was listed. The large pieces completed division in the usual time. Afterwards the injured cells replaced the missing parts. In one case (no. 69) where the cut was in the median longitudinal plane (long. B), division was completed in both fragments, but only those of the left side, the ones with micronuclei, regenerated. Experiment no. 84 is particularly interesting, for the cut was oblique E. This divided the animal so that when division was completed there was a small and a large cell originating from both the posterior and the anterior half. Neither of the small pieces regenerated. Both of the large pieces reformed the missing parts, but only one of them divided. Probably the micronucleus of the anterior half was destroyed, for the cut was about in its region.

Cells cut in late division

Summary of table 9. Six individuals of U. setigera were cut in late division stages. In general the results correspond to those obtained from cells cut slightly earlier. When cut transversely division proceeded normally and the injured cells regenerated after division. This transverse cut was made in two experiments (nos. 20 and 35). The amicronucleate parts cut off died very soon, in one case without any, in the other with slight regeneration. Two individuals were cut in the median longitudinal plane (long. B). In one case, experiment no. 113, division was completed in both parts. Those with micronuclei regenerated in ten to eighteen hours and later divided. The other two pieces died without regenerating completely. In the other case where a longitudinal cut was made, the micronucleate part divided and regenerated, but the amicronucleate piece instead of dividing, doubled on itself and formed a monster which died in about twenty-four hours. In experiment no. 132 the cut was oblique (oblique B) and evidently passed in front of the micronuclei, for the anterior part which contained also a small piece from below the division plane divided and both pieces thus formed died without regenerating; while the posterior part divided and both cells regenerated completely and later divided. In experiment no. 77 the animal was cut much as was the one in experiment no. 84, table 8. The results were the same. The anterior part divided and the anterior cell thus formed lived, regenerated, and later divided while the posterior one died with but slight regeneration. The posterior fragment of this experiment also divided, but the resulting anterior part died, while the posterior part lived and later divided.

Abnormal regeneration

Abnormal regeneration—table 10. It sometimes happened that an attempt to cut a cell was not successful and that the incision did not separate the two halves. Usually in such cases the wound healed in a few hours and the animal was normal. Individual no. 114 was in an early stage of division when it was injured while attempting to cut it. In this case the plane of the cut was transverse in the center of the body. Instead of dividing normally, the growth of the precocious cirri became abnormal and the cut surfaces fused together so that a monster was formed. This lived for five days and on the fourth had the appearance shown in figure 25. In all probability, in this instance the micronucleus was injured by the cut. Two other experiments (nos. 32 and 72) in which abnormal development occurred may be accounted for by this explanation. The first of these was a cell cut twenty-three minutes after division and the other was operated on at a mid-division stage. In the former the cut was made in a longitudinal plane between the posterior cirri and extended a little to the left. The pieces were united by a strand of protoplasm and seemed to rotate on each other, so that when reunited they had the appearance as shown in figure 26.

In an attempt to determine whether a cut in the region of the micronucleus would produce abnormalities, sixteen cells were operated on. Four of these were successful. Two were cut in early division. One of these did not complete division, but developed as has been described for individual no. 114.

TABLE 10

Uronychia setigera. Abnormal regeneration due to incomplete cutting

EXPERI- MENT NUMBER	TIME OF CUTTING	PLANE OF CUT	RESULTS
114	Early div.	½ Trans B. from left side	Did not complete div. but formed a monster
32	23 min. after div.	⅓ Long. B. from post.	Lived three days but was never normal
72	Mid-div	Trans. B.	Ant. cell died. Post. abnormal, never div.
146	Early div.	½ Obl. D.	Div. ant. cell died, post. abnormal
145	Early div.	½ Obl. A.	Did not divide. Abnormal
147	Late div.	½ Obl. D.	Div., ant. abnormal, post. normal
148	Mid-div.	½ Obl. A.	Div., ant. normal, post. abnormal

The other completed division, but the anterior cell soon died and the posterior became abnormal, although it lived for four days. In the case of a cell operated on in mid-division the posterior cell was injured, and although the anterior cell was normal, the cirri of the injured part developed to at least twice their normal length. In the fourth case the individual had almost completed division when the anterior part was injured. Division was soon finished, but the cirri of the injured cell were long and very irregular.

It is doubtful whether abnormalities resulting from incomplete cuts near the center of the body can be accounted for on the basis of an injury to the neuromotor system, even if we assume that such a system exists in Uronychia. If such an explana-tion is correct, any cut severing the fibers connecting the moto-rium with the cirri should produce the same results. This is not the case, for only when the cut was in the region where the micronucleus is usually found, did monsters develop.

TABLE 11

Experiments on starved cells of Uronychia setigera. Cells starved two days

EXPERI-MENT NUM-BER	TIME AFTER DIVISION		PLANE OF CUT	ANTERIOR OR RIGHT PIECE		POSTERIOR OR LEFT PIECE	
				Regenera-tion	Fate	Regeneration	Fate
	hrs.	*min.*					
170		10	Trans. A.	None	Died at once	Complete	Divided
171		15	Trans. B.	None	Died at once	Complete	Divided
172	1	00	Obl. D.	Slight	Died	Complete	Divided
173	3	45	Trans. B.	Slight	Died	Complete	Divided
174	5	30	Trans. A.	None	Died	Complete	Divided
175	10	00	Trans. B.	None	Died	Complete	Died
177	10	00	Trans. B.	Slight	Died	Complete	Divided
178	14	00	Trans. B.	Slight	Died	Complete	Divided
176	12	00	Trans. B.	Slight	Died	Complete	Died

TABLE 11A

Uronychia setigera starved four days

EXPERI-MENT NUM-BER	TIME AFTER DIVISION		PLANE OF CUT	ANTERIOR OR RIGHT PIECE		POSTERIOR OR LEFT PIECE	
				Regeneration	Fate	Regeneration	Fate
	hrs.	*min.*					
179		20	Obl. F.	Complete	Died	None	Died at once
180		30	Obl. E.	Complete	Died	None	Died at once
181		45	Trans. C.	Complete	Divided	None	Died at once
182	1	45	Trans. B.	Slight	Died	Complete	Died
183	2	30	Trans. C.	Complete	Died	None	Died at once
184	4	00	Trans. C.	Complete	Died	Slight	Died
185	4	00	Trans. B.	Partial	Died	Slight	Died
186	7	00	Trans. C.	Complete	Divided	Slight	Died

Starved cells

Experiments on starved cells—table 11.   When starved by being placed in a poor food medium, U. setigera grows smaller, but the protoplasm does not become vacuolated as does that of Paramecium.   Individuals will live for some time and even divide if put into filtered sea-water, although the interval between divisions is greatly increased.

Experimental cuttings were made on individuals starved two and four days, although not enough data were obtained to furnish absolute evidence of the effects of starvation.   It was soon found that if put back into plain sea-water cut cells died without regeneration, while if transferred to food media some would live and divide, so this method was used in the experiments performed.

In all, thirty-seven cuttings were made.   In twenty of these experiments both fragments died at once.   These are not listed in table 11.

Nine individuals were cut after two days of starvation.   Both pieces died in two of these experiments, although regeneration took place in the larger micronucleate fragments.   In the other seven the parts with micronucleus regenerated and divided. The amicronucleate parts died with no regeneration or with very little.

Eight individuals were cut after four days of starvation.   In five of these the small parts died at once or within a few hours without regeneration, while the larger micronucleate fragments regenerated even though they died without division.   In one case both fragments showed some regeneration, but died before it was completed.   In two experiments the larger pieces completed regeneration and divided, while the small parts died without regeneration.

Cutting cirri only

In experiment no. 46, listed in table 4, by accident the posterior cirri were cut off without injuring the body.   Later the cell was cut in the desired plane.   However, it was noticed that

the posterior cirri never regenerated, but were replaced by new
ones at the following division. This suggested that possibly
the cirri differed from the rest of the cell in their power to regenerate, and this led to further experimentation. In all,
eight attempts were made to remove part of the cirri without
injuring the cell. It was possible to tell whether or not the
body was injured, for when the cell was cut the cirri were held
together by the protoplasm removed with them. Six of the
eight attempts were successful. Without the posterior cirri
the animals swam about normally, but did not dart and jump.
In no case did regeneration of the missing cut cirri take place,
even though the cutting was done from twenty-two minutes
to four hours after division. Figure 27 represents a cell whose
cirri were removed two hours and forty-five minutes after division. The drawing was made from life twelve hours later.
In this, as in the other five of the six successful experiments,
the new cirri formed at the division period to replace the muti-
lated ones.

Stained preparations, both of total mounts and of sections,
of normal individuals show that the cirri have a well-developed
plate of basal granules imbedded in an area of dense protoplasm,
as Maier ('02) has demonstrated for the cirri of Stylonychia
histrio. If this basal plate is removed or injured, new cirri
will form, but if the cirri themselves are cut without touching
the basal granules no new growth takes place. The basal granules therefore do not have the power in themselves to reform the
cirri. New basal granules can be formed from which new cirri
will develop, but the old basal granules apparently have no
power to replace lost parts of cirri. Evidently the development
of the cirri is dependent on the activity of the basal granules
and once a basal plate has formed a cirrus, its power of causing
further growth ceases under normal conditions. In experiments in which the micronucleus was injured or destroyed during the division of the animal, as for instance experiment no.
114, figure 25, the cirri did not stop growing when they had
reached normal length and so became abnormally long. This
may indicate that the micronucleus has some influence over the
activity of the basal granules and that if it is injured during the

formation of new cirri, the growth will continue as long as the animal lives. If, however, the cirrus has completed its growth, the lack of a micronucleus does not result in any further changes.

Entz ('09) and Collin ('09) claim a nuclear origin for these basal granules. In the normal division of Uronychia the first evidence of their appearance is at the time of the formation of the precocious cirri, and these arise at the surface of the cell, quite close to the spot which they will occupy in the adult. If they do have their origin from the micronucleus it must be that the substance from which they are formed is set free during the process of normal cell activity and not just at the time of the formation of the cirri, for the above experiments show that new cirri with basal granules develop in amicronucleate fragments.

Collin claimed that in Anoplophyra branchiorum the basal granules developed in connection with the macronucleus. This may be the case in Uronychia, for no experiments were made which would be conclusive in regard to this point. However, the observations of normal growth as well as of regeneration after injury are in closer agreement with Maier's view that the basal granules arise as "cytoplasmic bodies at the surface of the cell." He believes, furthermore, that they serve merely as anchors or supports for the cirri and are not to be regarded as kinetic in function. Observations on Uronychia agree with Yocom's work ('18) on Euplotes patella. He describes a basal plate made up of basal granules imbedded in a dense substance which serves as a support for the cirri. He adds, "this is in agreement with Maier who considered the basal plate as the means of support for the cirrus but it is to be remembered that in Euplotes the function of support is to be attributed only to the dense opaque protoplasmic plate in which the basal granules are imbedded and that the basal granules themselves are given an entirely different function." In Euplotes patella Yocom demonstrated that the basal granules are in direct contact with the fibers of the neuromotor apparatus and he believes, therefore, that "this forms a basis of attributing to the basal granules the function of receiving stimuli. Such impulses received cause a contraction of the central contractile axis of each component cilium of the cirrus thus causing a lashing of the whole organ."

Cutting Uronychia binucleata

Experiments on Uronychia binucleata—table 12. It was noticed very early in the course of this work that regeneration occurred in both fragments of some individuals regardless of the time which had elapsed between division and cutting. Furthermore, both pieces divided. This was the first evidence showing that more than one species was being used. When it was discovered that two micronuclei were present in one species, while the

TABLE 12

Experiments on Uronychia binucleata

EXPERI-MENT NUM-BER	TIME AFTER DIVISION		PLANE OF CUT	ANTERIOR OR RIGHT PIECE		POSTERIOR OR LEFT PIECE	
				Regeneration	Fate	Regeneration	Fate
	hrs.	*min*					
150		5	Long. C.	Slight	Died	Complete	Divided
151		12	Obl. C.	None	Died	Complete	Divided
153		18	Obl. F.	Complete	Divided	None	Died
61		20	Obl. D.	None	Died at once	Complete	Divided
155		25	Long. C.	Slight	Died	Complete	Divided
159		45	Long. B.	Complete	Died	Complete	Divided
161	1	00	Obl. A.	Complete	Divided	Partial	Died
13	3	10	Obl. C.	Partial	Died	Complete	Divided

In the experiments listed below, the cut was transverse or but slightly oblique and near the center of the cells. The time after division varied from five minutes in the case of no. 149 to twenty hours after division in the case of no. 9. In all cases both fragments regenerated and divided. Nos. 149, 152, 2, 154, 156, 57, 48, 158, 10, 11, 160, 162, 163, 164, 165, 57, 50, 166, 13, 14, 15, 167, 95, 93, 169, 9.

other had but one, conflicting results obtained by cutting were explained.

Even before it was known that two species were being used for experimentation, a record was kept of the relative size of the individuals used and it was noticed that the large individuals showed greater powers of regeneration. Thirty-four experiments were performed on this species. Eight cases were recorded in which one part died, while the other fragment regenerated completely and divided. The amount of regeneration in the eight fragments depended on the size of the piece, the time after division of the cut, and the amount of mechanical injury done by the cut. For instance, the smaller fragment in experiment no. 61 was killed by crushing. In experiment no. 13 the cut was made three hours and ten minutes after division and the smaller parts constituted nearly one-fourth of the whole cell, yet regeneration was only partial. While in the case of the smaller part of no. 159, which constituted nearly half of the cell, regeneration was complete, even though the cut was made an hour after division.

Twenty-six individuals were cut through the middle of the cell in such a way that one micronucleus was present in each fragment. In all of these cases, regeneration was complete and division followed. It is not known what nuclear changes are involved in regeneration and subsequent division.

DISCUSSION AND CONCLUSIONS

The study of the three species of Uronychia found at Woods Hole, both of normal individuals and of those which were used for experimental purposes, seems to point to rather definite conclusions as to the function of certain cell organs. Because of the difference in nuclear structure in these species it is possible to use one as a control for the others.

Various functions have been ascribed to the micronucleus of the ciliates, including its activity in connection with regenera tion. Gruber ('85) thought that in Stentor, at least, the mi cronucleus was not as important for regeneration as the macronu cleus. He based this conclusion on the fact that in fragments of conjugating Stentor no regeneration takes place until one of the micronuclei takes on the form of the macronucleus. Stevens ('04), on the other hand, found that in Lichnophora no regeneration takes place unless both the macronucleus and micronucleus are present, and even then only slightly. Lewin ('10) agreed with Gruber and did not believe that the micronucleus is needed for regeneration or even for growth and division in Paramecium caudatum. He based his conclusions on the fact that he found a monster with the nuclear elements unequally distributed which, he claimed, produced on division a race of amicronucleate individuals. His published evidence, however, seems somewhat inconclusive. Calkins ('11) found that only a small percentage of Paramecia would regenerate regardless of the position of the plane of cutting. However, both nuclear elements were always present in all those which did regenerate. Evidently there is great variation in different species of Protozoa in regard to the ability to regenerate, and it is possible that the micronucleus functions differently.

The micronucleus has long been recognized as a diagnostic characteristic of the group of ciliates. In some instances the micronuclei cannot be found during the vegetative stages and becomes separated from the macronucleus only during conjugation, as Calkins ('12) demonstrated for Blepharisma undulans. In Opalina, Metcalf ('09) showed that when syngamy takes place the nuclei show two types of chromatin comparable to macro- and micronuclei. Dawson ('20) has described an Oxytricha with no micronucleus, and he has followed the life-cycle sufficiently to show that the sexual phases are abortive. Here evidently is a ciliate without one of the most important organelles. This Oxytricha is able to live and divide without a micronucleus, but such a case certainly is the rare exception.

In Uronychia it is clear that regeneration can and does take place under certain conditions without the presence of any micronucleus. Stained preparations fail to show that micronuclei have formed from the macronucleus, as Lewin ('11) suggested might be the case. Usually the amicronucleate pieces became abnormal if they lived for more than three or four days, so it might be said that for perfect regeneration the micronucleus is essential. These amicronucleate pieces apparently starved to death for, as far as could be discovered, no food was taken in or assimilated. In many cases no evidence was seen to indicate that a mouth was formed. Stained preparations do not show one, but as the mouth is not always demonstrable even in normal individuals, this is not conclusive.

The micronucleus is necessary for normal growth and division. In no case in U. setigera with its one micronucleus did both pieces divide, even though they did regenerate, while in binucleata, cut under the same conditions and differing only in the possession of two micronuclei, both pieces regularly did regenerate and divide.

It is not easy to cut Uronychia in such a way that either piece will be free from a part of the macronucleus.   In fact, it would be impossible to be sure that some of the macronucleus was not included, for there is so much variation in that structure.   These experiments do not offer any evidence of its function.

The fact that various monsters have been produced by injuring the cells in the region of the micronucleus indicates that it regulates the growth processes.   In those experiments in which the micronucleus was destroyed during the division process (cf. table 12) the process of regeneration of new cirri did not stop at the usual time and monsters resulted.   Even in the regeneration of amicronucleate pieces resulting from cuts not made during division, it was noted occasionally that the cirri continued growth throughout the life of the cell.   Usually the cell died from starvation before striking abnormalities were produced. This function of the micronucleus might be compared with the regulating action of the endocrine organs in the metazoa.

In summarizing his work on Uronychia transfuga, Calkins wrote as follows: "The results might be interpreted by the assumption of a specific substance, possibly enzymatic in nature which accumulates with the age of the cell until a condition analogous to saturation is reached.   With the formation of the new cell organs this substance, it may be further assumed, is exhausted and regeneration is impossible save with the full complement of cell organs."   These experiments seem to bear out such conclusions.   A comparison of these results with those obtained from the experiments on U. transfuga seems to show that this substance accumulates in the protoplasm of the cell at an earlier period in the American species.   For instance, Calkins found that regeneration of fragments without a micronucleus was seldom completed if the cutting was done before the animal had begun its division.   In U. setigera, however, after five or six hours had elapsed since division, amicronucleate pieces almost always regenerated the lost parts and assumed a more

or less normal form.   Therefore, such a substance is not formed
by the micronucleus and stored in it until the time of division
when, by the changes in the micronucleus, it is liberated.   Rather
it indicates that such an hypothetical substance formed while the
micronucleus is present accumulates in increasing amounts in
the protoplasm up to the time of division when it is used up in
the regeneration of those parts which have to be formed anew.
In starved cells it would appear that this substance is not formed
as rapidly as in normal individuals, indicating that it depends on
the taking in and assimilation of food.

Just what the nature of this hypothetical substance is is doubtful.   Meyer ('04) described a substance which he discovered
first in Spirillum volutans and which takes nuclear stains, although it is not chromatin.   He believes this substance, which
he calls volutin, to be a reserve of nuclear material.   This has
been found in many other bacteria, in yeast, and in the group of
the flagellates.   Reichenow ('09) demonstrated that volutin is
a nucleic-acid combination and that as the chromatin in the
nucleus increases, the volutin decreases in the cytoplasm; also
that when the nucleus is not growing the volutin increases.
I have not been able to demonstrate the presence of volutin in
Uronychia, but it or some similar substance would explain the
action of regeneration; that is, the nuclear reserve would be
used up during cell division, and as a result the ability to regenerate would be at a low ebb immediately after division.   As
the growth processes go on, a new supply of the reserve substance
would form, and thus the power of regeneration would increase.

Lund ('18) has carried out a series of experiments showing
that the resistance of Paramecium caudatum and Didinium
nasutum to KCN varies according to the age of the cell.   She
states that the resistance of Paramecium to KCN "when allowed
to feed on bacteria, showed a marked increase, and when fed
on yeast the resistance increases to a smaller degree, from the
time of division up to the following division."   "When Paramecium and Didinium are prevented from obtaining food the
resistance to KCN gradually decreases below its value at the
completion of division."   These differences in resistance are

explained by assuming that they are due to changes in permeability of the cell. There is clearly a close resemblance between the curve of variations in the resistance of Paramecium to KCN and the curve of regenerative power of Uronychia. Both are due to changes in the cytoplasm, and it may be that the substances which make the cells less susceptible to the action of KCN are the same as those which increase the regenerative power.

It is possible that the change is a physical one, as Lund suggests. Instead of a volutin-like substance being formed, the cytoplasm changes with cell age, the age at which the changes occur varying with different species, until a certain physical state is reached in which enzymes originating from the micronucleus are activated. These would be liberated as formed and would not be stored in the micronucleus to be set free only at division. The results on the whole seem to bear out the statement of Morgan ('01) when he said that "the nucleus supplies certain products of metabolism that must be present before the protoplasm can successfully carry out its innate tendency to complete the typical form."

In Paramecium Calkins ('11) found that "there is strong evidence of a division zone which lies in the center of the cell. If the cell is cut anterior or posterior to this zone the fragment divides in the original plane into a truncate abnormal form and a normal form. The truncate form may divide again not through its center but through the center of the cell were it perfect." Lewin ('10) finds a somewhat similar condition in his merotomy experiments on Paramecium, for he states that "since under normal conditions no two sister merozoites were found to divide there is a suggestion that possibly there exists in the cell a localized division center which passes on sectioning to one merozoite leaving the other incapable of division." According to these conclusions, the reason why division does not take place in parts of protozoan cells is not necessarily the lack of one of the cell components, such as the micronucleus, but rather that in a protozoan cell there is a potential plane of division, and when the cell is cut this zone can be present in only one of the fragments.   If this were the case, it would appear that a cell cut below the center would lead to division of the anterior part only, the posterior part not having the division plane present.   In Uronychia, however, in some cases the smaller posterior piece was the one to divide, while the anterior part with the so-called division zone died without division after it had regenerated.

That a division zone does exist in Uronychia is indicated by those cells which were cut from five hours after division up to the time of the next division.   In many cases the subsequent division, after regeneration, of these individuals resulted in the formation of two unequal cells, and in every case the smaller cell was from the side which had been injured by the cutting.   This division zone is not as marked in Uronychia as that which Calkins describes for Paramecium and it is not developed until a few hours after division.   This is shown by the fact that the size of the daughter cells resulting from a regenerated individual did not differ to any appreciable extent unless the cutting had been done five or more hours after division.

SUMMARY

1. Three species of Uronychia are found at Woods Hole, one of which has two micronuclei, while the others have but one.

2. In Uronychia all the old cirri are absorbed during the division of the animal and new ones are formed precociously.

3. The power to regenerate parts lost by cutting or other injury is not always dependent upon the presence of a micronucleus.

4. The ability to grow and divide is dependent on the presence of a micronucleus.

5. The power to regenerate lost parts varies with the age of the cell, being least shortly after division and increasing up to the next division, being best developed at the start of division.

6. In division, nuclear changes and cytoplasmic changes progress together, and it is difficult to tell which starts the process.

7. In Uronychia the large cirri are highly differentiated, in that they do not regenerate unless the body is injured.   The formative agency of the cirri lies in the body protoplasm, namely,

the basal bodies. Simply cutting the cirri without injuring the basal bodies does not result in regeneration.

8. Abnormalities may be produced in Uronychia if the micronucleus is injured. Such forms die without dividing.

9. There is evidently a division plane established at a fairly early period in the cell, and this is not altered by cutting the cell.

LITERATURE CITED

BUTSCHLI, O. 1885 Bronn's Klassen und Ordnungen des Thier-Reichs. Bd. 1, Abth. III.

CLAPARÈDE ET LACHMANN 1856-57 Études sur les Infusoires et les Rhizopodes. Inst. Nat. Génevois, T. 4-5, pp. 184.

CALKINS, G. N. 1904 Studies on the life history of Protozoa. IV. Death of the A series. Jour. Ex. Zool., vol. 1, p. 422.
  1911 Regeneration and cell division in Uronychia. Jour. of Ex. Zool., Feb., p. 95.
  1911 Effects produced by cutting Paramecium cells. Biol. Bull., vol. 21, p. 36.
  1901 Marine Protozoa of Woods Hole. U. S. Fish Commission Bulletin.
  1912 The paedogamous conjugation of Blepharisma undulans St. Jour. Morph., vol. 23.

COLLINS, B. 1909 La conjugaison d'Anoplophyra branchiarium. Arch. de Zool. Exp. et Gen., T. 1, ser. 5.

DAWSON, J. A. 1920 An experimental study of an amicronucleate Oxytricha. Jour. Exp. Zool., vol. 30.

DOFLEIN, F. 1916 Lehrbuch der Protozoenkunde, Vierte Auflage. Jena.

ENTZ, G. 1909 Studien uber Organization und Biologie der Tintinniden. Arch. fur Prot , Bd. 15.

GRUBER, A. 1883-85 Ueber kunstliche Theilung bei Infusorien. Biol. Centralb., Bd. 3, S. 389; Bd. 4, S. 717; Bd. 5, S. 137.

HOLMES, S. J. 1907 The behavior of Loxophyllum and its relation to regeneration. Jour. Exp. Zool., vol. 4, p. 399.

KENT, W. S. 1881 Manual of the Infusoria.

LEDANTEC La régénération du micronucleus chez quelques Infusoires ciliés. C. R. Acad. Sci. Paris. T. 125.

LEWIN, K. R. 1910-12 Nuclear relations of Paramecium caudatum during the asexual period. Camb. Phil. Soc. Proc , vol. 16, p. 39.
  1911 The behavior of the infusorian micronucleus in regeneration. Roy. Soc. Proc., Ser. B., vol. 84, p. 332.

LUND, B. L. 1918 The toxic action of KCN and its relation to the state of nutrition and age of the cell. Biol. Bull., vol. 35, p. 207.

MAIER, H. N. 1903 Der feinere Bau der Wimperapparate der Infusorien. Arch. fur Prot., Bd. 2, S. 73.

MAUPAS, E. 1883 Contributions a l'étude morphologique et anatomique des Infusoires ciliés. Arch de Zool. Exp , 2 Ser., T. 1.

METCALF, M. Opalina: its anatomy and reproduction with a description of infection experiments and a chronological review of the literature. Arch. fur Protistenk., Bd. 13.

MEYER, A. 1904 Orientierende Untersuchungen über Verbreitung, Morphologie und Chemie des Volutins. Botan. Ztg., Bd. 62, S. 113.

MINCHIN, E. A. 1912 An introduction to the study of the Protozoa. London.

MORGAN, T. H. 1901 Regeneration of proportionate structures in Stentor. Biol. Bull., vol. 2, p. 311.

REICHENOW, E. 1909 Untersuchungen an Haematococcus pluvialis nebst Bemerkungen über andere Flagellaten. Arb. a. d. Kais. Ges.-Amte, Bd. 33 S. 1.

STEVENS, N. M. 1904 Further studies on the ciliate Infusoria Lichnophora. Arch. fur Protist., Bd. 3.

STEIN, F. 1859 Der Organismus der Infusionsthiere, I. Abtheilung.

WALLENGREN, H. 1902 Zur Kenntnis des Neubildungs- und Resorptions processes bei der hypotrichen Infusorien. Zool. Jahr. Abt. fur Anst. der Thiere, Bd. 15, S. 1.

YOCOM, H. B. 1918 The neuromotor apparatus of Euplotes patella. Univ. of Cal. Pub. in Zool., vol. 18.

PLATE 1

EXPLANATION OF FIGURES

1 Uronychia setigera. From life. Dorsal view.

2 U. setigera. From life. Ventral view.

3 U. setigera. From life. Side view.

4 U. binucleata. From life. Ventral view.

5 U. binucleata. From life. Side view.

6 U. binucleata. From life. Dorsal view.

7 U. transfuga. Dorsal view of a stained preparation.

DONNELL BROOKS YOUNG

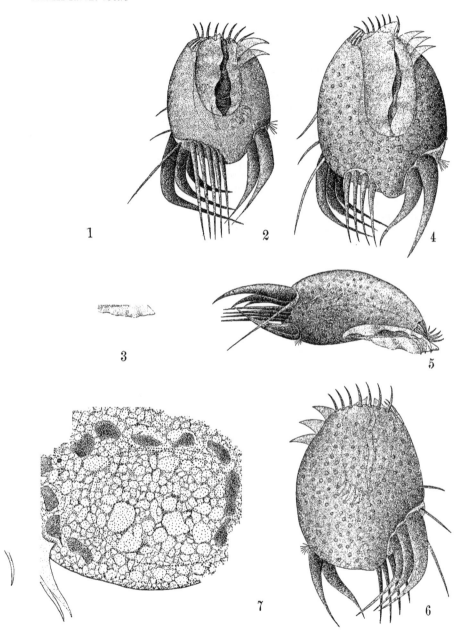

PLATE 2

8 U. transfuga. From Wallengren, "Zur Kenntnis des Neubildungs- und Resorptions-processes bei der Theilung der hypotrichen Infusorien," fig. C.

9 U. transfuga. From Bütschli, plate LXXII, fig. 4a.

10 U. transfuga. From Bütschli, plate LXXII, fig. 4b.

11 U. setigera. From Calkins, "Marine Protozoa of Woods Hole," fig. 55.

12 U. transfuga. From Calkins, "Regeneration and Cell Division in Urony-chia," fig. 1.

13 U. binucleata. Dorsal view of a stained preparation.

14 U. setigera. Side view of a stained preparation.

15 U. setigera. Ventral view of a stained preparation.

16 U. setigera. Ventral view of an early stage in division. From a stained preparation.

17 U. setigera. Ventral view of the first stage in division. From a stained preparation.

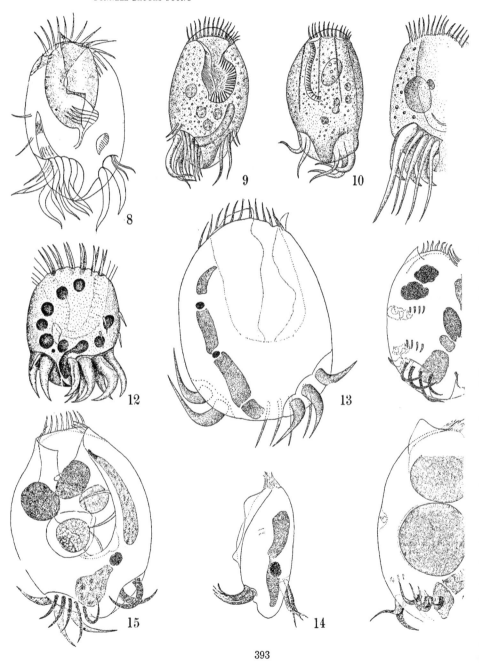

8

9

10

12

13

15

14

PLATE 3

18 U. setigera. Dorsal view of about the same stage as shown in figure 17. From a stained preparation.

19 U. setigera. Dorsal view of a mid-division stage. From a stained preparation.

20 U. setigera. Ventral view of a mid-division stage. From a stained preparation.

21 U. setigera. Ventral view of the daughter cells just ready to separate. From a stained preparation.

22 U. setigera. An individual which was cut in such a way that the micronucleus was destroyed. The specimen was stained thirty-six hours after cutting.

23 U. setigera. An individual which was cut in a median transverse plane at an early division stage. The micronucleus was evidently destroyed. Regeneration was complete. Experiment no. 3.

24 U. setigera. Another individual cut as in the case of experiment no. 3. Experiment no. 65.

25 U. setigera. A monster resulting from a cell injured in an early division stage. Experiment no. 114.

26 U. setigera. An abnormal cell which developed as the result of a cut which almost separated the two pieces. They rotated on each other and fused as shown.

27 U. setigera. A cell which had the posterior cirri cut and which did not regenerate them until division. This sketch was made twelve hours after the cirri were cut.

28 U. setigera. A monster which developed from one of the amicronucleate cells of experiment no. 131.

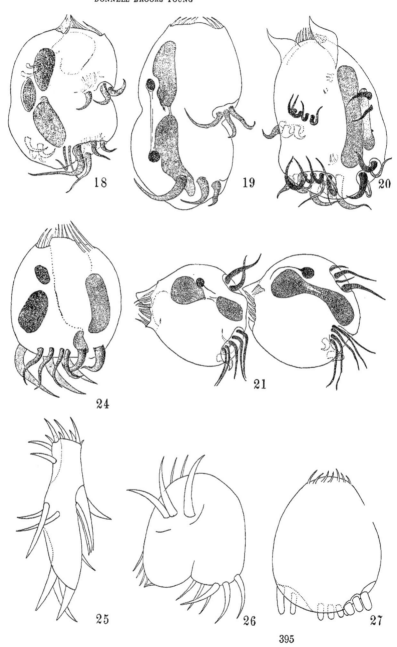

PROMPT PUBLICATION

The Author can greatly assist the Publishers of this Journal in attaining prompt publication of his paper by following these four suggestions:

1. *Abstract.* Send with the manuscript an Abstract containing not more than 250 words, in the precise form of The Bibliographic Service Card, so that the paper when accepted can be scheduled for a definite issue as soon as received by the Publisher from the Editor.

2. *Manuscript.* Send the Manuscript to the Editor prepared as described in the Notice to Contributors, to conform to the style of the Journal (see third page of cover).

3. *Illustrations.* Send the Illustrations in complete and finished form for engraving, drawings and photographs being protected from bending or breaking when shipped by mail or express.

4. *Proofs.* Send the Publisher early notice of any change in your address, to obviate delay. Carefully correct and mail proofs to the Editor as soon as possible after their arrival.

By assuming and meeting these responsibilities, the author avoids loss of time, correspondence that may be required to get the Abstract, Manuscript and Illustrations in proper form, and does all in his power to obtain prompt publication.

THE JOURNAL OF EXPERIMENTAL ZOOLOGY, VOL. 36, NO. 4
NOVEMBER, 1922

Resumen por el autor, W. W. Swingle.

Experimentos sobre la metamorfosis de los anfibios
neoténicos.

1. El autor ha alimentado Necturus adultos con grandes
cantidades de tiroides, en plena actividad fisiológica, y lóbulo
anterior de la glándula pituitaria, transplantando simultánea-
mente la tiroides de la rana. Los resultados han sido negativos al
cabo de cuatro meses, a pesar de la enorme cantidad de tiroides
ingerida. 2. Las tiroides de Necturus normales fueron trans-
plantadas en larvas jóvenes de Rana clamata, desprovistas aún
de miembros. Al cabo de diez a catorce días aparecieron los
síntomas de hipertiroidismo, tales como el desarrollo de los
miembros y atrofia casi completa de la cola. A pesar de sus
caracteres larvarios, Necturus posee glándulas tiroideas activas
y parece haber perdido su capacidad de transformarse bajo el
estímulo de la alimentación tiroídea. 3. La tiroides de un ajolote de 14.25 pulgadas de longitud y por lo menos de cuatro
años de edad fué transplantada en larvas de R. clamata, desprovistas aún de miembros. La glándula era grande, vascular
y las vesículas estaban distendidas por el coloide. Se cortó
en seis pedazos, que fueron injertados en otras tantas larvas.
Un renacuajo murió; los restantes presentaron la reacción
hipertiroidea típica a los ocho dias. Dos semanas después del
injerto las larvas presentaban patas y marcada reabsorción de
la cola. De este modo, una sola tiroides de ajolote contiene
bastante cantidad de hormón activo para metamorfosear prácticamente a cinco renacuajos, pero cuando se la deja persistir
dentro del ajolote no puede iniciar la metamorfosis. La neotenia
del ajolote parece depender aparentemente de la incapacidad de
la glándula tiroides para verter su hormón, completamente
formado y fisiológicamente activo, en la sangre. Un factor que
permite la excrección falta en este caso. Los ajolotes se met-
amorfosean rápidamente cuando se les alimenta con grandes cantidades de tiroides. Experimentos semejantes fueron llevados a
cabo con la tiroides de anuros neoténicos.

Translation by José F. Nonidez
Cornell Medical College, New York

EXPERIMENTS ON THE METAMORPHOSIS OF NEOTENOUS AMPHIBIANS

W. W. SWINGLE

Osborn Zoölogical Laboratory, Yale University

TWO TEXT FIGURES AND TWO PLATES (EIGHT FIGURES)

The discovery of the causal relationship existing between certain endocrine secretions and amphibian metamorphosis has served to stimulate anew the interest of investigators in the problem of neoteny and paedogenesis as presented by various species and genera of this vertebrate group.

The retention of the larval form either permanently or for periods far beyond the normal time required for metamorphosis is known as neoteny or neotenie (Kollmann, '82). This author, who has studied the problem carefully, distinguishes between partial neoteny, where the animal is simply retarded in metamorphosis beyond the normal time and passes the winter as a tadpole, and total neoteny, in which case the animal retains its gills and other larval characters becoming sexually mature in this condition.

Partial neoteny is quite common in such frog species as R. clamata and R. catesbeiana as there is a pronounced tendency of these forms to prolong the larval life an extra six or eight months beyond the usual one- and two-year period. Indeed, the larval life of Rana catesbeiana may extend to the period of sexual maturity in some tadpoles in so far as the possession of ripe spermatozoa is concerned, and this appears to be the explanation of the second larval sexual cycle described by the writer in a previous paper ('21).

The classical example of total neoteny is axolotl, the paedogenetic larva of Amblystoma tigrinum. The researches of Duméril ('65), Chauvain ('75) and others on the metamorphosis of axolotl have resulted in the more or less current belief that the peculiar group of aquatic amphibians known as perennibranchiates (Necturus, Proteus, Typhlomolge, etc.) are permanent larval forms capable of reproducing the species. The group is supposed to represent a sort of retrograde evolution from an originally terrestrial life to permanent aquatic existence by suppression of metamorphosis. The evidence for this view is suggestive; briefly stated it is somewhat as follows:

1. It is .generally conceded that the perennibranchiates do not form a natural group, but are to be regarded as a heterogeneous assembly; various genera are undoubtedly represented in the group. These animals probably became neotenic at a phylogenetically old stage and are hence the oldest and not the youngest members of the present-day urodeles. ·

2. Various anatomical features of the group, such as the pentadactyloid limb, presence of lungs, suppression of internal gills, and connection of the pelvic girdle with the vertebral column, point to a terrestrial existence somewhere in the history of the group.

3. Perhaps the most suggestive line of evidence for the view that the perennibranchiates are permanent larval forms is the occurrence of neoteny and paedogenesis as aberrations of development in semiterrestrial species of urodeles. For instance, it has long been known that the larvae of certain European salamanders fail to undergo metamorphosis and occasionally attain the size of 80 mm., whereas the normal size at transformation is 40 mm. Larvae of Triton have been reported 80 to 90 mm. long with functional gills and sexual organs fully developed. DeFilippi ('61) found in one locality in Lombardy sexually mature larvae. According to him, such gill breathing, sexually mature specimens occur constantly in a small lake in the province of Ossola in the Italian Alps. Many other cases have been reported, and the classical example of the axolotl is well known. It will be recalled that this creature was classified by systematists as a distinct species of perennibranchiata until Duméril described its metamorphosis into Amblystoma tigrinum.

4. Lastly, it has been repeatedly stated in the literature that one of the perennibranchiates, Typhlomolge, is hereditarily

lacking in the thyroid apparatus, which, if true, accounts for the suppressed metamorphosis. It will be recalled that it is possible to induce neoteny in anuran larvae by thyroid extirpation.

This evidence is suggestive, and taken in conjunction with what we know of the thyroid gland and its relation to metamorphosis suggests that the thyroid mechanism of forms such as Necturus, axolotl, and neotenous anuran larvae is defective and incapable of bringing about transformation.

EXPERIMENTS ON PERENNIBRANCHIATES (NECTURUS MACULATUS)

A group of adult necturus were obtained from the Ohio Valley and repeatedly injected with thyroid extract, and at the same time forcibly fed large quantities of the desiccated commercial preparation by means of a long glass pipette thrust down the throat. This procedure was repeated several times with negative results, despite the fact that the thyroid dosage was relatively enormous. The physiological activity of the thyroid preparation was tested by feeding small amounts to larvae of R. clamata, averaging 50 mm. total length, but without hind legs. These animals promptly showed marked indications of metamorphosis within eight days from the date of feeding (fig. 7). The iodine content of the desiccated tissue was given on the label of the bottle as 0.21 per cent by weight and the analysis made by the chemists of Parke, Davis & Company, Detroit. It is obvious that the thyroid preparation used in the experiment was not responsible for the negative results. One animal was injected twice intraperitoneally with 10 mg. of thyroxin iodine obtained from the laboratories of E. R. Squibb & Sons, and then forcibly fed large quantities of desiccated thyroid and anterior pituitary lobe substance. The physiological activity of the thyroxin iodine was tested by placing two young specimens of. Amblystoma punctatum in a 1 to 50,000 solution, whereupon they metamorphosed within two weeks. None of these agents, singly or taken together, produced the slightest indications of metamorphosis, nor, I may add, appeared to harm the Necturus in any way. As a last resort, three thyroid glands of newly metamorphosed

Rana clamata frogs were engrafted subcutaneously in one animal. Four months later, no metamorphic changes had appeared. The experiment was abandoned, with the conviction firm in the writer's mind that, although adult Necturus may possibly be induced to metamorphose, thyroid tissue alone is not the agent that will accomplish the transformation. Axolotls readily respond to thyroid administration by metamorphosis according to Laufberger ('13) (cited by Adler, '16), Huxley ('20), and others.

Examination was made of the thyroids of untreated Necturus, but nothing unusual was observed, except that in some animals the glands are small and the vesicles more or less isolated from each other, while in other animals the glands may be large. The blood supply appeared normal in those individuals with compact glands. The thyroid of one animal consisted of but four to six extremely large follicles on either side, in others the gland was larger and comprised ten to sixteen large follicles, while in one animal the glands consisted of twenty-one or twenty-two follicles on each side. In the last-mentioned case the animal had been treated with thyroid and thyroxin iodine several weeks previous to examination.

In order to test the physiological activity, a series of heteroplastic thyroid transplantation experiments were made. In the first set of experiments one-half of the thyroid gland (from one side only) was engrafted intraperitoneally into immature Rana clamata tadpoles averaging 52 mm. total length, with hind-leg buds 2.7 mm. long, but undifferentiated. Within eight days the tadpoles which received the graft showed all the symptoms of hyperthyroidism. The animals were greatly emaciated, with protruding eyes; the hind legs increased in length from 2.7 mm. to 8 mm. with complete differentiation. In two animals the right fore leg had appeared and the tail atrophied to half its original length. At the end of the eighth day from the date of transplantation the larvae measured 26 mm. total length, whereas the week previous the same individuals averaged 52 mm. It was impossible to keep the tadpoles alive until they completely metamorphosed, so great was the acceleration of metabolism and metamorphic change. Death in most cases was apparently due to respiratory difficulties. Several tadpoles were kept alive until tail resorption was nearly complete by placing them in shallow containers and passing a stream of compressed air through the water.

In a later series of experiments the thyroids were cut into small pieces and each part transplanted separately into immature tadpoles without hind limbs. One animal received three large colloid-filled vesicles dissected out of the gland; another received four follicles; the remainder received seven large follicles each. The grafts were made July 26th, and on July 30th, when examined, several engrafted individuals showed evidences of hyperthyroidism, such as emaciation and limb development. By August 2nd all of the animals showed a marked reaction to the graft; emaciation was very marked, the eyes protruded, and the legs had greatly increased in length. One animal had the right fore leg through the skin, and the remainder showed autolysis of the skin in the region where the fore limb later appears. The control animals remained unchanged (figs. 3 to 6). August 7th, when the experiment was discontinued, all engrafted animals were in advanced stages of metamorphosis. During the course of the experiment the animals were fed quantities of Spirogyra. They fed very little after the first three or four days, and not at all following the onset of marked metamorphic change.

This experiment shows clearly that the thyroid glands of adult Necturus are highly active metamorphosis-inducing agents. It is reasonable to assume that if they are capable of producing marked metamorphic changes when transplanted into immature anuran larvae within eight days, they are potentially competent' of doing likewise in Necturus if this animal still retained any capacity to transform. It will be shown later that the relation of the thyroids to metamorphosis in Necturus is quite different from the situation existing in axolotls and neotenous anurans.

While writing this paper the writer came upon a statement by Uhlenhuth ('21) that Jensen ('14) had subjected Proteus to the action of thyroid substance, but did not get any demonstrable results. Uhlenhuth's comment upon this experiment is interesting. He says: "Many causes may have been responsible for this failure (i.e., Jensen's), in particular the fact that the animals were too old when they were subjected to the thyroid feeding." And elsewhere the same author states "that nothing is known of the endocrine system of Proteus."

The writer ventures to predict that if the thyroid glands of an untreated Proteus are engrafted into anuran larvae, the latter will react similarly to those engrafted with Necturus glands. These experiments negative any such assumption as Uhlenhuth's ('21, page 201) that, "if the thyroid substance is capable of causing the development of the characters of a terrestrial amphibian, the administration of thyroid substance should cause the metamorphosis of Proteus anguineus." This writer apparently holds the view that any perennibranchiate will, if fed thyroid, metamorphose, and that retention of larval characters in these forms is due to absence or defect of the thyroid mechanism. The present experiments lend no support to any such hypothesis. In regard to the statement that the endocrine system of Proteus is unknown, it is interesting to note that Franz Leydig in 1853 (p. 62), describes the thyroids of this animal. According to his description, they apparently do not differ greatly from those of Necturus.

Uhlenhuth lays great emphasis upon the fact that one of the perennibranchiates, Typhlomolge rathbuni, is said to lack the thyroid gland, and states in effect that the reason for the retention of the larval characters is due to the thyroid absence. Perhaps this is true, but it strikes the writer as being rather odd that the retention of larval characters of this perennibranchiate should depend upon the absence of the thyroid function, and that such characters are retained in another form (Necturus) in spite of the presence of a most potent and active thyroid, in spite of feeding, injecting, and engrafting of thyroid substance.

There is another point of interest in connection with the discussion of Typhlomolge, and that is the apparent absence of thyroid glands in adults of this peculiar animal. Emerson ('05) studied the general anatomy of two specimens preserved in 4 per cent formalin, and merely mentions in the course of her discussions, "Sections of the head reveal the presence of a thymus gland, but I do not find thyroids." This is the only mention made of the thyroids in her paper. Recently, through courtesy of Prof. H. H. Wilder, the writer had an opportunity of examining Miss Emerson's material, consisting of serial sections through the head of one animal. No trace of a thyroid was observed, but it should be stated that some of the epithelial structures had disappeared.

The writer has carefully examined three adult specimens of Typhlomolge and failed to find any trace of a thyroid. The entire lower-jaw region, back to and including the heart, was dissected under a high-power binocular microscope, and some tissue sectioned, but with negative results. However, the failure to find the glands does not necessarily mean that they were not present or had not been present at some period, because the animals had been preserved over fifteen years in alcohol and many of the epithelial structures had undergone disintegration during this interval.

If the thyroid mechanism of Typhlomolge is congenitally lacking, then this amphibian is the only vertebrate known in which the gland is normally absent. It has been stated in the literature that Typhlomolge is only the neotenic larva of Spelerpes, but it is well known that Spelerpes larvae possess thyroids. It seems probable that the Texas cave salamander has a thyroid, but that it develops as a diffuse aggregation of follicles, somewhat similar to the condition known to exist in teleosts. At any rate the question of the presence or absence of the gland deserves further investigation before it can be accepted as an established fact, because the history of vertebrate morphology is replete with descriptions of forms supposedly anomalous for the lack of certain structures, only to be later shown to possess them.

To sum up, it may be said that these experiments indicate that Necturus and probably other perennibranchiates have permanently lost their ability to metamorphose into terrestrial forms under the stimulus of thyroid administration alone: our experiments indicate that the thyroid apparatus of these animals is highly active and potent despite their larval characters.

These experiments, however, do not rule out the possibility of inducing the metamorphosis of perennibranchiates by other means than that of thyroid feeding or transplantation. The cause of the non-metamorphosis of these forms may be pluriglandular in origin, and a result of defective interrelation of various endocrine glands. It should be added that the writer fed one Necturus small quantities of desiccated ovarian, testicular, adrenal, and anterior pituitary lobe tissues, along with large quantities of thyroid extract, but without avail. It is probable that transplants of these various glands simultaneously would have had more effect than feeding the desiccated substances, in case the animal possessed the capacity to transform. Administration of endocrine secretions, no matter in what quantity given, can give positive results only when acting upon an appropriate hereditary substratum. The indications are that the hereditary factors concerned in the metamorphosis of Necturus have become so modified that the appropriate substratum is lacking, thus rendering the thyroid hormone powerless. It is obvious that hereditary conditions in the perennibranchiates are quite different from those in axolotl, in regard to metamorphosis, since the latter readily respond to thyroid feeding and the former do not.

EXPERIMENTS WITH AXOLOTL THYROIDS

Through the courtesy of Prof. Henry Laurens, of the Department of Physiology, the writer obtained a very large specimen of Axolotl mexicanum (neotenic larva of Amblystoma tigrinum) for thyroid transplantation work. The animal was a very large one, measuring 14.25 inches from snout to tail-tip. The exact age of the specimen is unknown, as it was obtained, along with several others, from Albuquerque, New Mexico. When first brought to the laboratory the animal was about 8 inches long, and hence presumably about two years of age at the time; it was kept under laboratory conditions for two more years, thus making four years the animal's approximate age when used by me. This specimen was the only one of the lot that failed to metamorphose within a few months following removal from its native habitat to New Haven.

The thyroid glands were compact and large—larger than the glands of newly metamorphosed R. clamata frogs—and made up of large and small follicles filled with colloid. The blood supply to the glands was rich, apparently much more so than is the case with larval anurans; only a superficial examination was made to test this point because of the lack of sufficient material.

The pituitary gland was also examined; the pars anterior seemed normal or possibly rather small when the relative sizes of all the lobes are considered.

The thyroids were dissected out and each gland cut into three pieces of approximately equal size. Each piece of tissue was then transplanted intraperitoneally into immature Rana clamata larvae averaging 51.5 mm. total length with hind-leg buds 2.2 mm. undifferentiated. A few hours following grafting one tadpole jumped out of the container and was found dead, leaving but five transplanted tadpoles. Eight days after transplanting the pieces of axolotl thyroid, the engrafted tadpoles showed all the characteristic features of hyperthyroidism, such as cessation of growth, marked acceleration of limb development, tail atrophy and resorption, and body emaciation.

August 20th, or twelve days after transplantation, four of the engrafted animals were found dead (figs. 8 to 10). The photographs show very well the marked tail atrophy and resorption and the fore-leg development. The early death of the animals was due to the great acceleration of metabolism and metamorphic change. Undoubtedly, smaller pieces of the axolotl thyroid would have had the same effect upon metamorphosis without the too destructive rise in katabolic activity. The amount of tail resorption can be judged by the fact that during the twelve days of the experiment the average total length of the tadpoles decreased from 51.5 mm. to 29.6 mm. The control animals remained unchanged.[1]

[1] Since this was written one hundred and nine large axolotls were obtained by Professor Harrison from Albuquerque, New Mexico, and given to me for experimentation. This experiment was repeated on a large scale with identical results. Mr. Carl Mason. of this laboratory, metamorphosed thirteen normal, thyroidless, and pituitaryless R. sylvatica tadpoles by transplanting pieces of the thyroid of a single 14-inch axolotl.

The pituitary gland of the axolotl was also engrafted into an immature larva of R. clamata, measuring 45 mm. total length, hind limbs 2.5 mm. long and differentiated into thigh and shank, but without toe points. All three lobes of the gland were transplanted together. The success of the graft was attested by the color change in the larva induced by the expansive action of the pars intermedia secretion upon the melanophore system.

Eighteen days after the graft was made the animal had not changed in any way either in regard to limb development or growth, so the experiment was abandoned. Examination of the implanted gland showed it to be mostly resorbed.

The negative result following grafting of the axolotl pituitary is in striking contrast to that obtained when the pituitary of a newly metamorphosed frog is transplanted into immature anuran larvae. However, it must be remembered that in the latter case we are dealing with a homoplastic graft and in the former with a heteroplastic one, and a single transplant at that. It is unfortunate that a larger amount of axolotl material was not available, for it is of importance to know whether or not the pituitary gland of the axolotl is active, and some idea of its potency can be obtained by testing its effect upon limb development of anurans. If the results are consistently negative, then it is probable that the gland is defective in so far as its relation to the thyroid mechanism is concerned.

The results obtained by grafting portions of axolotl thyroids, are clear-cut and admit of but one interpretation: namely, that the thyroid apparatus of this animal is highly active and potent in inducing marked metamorphic change when transplanted into immature anurans, but is apparently incapable of initiating metamorphosis when left unmolested in its normal place. This experiment seems to rule out the idea that the axolotl's thyroid secretion is defective. If the thyroid glands of a single axolotl when cut into six fragments are capable of initiating metamorphosis in five anuran larvae grafted with a single fragment (the sixth animal died) within ten to fourteen days, surely we may safely assume that the same glands, entire, contain enough of the active hormone to initiate metamorphosis in the single axolotl of which they originally formed a part.   The metamorphosis of an anuran larva involves much more fundamental transformation and reorganization of tissues and organs than the same process in axolotl.

Our experiment suggests four possible factors to account for this anomalous situation: 1) Possibly the blood supply taking the hormone away from the gland is defective and the thyroid consequently unable to release its secretion.   2) The thyroid is able to collect, store, and transform the incoming iodine taken from the food and water into the physiologically active hormone, but owing to defective nervous stimulation the gland is unable to release the secretion into the blood stream.   3) The secretion, though perfectly formed, is unable to escape from the gland, owing to some defective interrelation between the pituitary and thyroid, or possibly some other endocrine gland which supplies the necessary stimulus to the thyroid, thereby acting as the releasing factor.   4) The blood and tissues of neotenous forms may contain substances that neutralize or render impotent the metamorphosis-inducing agent of the thyroid hormone.

If it is true, however, that the thyroid hormone is unable to escape because of defective outlet through the blood stream, or because of defective interrelationship between various components of the endocrine system, how is it that axolotls usually promptly metamorphose when taken from their native habitat or when subjected to sudden environmental changes, such as a change from New Mexico to New Haven?[2]   Furthermore, aside from the nervous system, there are few anatomical or physiological mechanisms which hold such power over other structures as to permit a gland like the thyroid to manufacture and store in large quantities a highly complex substance, but apparently prohibits its release.   When we consider the thyroid glands of an axolotl filled to capacity with highly active secretion, as our experiments clearly show, yet apparently unable to release the

[2] Of the 109 axolotls received from Albuquerque, only those that were thyroidectomized failed to metamorphose spontaneously a few weeks after removal from their native habitat.   Left unmolested in the New Mexican environment, the animals may remain permanent larvae and grow to a length of 14 inches.

hormone into the network of capillaries surrounding the gland in sufficient quantities to induce transformation, we are led to the conclusion that in the last analysis the crux of the problem is defective stimulation (perhaps inhibition) of the gland by that portion of the animal's nervous system responsible for the flow of secretion under normal conditions.

How else can one reduce to harmony the multiplicity of factors that have been invoked to explain axolotl metamorphosis, save by reducing them all to a common factor: i.e., agents which produce their effect by subjecting the organism to more or less violent changes of the environment thus acting as a constant nervous stimulant? For example, a few of the agents (aside from thyroid or iodine feeding)[3] that have served to initiate axolotl metamorphosis are: sudden changes in food supply, drying of swamps or pools in which the animals live, changes in the temperature of the water (Shufeldt, '85); forcing the animals to breathe air, insufficiently aerated water (Chauvin, '75, '77); administration of salicylic acid (Kaufman, '18); shifting of the animal from its normal habitat to other districts, such as from New Mexico to New Haven.

A glance at this list of factors indicates that their varied nature alone negatives the idea that any one of them can be the real causative factor in axolotl metamorphosis. However, all can be classed as shocks to the organism, and it may possibly be that such more or less constant excitation may bring about nervous stimulation to the thyroid sufficient to overcome the inhibiting influence and release the stored secretion, thus initiating metamorphosis. The nature of the inhibiting factor is of course the crux of the problem, and in the last analysis is probably of endocrine origin acting through the intermediation of the nervous

[3] Apropos of Uhlenhuth's claims that iodine has nothing to do with axolotl metamorphosis, the recent papers of Jensen and Hirschler are of interest. Jensen (Compt. Rend. Soc. de Biol., T. 85, 1921) metamorphosed axolotls by injections of iodocasein, iodoserumglobulin, and iodoserumalbumin. Also by feeding with an organic iodine compound—iodo-thyrosin. Hirschler (Arch. Entw. Mech., 1922) metamorphosed axolotls and anuran tadpoles by feeding elemental iodine in various forms. These investigators worked on strains of axolotls which do not spontaneously transform.

system. On theoretical grounds, the writer believes that electrical stimulation, thyroid puncture, and extirpation of the thyroid and reimplanting it into the same individual will metamorphose axolotl, but to date has been unable to obtain sufficient animals for experimentation along the lines indicated.

If such procedure should cause metamorphosis, then it is clear that the physiologically active hormone is not released from the thyroid in sufficient quantity to induce transformation. From the evidence at hand it seems to the writer that such is probably the case. Axolotls readily respond to thyroid feeding or to injections of iodothyrine by transforming, and the amount of thyroid substance required is not excessive. If the assumption were correct that the blood and tissues of this neotenic form contained substances which neutralized or rendered impotent the thyroid hormone, thus preventing metamorphosis, why should small amounts of thyroid substance, when fed, be able to produce an effect? The evidence obtained from thyroid transplantation experiments with neotenous anurans is interesting in this connection.

EXPERIMENTS WITH THE THYROIDS OF NEOTENOUS ANURANS

The larvae of the green frog, Rana clamata, have a larval period of approximately one year: i.e., 370 to 400 days from the date of egg deposition. The animals attain a length of about 65 mm. at metamorphosis, which occurs in late July and early August. However, it has been repeatedly observed by the writer that many larvae fail to transform at the usual time and remain an extra year as tadpoles. Such animals are typically neotenous forms and, as they continue growing throughout the larval period, they ultimately reach a size considerably in excess of that generally exhibited by the species at metamorphosis. Larvae measuring 75 to 90 mm. total length, with differentiated hind legs varying from 4 to 20 mm., have been captured in the months of November, December, and January from various pools in the vicinity of New Haven. The tendency of the species is to prolong rather than curtail the span of larval existence. Because of this fact, a series of thyroid-transplantation experiments were carried out with these neotenous individuals, in the hope of determining the endocrine locus responsible for the failure to metamorphose at the proper time. The first procedure was to test the physiological activity of the thyroid apparatus of such animals by transplanting the glands into immature larvae of the same species with undifferentiated legs and noting the effect upon metamorphosis.

December 17, 1920, three immature larvae, averaging 33.5 mm. total length, without hind limbs, except undifferentiated epithelial buds, were engrafted with the thyroid glands of 80-mm. neotenous larvae with hind legs 11.6 mm. long. The engrafted forms were the smallest obtainable at this season of the year and not over six months of age. It is unfortunate that a larger number could not have been used, but the results are clean-cut, as will be seen later, when the experiment was rechecked with another group of animals in midsummer.

One animal was in advanced stages of metamorphosis on January 17th, just one month from date of grafting. The right fore leg was through the skin and the left fore leg appeared the day following (January 18th). For a week previous to the first appearance of the fore limbs the engrafted animal showed marked signs or hyperthyroidism, such as emaciation, protrusion of the eyes, slight tail atrophy, and autolysis of the skin over the region of the fore legs. The control larvae at this time averaged 34 to 38 mm. in length and showed no change in regard to leg development from the condition when the experiment was started. The other two larvae of the engrafted culture developed fore limbs January 22nd and typical frog mouth. All of the animals died before tail resorption was complete. Figure 1 is a drawing of an engrafted larva and its control. Figure 2 shows the neotenous type of tadpole from which the thyroid glands were taken. Generally the limb development is less marked than the drawing indicates. When the drawings were made the animals were nearly a year past their normal time of metamorphosis.

Briefly summarized, the experiment shows that transplantation of the thyroid glands of neotenous larvae 80 mm. total length, with differentiated hind limbs 11.6 mm., into immature tadpoles

33.5 mm., without limbs, brings about very marked metamorphic changes within thirty days, although the animals could not be reared to the stage of complete tail resorption.

It was observed that in the engrafted animals autolysis of the skin over the region of the fore limbs occurs independently of limb development as a distinct phenomenon of anuran metamorphosis.   Years ago Braus ('06) described similar phenomena in developing tadpoles after extirpation of the limb bud.   I mention it here because of the remarkable autolysis which is sometimes observed in transplanted larvae; the fore limbs may be small, whereas the skin area destroyed may be very large indeed compared with limb size.   It should be remembered, however, that in anurans the fore and hind legs tend to keep pace with one another in development, only the fore legs are not visible because of the opercular covering.

The chief point of interest, however, is the odd fact that much greater metamorphic change follows transplantation of the thyroids of 80-mm. neotenous larvae with hind legs 11.6 mm. into immature larvae without limbs than occurs in control animals 80 mm. with legs 11.5 mm.   In other words, the grafted glands wrought far greater changes in the same time interval (approximately one month) when transplanted into immature larvae than when left undisturbed in the mature forms.

This result, so curious and at variance with what one might expect, led to a repetition of the same experiment the following summer, from a different angle.   Immature larvae averaging 40 mm. total length with hind leg buds 0.5 mm. were transplanted with the thyroids of mature though not neotenous tadpoles 68 mm. total length with hind legs 11.5 mm.   The results were similar to those of the earlier experiments, though not so striking as one would expect, because the mature control animals (not neotenous) in this experiment were approaching metamorphosis at the end of the experiment, whereas in the previous experiment the neotenous controls showed no change, and in fact passed the winter in the laboratory as tadpoles.

The experiments indicate that the thyroids of extra-season, neotenous anuran larvae with hind limbs 11 mm. long are physio-

logically active and capable of inducing metamorphosis, provided their contained secretion could get into the blood stream.
Like the thyroid mechanism of axolotl, the glands of these larvae
seem to be rendered more or less functionless by an inhibiting
factor which prevents secretion of the hormone into the circula-
tion. Following transplantation, the inhibition apparently is
overcome by the acquisition of a new blood and nerve supply

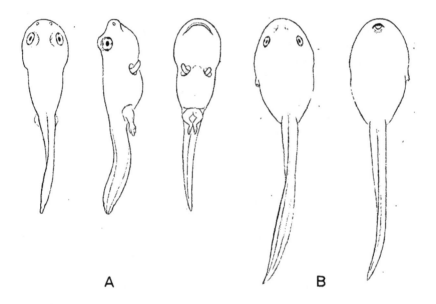

Fig. 1 Immature R. clamata larvae engrafted with the thyroids of extra large
neotenous tadpoles with hind limbs 11.6 mm. long, B, immature control animals
of A groups. × 2.

in the new environment, because the absorbed secretion induces
metamorphosis.

A modification of the experiment just recorded was attempted;
the thyroids of large neotenous larvae were engrafted intraperitoneally into other extra-season animals of the same size
and developmental stage. The idea was, that since the glands
of such individuals are physiologically active and capable of

inducing marked metamorphic changes when transplanted into immature animals, they would probably produce similar effects in neotenous larvae while undergoing resorption in the foreign

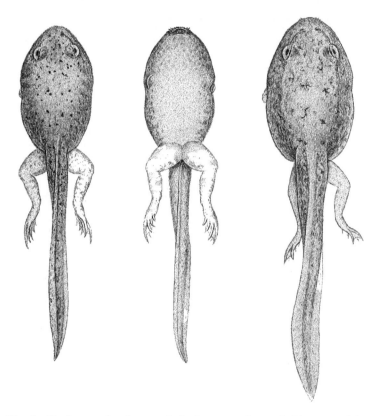

Fig. 2   Neotenous R. clamata tadpoles.   Numbers of this type of larvae, measuring 75 to 90 mm. total length, with hind limbs 5 to 25 mm. long, pass an extra year beyond their usual period of metamorphosis in this stage.   Drawing natural size.

environment because inhibiting factors could not prevent the release of the contained secretion.   The experiment is briefly presented in the following tables:

TABLE 9

December 22, 1920

ENGRAFTED		ANIMALS FROM WHICH GLANDS WERE TAKEN		CONTROLS	
Total length	Hind legs	Total length	Hind legs	Total length	Hind legs
mm.	*mm.*	*mm.*	*mm.*	*mm.*	*mm.*
80	8.5	81	8.5	81	8.5
82	10.0	82	8.5	82	9.0
83	9.0	81	9	80	8.0
80	8.5	80	10	79	8.5
81	8.5	83	8	81	10.0
82	9.0	79	8.5	83	8.5
81.3	8.9	80.7	8.8	81	8.8

TABLE 10

January 30, 1921

ENGRAFTED		CONTROL	
Total length	Hind legs	Total length	Hind legs
mm.	*mm.*	*mm.*	*mm*
80	25	80	9.5
82	24	81	10.0
80	24	82	10.0
83	23	81	10.5
82	26	83	10.0
81.4	24.4	81.4	10.0

Note: One engrafted animal died January 23rd.

TABLE 11

March 1, 1921

ENGRAFTED		CONTROL	
Total length	Hind legs	Total length	Hind legs
mm.	*mm.*	*mm.*	*mm*
83	31	82	13
82	32	83	13
82	29	81	14
84	30	82	12.5
82.7	30.4	82	13.4

One engrafted animal died February 25th. The experiment was abandoned March 10th, as some of the transplanted individuals had fore limbs. The data clearly indicate that the thyroid glands of neotenous tadpoles with hind legs 8 to 10 mm. are physiologically active and capable of inducing acceleration of limb growth and development when transplanted into other neotenous animals of similar size and developmental stage. However, metamorphic change following grafting is not so rapid or marked as when the same type of gland is implanted in immature (non-neotenous) larvae. But in either case the absorption of the secretion of the grafted gland induced a greater reaction than the same gland is capable of producing (in the same time and under similar conditions) when left unmolested in the neotenous animal of which it was originally a part.

SUMMARY OF CONCLUSIONS

1. Adult Necturus were fed, injected, and engrafted with physiologically active thyroid substance with negative results. The injection of 20 mg. of thyroxin iodine had no effect upon metamorphosis, whereas a 1 to 50,000 solution of this compound readily metamorphosed immature larvae of Amblystoma. The experiment indicates that perennibranchiate amphibians are permanent larvae and have lost the ability to transform under the stimulus of thyroid treatment.

2. Necturus differs markedly from the axolotl in its reaction to thyroid administration, as the latter readily metamorphoses when fed or injected with this substance.

3. The thyroid glands of the perennibranchiate Necturus are quite variable in size, and in some animals may occur as large vesicles more or less isolated from one another. In other individuals the glands are rather small and may consist of but four to six extremely large vesicles. In Necturus the thyroids are generally located near the apex of the triangle formed by the geniohyoid and external ceratohyoid muscles.

4. Despite its larval characters, Necturus possesses thyroid glands of great physiological activity, as shown by heteroplastic transplantation into immature anuran larvae.

5. A giant axolotl, 14.25 inches long, several years of age, was found to have a highly active metamorphosis-inducing thyroid apparatus. The thyroid of a single specimen cut into six pieces and transplanted metamorphosed five immature anuran tadpoles within two weeks. The sixth animal died following the operation.

6. The axolotl's thyroid is normal in appearance and of large size, consisting of numerous large vesicles filled with colloid. The gland is surrounded by a rich network of capillaries.

7. The failure of the axolotl to metamorphose appears to be due to the inhibition or the defective development of some unknown factor which normally serves to release the fully formed hormone from the thyroid into the blood stream. It is suggested that defective nervous stimulation or perhaps inhibition is the immediate cause of retention of the secretion within the thyroid vesicles, but that in the last analysis some defect of interrelation of the various components of the endocrine system is probably responsible for the nervous inhibition or lack of normal stimulation.

8. Experiments on large neotenous anuran tadpoles indicate that the failure of these animals to metamorphose at the proper time probably is due to the same causes responsible for axolotl neoteny: i.e., the thyroid glands apparently do not secrete their fully formed hormone into the blood stream because of some unknown inhibiting influence. The thyroid inhibition seems to be less marked in anurans than in the axolotl, since neotenous tadpoles eventually metamorphose if given sufficient time.

9. The next step in the analysis of amphibian neoteny is to determine the nature of the factor responsible for the failure of the thyroid to release its hormone (or at any rate to render it impotent in so far as metamorphosis is concerned). Is this unknown factor hormonal, or nervous, or both?

10. In the older work on amphibian neoteny too much stress was laid upon the exogenous factors as causative agents, and too little, if any at all, upon endogenous factors, and heredity.

ADDENDUM

Uhlenhuth's claim that urodeles differ from anurans in that their metamorphosis is independent of iodine and influenced only by the thyroid hormone itself is rendered invalid by the following experiment: The thyroid glands of large axolotls (seven inches total length) were extirpated and the animals kept for five months in the laboratory following the operation, then injected twice with strong doses of tyrosine in which two atoms of iodine had been substituted for two hydrogen atoms of the molecule forming the compound 3-5 diiodotyrosine. The animals metamorphosed within seventeen days following injection. Control thyroidless axolotls injected with equal quantities of pure tyrosine and later with large amounts of 3-5 dibromtyrosine showed no evidences of metamorphosis.

Further evidence that Uhlenhuth's claim is not valid is furnished by Huxley and Hogben who metamorphosed Salamandra and Triton larvae by immersion in dilute solutions of iodine (Proc. Roy. Soc., vol. 93, 1922); by Hirschler, who metamorphosed axolotls and tadpoles by administration of elemental iodine and iodoform (Arch. Entw. Mech., 1922), and lastly by Jensen who metamorphosed axolotls by injections of iodized casein, iodized serum globulin and iodized serum albumen (Compt. Rend. Soc. de Biol., 85, 391–392, 1921).

BIBLIOGRAPHY

ADLER, LEO 1916 Untersuchungen über die Entstehung der Amphibieneotenie. Pflüger's Archiv, Bd. 39.

BRAUS, H. 1906 Vordere Extremitat und Operculum bei Bombinatorlarven. Morph. Jahrbuch, Bd. 35, Heft 4.

v. CHAUVIN, MARIE 1875–76. Über die Verwandlung des mexikanischen Axolotl in Amblystoma. Zeitschr. f. wissensch. Zool., Bd. 25, Suppl., und Bd. 27.

DE FILIPPI 1861 Sulla larva del Triton alpestris. Arch. per la Zool. e per l'Anat. Comp. Genova (quoted from Gadow).

DUMÉRIL, AUGUST 1865 Nouvelles observations sur les axolotls nés a la menagerie. Comp. Rend., T. 61.

EMERSON, E. T. 1905 General anatomy of Typhlomolge rathbuni. Proc. Soc. Nat. History, Boston, vol. 32.

GADOW, H. 1909 The Cambridge Natural History, vol. 8.

HUXLEY, J. 1920 Metamorphosis of axolotl by thyroid feeding. Nature, vol. 104, no. 2618.

JENSEN, C. O. 1916 Ved Thyroiden—praeparater fremkald Forwardlung tros Axolotll'en. Oversigt. Klg. Danske Vidensk., Selsk. Forhandl., Copenhagen (cited by Uhlenhuth, '21).

KAUFMAN, L. 1918 Researches on the artificial metamorphosis of axolotls. Bull. Acad. Sc. Cracon, Ser. B., 32 (cited by Uhlenhuth).

KOLLMANN, J. 1884 Das Überwintern von europäischen Frosch und Tritonlarven und die Umwandlung des mexikanischen Axolotl. Verhandl. d. Naturh. Gesellsch. Basel.

LAUFBERGER, V. 1913 Ovzbwzeni metamorfos axolotln Krmenim zlazon stitnon. Biologicke Lysty (cited by Adler, '16).

LEYDIG, FRANZ 1853 Anatomisch-physiologische Untersuchungen über Fische und Reptilien. Berlin.

SHUFELDT, R. W. 1885 Mexican axolotl and its susceptibility to transformation. Science, vol. 6.

SWINGLE, W. W. 1921 The germ cells of anurans. I. The male sexual cycle of Rana catesbeiana larvae. Jour. Exp. Zoöl., vol. 32, no. 2.

1922 The thyroid glands of the perennibranchiate amphibians. Anat. Rec., vol. 23, no. 1, p. 100, Proc. Am. Soc. Zool.

1922 Experiments with necturus and axolotl thyroids. Anat. Rec., vol. 23, no. 1, p. 106, Proc. Am. Soc. Zool.

UHLENHUTH, E. 1921 Internal secretions in growth and development of amphibia. Am. Nat., vol. 55, no. 638.

PLATE 1

EXPLANATION OF FIGURES

3   Larva grafted ten days with small piece of Necturus thyroid, and control.

4 and 5   Larvae engrafted eight days with pieces of Necturus thyroid.

6   Larva grafted six days with Necturus thyroid, and normal control.

7   Larva fed mammalian thyroid tissue (desiccated) eight days. Control animal same as in figure 6.

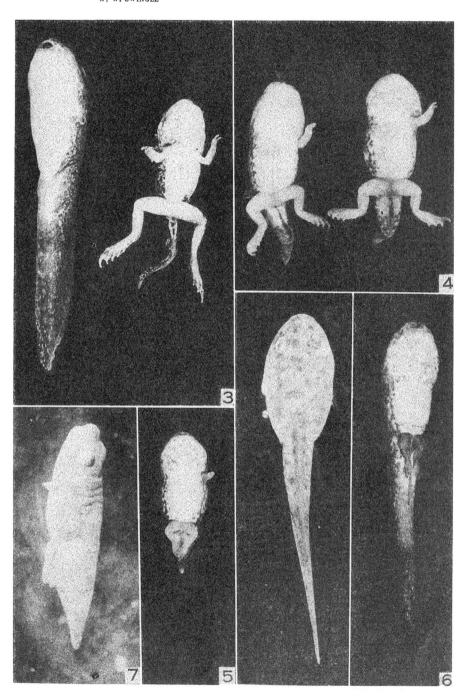

PLATE 2

EXPLANATION OF FIGURES

8 Immature R. clamata larvae ten days following transplantation small pieces of axolotl thyroids.

9 Same as figure 8. Ventral view.

10 Control larva for animals shown in figures 8 and 9.

420

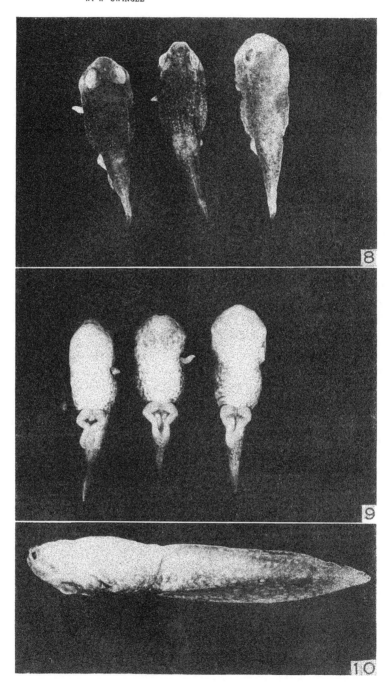

Resumen por el autor, H. P. Kjerschow Agersborg.

Algunas observaciones sobre los estímulos cualitativos químicos y físicos en los moluscos nudibranquios, con
especial mención del papel de los "rinóforos."

Hermissenda responde a los estímulos tactiles aplicados sobre
cualquier parte del cuerpo. Los tentáculos dorsales producen la
respuesta más efectiva. La cabeza y los tentáculos dorsales son
más sensitivos a los ácidos y sales en solución que cualquier
otra parte; el extremo de los tentáculos dorsales produce la
respuesta más efectiva. Los tentáculos orales son casi tan sensitivos como los dorsales a la acción de los diferentes estímulos
químicos; pero los orales poseen en adición una función selectiva
en el sentido de que cuando reciben un estímulo de algún alimento sabroso puede obligarse al animal a moverse en la dirección del estímulo. Los tentáculos orales producen una reacción
positiva definida hacia los estímulos alimenticios; la reacción
hacia los estímulos de distancia es menos definida que la de los
estímulos de contacto. Los tentáculos dorsales ("rinóforos")
no dan ninguna muestra de sentido olfatorio. Las reacciones
de Dendronotus no son marcadas, en general. Melibe selec
ciona su alimento con ayuda de sus tentáculos orales (cirros);
los tentáculos dorsales, aunque están ricamente inervados por
nervios de la región anterior del cerebro ("ganglios olfatorios")
no parecen funcionar como órganos olfatorios. La reacción
hacia las corrientes de agua no se altera cuando se extirpan los
tentáculos dorsales.

Translation by José F. Nonidez
Cornell Medical College, New York

SOME OBSERVATIONS ON QUALITATIVE CHEMICAL AND PHYSICAL STIMULATIONS IN NUDIBRANCHIATE MOLLUSKS WITH SPECIAL REFERENCE TO THE RÔLE OF THE 'RHINOPHORES'[1]

H. P. KJERSCHOW AGERSBORG

University of Nebraska, Lincoln, Nebraska

TWO FIGURES

INTRODUCTION

The purpose of the investigation upon which this paper is based was to determine the exact function of the dorsal tentacles which have come to be considered as organs of smell, and are generally called 'rhinophoria.'

Early writers on nudibranchiate mollusks, Alder and Hancock ('45, p. 19), Hancock and Embleton (':52, p. 242), Jeffreys ('69), ascribed to the dorsal tentacles the function of olfaction, and Bergh ('79), agreeing with these authors, employed the term rhinophoria. In fact, Tapparone-Canefri ('76) suggested this term for Melibe papillosa De Filippi, calling the 'tentacula' rhynophoria. Also later writers, Fischer ('87), Pelseneer ('06) (Prof. E. Ray Lankester's "A Treatise on Zoology"), seem to agree on this point. Hescheler ('00), however, uses the term 'Kopftentakel' (Prof. Arnold Lang's "Lehrbuch der vergleichenden Anatomie der wirebellosen Thiere"). Copeland ('18) thinks that the snails Alectrion obsoleta and Busycon canaliculatum are as successfully directed toward distant food by means of an olfactory apparatus consisting of a single organ of smell, associated with a siphon terminating in a shifting 'nostril' for sampling the surrounding water and its contents, as animals with paired olfactory organs and fixed nostrils. But Arey ('18) disagrees with

[1] From Puget Sound Biological Station, Friday Harbor, Washington; and contribution from the Zoological Laboratory, of the University of Illinois, under the direction of Henry B. Ward, no. 206.

423

Copeland on this point. He finds that the 'rhinophores' of Chromodoris elegans and C. zebra have nothing in connection with olfaction. Alder and Hancock, Hancock and Embleton, Jeffreys, Tapparone-Canefri, Bergh, Fischer, et al., based their opinion, relative to the function of the dorsal tentacles, on morphological data. The findings of Arey are somewhat in agreement with my own observations (vide infra). And, as we will see, it may not be well to designate a specific function to these organs, even though their response may be similar to that of vertebrates with definite localized and known sense organs relative to their specific function. Thus, even though the so-called rhinophorium is highly innervated with nerves from the anterior part of the brain (Hancock and Embleton), it may not be a good criterion at all by which to judge its function, for the simple reason that the brain of invertebrates is not analogous to the brain of vertebrates. In fact, Minnich ('21) has shown that the organs of taste in the butterflies Pyrameis atalanta Linnaeus and Vanessa antiopa Linnaeus are located in the tarsi. He demonstrated that tarsal chemoreceptors are present in all four tarsi of the walking legs. The removal of the antennae labial palpi, and rudimentary fore legs in Pyrameis does not affect in any significant way the responses produced through contact chemical stimulation of the tarsi. From the anatomy of Lepidoptera it is seen that the pedal nerves are innervated from the thoracic ganglia, and not from the brain. Chemo-receptors, according to Minnich, may be divided into two classes: first, those affected in general by volatile materials, the source of which may be more or less remote from the receptive surface; second, those affected in general by non-volatile materials, the source of which must be in intimate contact with the receptive surface. The former serve as distance chemoreceptors; the latter, as contact chemoreceptors. The above distinction, how-ever, is far from being an absolute one, but is merely useful as the best single condition by which the two groups of sense organs may be conveniently differentiated. For, it is contended, in the last analysis both are stimulated by a solution of the exciting material, the solvent consisting, at least in part, of the secretion present on the sensitive surface.

DISCUSSION

The following data were collected from experimental studies of Hermissenda opalescens Cooper, Dendronotus giganteus O'Donoghue, and Melibe leonina Gould.

Type 1—Hermissenda opalescens Cooper

This species has been accurately described by O'Donoghue ('21); it is, however, necessary to give a brief account of certain morphological structures because of the variance in the nomenclature employed by many writers relative to certain points. Gasteropods have commonly one or two pairs of tentacles on the head. The anterior pair is frequently spoken of as oral tentacles or buccal or inferior (Fischer), vordere Fühler (Claus and Grobben, '10), or Mundtentakeln (Lang and Hescheler), and the posterior as dorsal tentacles, hintere Fühler (Claus and Grobben), or Kopftentakeln (Lang and Hescheler), or rhinophores by other writers, notably by Bergh. In some species (Melibe Rang, Dendronotus Ald. & Hanc.) the oral tentacles may be lacking, as a distinct pair, or may consist of one or more rows of varied sized cirrhi around the fringe of the hood (Kjerschow Agersborg, '19, '21, '21 a, '22), and in that way the oral tentacles sometimes may not be readily recognized. The dorsal tentacles may also be modified so as to make them at first sight quite indistinguishable from the papillae (dorsal cerata), as in Dendronotus. In this paper I will treat of the rhinophores as dorsal tentacles (not 'head tentacles' (Kopftentakeln), as they are sometimes situated on the neck, e.g., Hermissenda, etc.), and the cerata as papillae (Kjerschow Agersborg, '22). The anterior or buccal tentacles may most appropriately be called oral tentacles not only because of their proximity to the mouth, but also because of the common usage of the term in this connection.

The oral tentacles (O'Donoghue, '21) in Hermissenda opalescens consists of one pair, situated anterolaterally on the head. They are lanceolate, tapering gently to a point. In a specimen which measured 40 mm. from the anterior end to the tip of the foot, they were 12 mm. or a little less than twice the length of the dorsal

tentacles, the anterior quarter being white.   In normal specimens
they are in constant motion, being generally directed from the
head at an angle to each other of about 65°.   When the animal
moves, they are directed forward and sideways in a lashing fashion, the tips being the most motile parts.

The dorsal tentacles are situated on the top of the neck, one
on each side of the red median dorsal line, and 1.5 to 2 mm.
back of the anterior border of the foot.   They are about one-half
the length of the oral tentacles and they are directed almost
vertically from the body, or sometimes at an angle to each other
from 0° to ca. 23°.   In external structure, they differ from the
oral tentacles, which are smooth, in that they are supplied with
a series of prominent annulations.   Like the former, they are
white at the ends.   For experimental purposes, I selected at
random a number of large and small individuals, which I paired
into a number of sets, each consisting of a small and a large
individual (vide ut tabula).

Tactile stimulation.   The entire surface, including the papillae,
the parts between the papillae, the foot including the lanceolate
posterior prolongation which extends back of the papillae commonly up to one-half the length of the body, the oral and dorsal
tentacles, and the anterolateral prolongations of the foot, is
sensitive to touch.   The tentacles are more sensitive to touch or
tactile stimuli than are any other parts of the body.   It is hard
to tell which of the tentacles are the most sensitive to tactile
stimulation, because the oral tentacles are frequently in constant
motion, but judging from the response obtained when the two are
touched similarly, it appears that the dorsal tentacles are the more
sensitive of the two, at least to tactile stimuli.   That is, an
equally gentle touch to them gives a more effective response in
the dorsal tentacles, in that they not only contract by shortening
relatively more, but also by remaining retracted longer than the
oral tentacles.   The oral tentacles may shorten a little, but their
main response to tactile stimulation consists in being jerked
away from the stimulus and then be put back at once to their
former or similar position.

Chemical stimulation. As shown per the accompanying tables (pp. 433–438), a number of chemical solutions of different concentrations were applied to the various parts of the body by the capillary-pipette method. This pipette was sealed at one end and insulated around the middle by a cork, for manipulation purposes. The same amount of the particular chemical used (unless stated differently) was applied each time. The pipette was rinsed externally each time after refilling to avoid diffusion of the chemical solution in the medium of the specimen experimented on and in order to secure as nearly as possible the normal condition of the sea-water. Sixty-two animals were used. Each one was placed in a finger-bowl half filled with fresh sea-water, and the pipette then applied with the chemical solution so as not to touch the body with the pipette, but to allow only the chemical solution in question to pass onto the desired part of the body. Solutions heavier than sea-water were allowed to flow from above; those lighter than sea-water were applied to the animal when it was crawling on the surface film of the water or on the side of the vessel.

The following signs represent the relative degrees of modes of response to the stimuli: the negative sign represents an attempt to avoid the stimuli. There are five of this kind, three positive and three others as explained below:

1. −g, means general but a slow reaction with an attempt to avoid the stimulus.

2. −, means a more definite negative response.

3. , means a still more definite negative response.

4. −, means explosive negative response.

5. − −, means explosive negative response, violent contractions and twistings of the body or parts of it.

6. The plus signs, +, + +, and + + +, mean various degrees of positive response with an attempt to move in the direction of, or toward the stimuli.

7. 0, means no reactions.

8. ⊕, means indefinite reaction which shows no particular effort to avoid the stimulus.

9. ⊖, means indefinite reaction but rather with opposite direction to ⊕.

In general, the head is the most sensitive part of the body both to tactile and chemical stimulation. The effect of acid gives an almost constant result. Comparatively, the dorsal tentacles are more sensitive than the oral tentacles to acids, and the tips of the dorsal tentacles respond more definitely to this stimulus than do the tips of the oral tentacles. Hermissenda opalescens gives a positive reaction toward 2 M ethyl alcohol and a negative reaction to 2 M methyl. That is, the organism would even suck the ethyl-containing pipette, but avoid the methyl-containing one. Specimens would eat readily cucumarian gonads treated with ethyl alcohol and show no ill effects. One specimen which ate 10 mm. of cucumarian gonads treated for 40 min. in 2 M methyl, died before the next day. It was found that this species would eat very readily various kinds of animal matter: jelly-fish tentacles, gonads of sea-urchins (Strongylocentrotus dröbachiensis), sea-cucumber (Cucumaria japonica), and of the sand-dollar (Echinarachinus excentricus), etc. The gonads of Cucumaria are filamentous in nature and are readily measured quantitatively.[2] I treated, therefore, strings of gonads of Cucumaria japonica in various chemical solutions: M 0.25 lithium chloride; M to M 0.5 sodium chloride; M 0.05 magnesium sulphate; M 0.10 sodium salicylate; sat. sol. quinine sulphate; 2 M glycerol; 2 M ethyl, and methyl from 5 min. to ca. 6 hrs. Hermissenda showed a marked difference in response to food thus treated in comparison of the reaction to normal food, i.e., food which had not been treated. And this difference corresponded somewhat to the reaction when these chemicals were applied onto the surface of the body. For example, the response to M 0.25 LiCl. is 0 to all parts of the body. When presenting food to the animal treated in this solution, it was found that the animal took to it as readily as to untreated food. The response to M NaCl is—(see explanation of sign); food treated in this solution was taken by the animal from 0 to 50 mm. in one minute. The exact results were: Out of ten specimens, five of which were given food treated ten

---

[2] These strings (filaments) were on the whole ca. 1 mm. in diameter; the exact volume was not computed. The amount later is indicated in mm. of the gonadic filament.

to fifteen min. ate as follows. Ten min. treatment, 25 mm.; 15 min, treatment: 17, 19, 20 and 25 mm. The other five had their food treated for 1 hr. and 40 min. in the same solution and the amounts these ate, respectively, were: 0, 12, 17, 20 and 50 mm. The same ten specimens, on another day, ate as follows, the food being treated for 15 min. in M 0.05 MgSO$_4$:

	mm.	mm.	mm.	mm.	mm.	mm.	mm.	mm.	mm.	mm.
At 10 a.m.............	50	20	30	50	50	50	50	30	20	20
At 3 p.m.............	15	15	20	35	22	50	30	10	30	32
Total in day.........	65	35	50	85	72	100	80	40	50	52

It was noted that the response of the organism when stimulated with M magnesium sulphate was 0. That is, it notices the stimulus, but does not attempt to avoid it definitely. One might, therefore, expect a result as noted above which is nearly similar to the results obtained from the control (see the table below). Fed on gonads treated with 0.1 M sodium salicylate for 10 min. the result was that all moved away from it at first. Three min. later, one ate 30 mm. which had been treated for 30 min., one nibbled at it and bit it into parts, but did not eat it. The others moved away from it. The reaction of the organism to M sodium salicylate is from $--$ to $----$; it was, therefore, no surprise that the organism did not want to feed on this food even though it was treated in a much more dilute solution as compared with those mentioned above. The Nudibranch ate food treated for 6 hrs. in glycerol: 40, 40, 40, 40, and 50 mm. The reaction to glycerol was 0. The same individuals at the same hour also ate, respectively, 50, 17, 25, 17, and 50 mm., treated in 2 M ethyl alcohol. The average amount of food taken at this time was: 90, 57, 65, 57 and 100 mm. The response to 2 M ethyl was $\oplus$ or nearly $+$. The other five would not take food treated in 2 M ethyl, but when it was treated in M ethyl they ate: 50, 50, 50, 50, and 40 mm., respectively. The amount of food taken, which was treated in sat. sol. of quinine sulphate, was as follows: a, bit at it and spat it out; b, do. do.; c, ate 25 mm.; d, ate 50 mm. at once, and e, avoided it altogether.

The chemical response to quinine sulphate is shown in the table to be from −g to   . The same individual which ate 50 mm. treated with quinine, ate 20 hrs. earlier 25 mm. food treated with 2 M NaCl for 15 min.   That is, my data show this animal ate on the average for six days 51 mm. per day, the food being treated in various ways; or, it ate ca. 3.23 mm. more than the average of five individuals, but its records show it ate some each day, i.e., from 25 to 85 mm. each day during the week.   It may be, there-fore, that specimen d was in a better physiological condition at the time when it took a full meal of food, e.g., 50 mm., which the others would not do.   An individual which in a morning did not eat from the food treated with M ethyl ate 10 mm., in the after-noon, of food treated with 2 M methyl; on the next morning it was dead, however.   The chemical response to methyl was (vide ut infra tabula).   It is then seen there is a definite relation between the sensitivity of the external parts of the organism to certain chemical solutions, and the taste of the organism, e.g., quinine sulphate gives a response from −g to −, the organism is careful about eating food treated in such a solution.   If hungry, it may eat it at once; if not hungry, it may bite at it (taste it?) and then either eat only a little, void it, or avoid it altogether. (N. B.—It is to be noted that the food was left with the individual up to one minute, after which time it was removed in all cases.)

The most interesting results were obtained by using broth of various animals, e.g., of jellyfish, sea-urchin's gonads, etc., and oils.   I also used an emulsion prepared from raw jellyfish and from the gonads of sea-urchins.   The reaction was positive for all of these, but more so for the cooled broth.   Applying the emulsion or broth with a pipette on the dorsal tentacles, Her-missenda turns round apparently in search for the stimulus; applying the stimulus in front, it makes progressive movements toward the stimulus and bites in the pipette and works the oral apparatus continuously.   When broth of the sea-urchin's gonads is applied onto the top of the head or on the dorsal ten-tacles the response is not really definite, the organism seems to be confused; but when the same amount of broth is let a short distance (5 mm.) in front, it moves toward the same and starts working the oral apparatus sucking in the mixture. Applying broth on the back it turns around, as if searching for the stimulus. The reaction for broth of sea-urchin gonads is twice as strong as for that of jellyfish. The response was from + to + + +, but most effective for the sea-urchin's. The oral tentacles showed a most definite positive reaction toward the stimulus. Touching the tentacle on the left or right side, I could lead the animal at will from one side to the other; it always turned after a short interval of time toward that side which tentacle was stimulated. This is equally definite when using a piece of solid food which it eats readily, e.g., gonads of sea-cucumber or sea-urchin. The reaction as shown by Hermissenda toward distance stimuli is less definite than that toward contact stimuli. Receptors for distance stimuli seem to be present in the oral tentacles; but receptors for contact stimuli, judging by the mode of response, are more specific than are the distance receptors. Contact receptors are present all over the body, perhaps also chemoreceptors, but the latter are specialized in the oral tentacles, as they seem to be used in discriminating between foods. They may, therefore, be gustatory in function, because the animal may be led about, from side to side, in the dish by touching the oral tentacle by some palatable food; that is, a food which the organism feeds on readily. No such results are effected by treating the dorsal tentacles, the so-called rhinophores, in the same way. The head, in general, seems to be most sensitive to stimuli, tactile or otherwise. When food stimuli are applied to the head, the animal turns in various directions so as to search for them. The dorsal tentacles do not respond to distance stimuli; their response to a food stimulus is similar as that given to a tactile stimulus, e.g., a clean glass rod. Hermissenda does not have the ability to locate solid foods; it really seems that the animal comes upon the foods accidentally. It is apparent that the animal can taste food in solution with most parts of its external surface, and particularly that of the head and the oral tentacles; but it does not look as though the organism is capable of scenting its food. I am not able to say that the dorsal tentacles are used for this purpose. I repeatedly found hungry animals within ca. 1 cm. from food they readily ate when the same was placed in front of them so as to be touching the lips. In fact, the animals would then make a swift nip for such foods with the jaws. If an animal, which had not fed for twelve hours, was at rest next to a piece of gonads (the food it ate the most readily) it would immediately become active by being touched on the oral tentacles by the food, and then begin to crawl about. If it came upon the food, feeding started at once. Touching the animal similarly gently on other parts of the body did not result in the active moving about. This may show that the chemoreceptors, as stated above, are better organized or specialized in the oral tentacles. During such forced movements, the animal moved practically in a straight line (in a circle in the finger-bowl), but it bent the head, now and then, from side to side, and was constantly lashing the oral tentacles as if it were feeling its way.

I repeated Arey's experiment by holding a drop of various kinds of oils between the 'rhinophores' without allowing the oils to come in contact with them, but I did not get any response. Allowing a drop of oil to come onto the body, the oral tentacles only gave definite response. The response was − to saffrol, and   to bergamot oil. The dorsal tentacles, for the same oils gave 0. The rest of the body gave   , or ⊖. The animal does not show any awareness to clove oil when a drop of it is suspended free between the dorsal tentacles. Allowing the same to touch any part of the body the response is −. The response to organnum oil is ⊖ for the head and 0 for the rest of the body (vide ut infra tabula). To cedar-wood oil, the response is either ⊖ or 0. And to orange flavor it is 0.

The following tables show the exact data relative to the number of specimens used, the size and condition of the specimens, the stimuli employed, the parts stimulated, the relative response, etc., of Hermissenda opalescens Cooper (tables pp. 433–438).

Arey quotes Crozier as thinking that the dorsal tentacles of Chromodoris are rheotropic, their prime importance being in effecting orientation to the water current. But the dorsal tentacles of Hermissenda do not seem to have a rheotropic function, because specimens with one or both of the dorsal tentacles removed oriented as easily and moved against the current as did

TABLE I

NUMBER OF INDIVIDUAL AND THE DATE	SIZE AND CONDITION	THE STIMULI USED	PARTS STIMULATED								REMARKS For explanation of signs, see the text
			Head	Oral tentacles	Dorsal tentacles	Papillae	Body	Anterior foot	Middle of foot	Tip of foot	
No. 1 8/3/'21	Good 30 mm.	2 M, HCl	—	—	—	— g	— g	— g	— g	— g	To M solution, no response
No. 2	G, 25 mm.	2 M, NaCl	—	—	—	— g	— g	—	— g	— g	To M solution, no response
No. 3	G, 20 mm.	2 M, NH4Cl	—	—	—	—	—	—	— g	—	To M solution, no response
No. 4	G, 20 mm.	2 M, NH4Cl	—	—	—	—	—	—	—	—	To M solution, no response
No. 5	G, 20 mm.	2 M, HCl	—	—	—	—	—	—	—	—	To M solution, no response
No. 1'	G, 20 mm.	M, NaCl	—	—	—	—	—	—	—	—	
No. 1'	G, 20 mm.	2 M, NaCl	—	—	—	—	—	—	—	—	
No. 2'	G, 25 mm.	M, aCN	—	—	—	— g	—	—	—	— g	
No. 2'	G, 25 mm.	2 M, NaCl	—	—	—	—	—	—	—	—	
No. 3'	G, 25 mm.	M, NaC	—	—	—	—	—	—	— g	—	Applied at posterior tip of foot, excellent forward ... at ater or oot stops ...
No. 3'	G, 25 mm.	2 M, NaCl	—	—	—	—	—	—	—	—	
No 4'	Sluggish, 30 mm.	M, NaC	—	—	—	—	—	—	—	—	
No. 4'	Sluggish, 30 mm.	2 M, NaCl	—	—	—	—	—	—	—	—	
No. 5'	G, 20 mm.	M, NaCl	—	—	—	—	—	— g	—	—	
No. 5'	G, 20 mm.	2 M, NaCl	—	—	—	—	—	—	—	—	

TABLE 2

NUMBER AND DATE	SIZE AND CONDITION	STIMULI	Head	Oral tentacles	Dorsal tentacles	Tips oral tentacles	Tips dorsal tentacles	Body	Anterior foot	Middle of foot	Tip of foot	Papillae	REMARKS
A	Normal 35 mm.	0.25 M LiCl	0	0	0	0	0	0	0	0	0	0	Tail = posterior tip of the foot which butts past the main part of the body
A	Normal 35 mm.	M, NaNO₃	—	—	—	‖	‖	—	— g.	—	—	—	‖ ‖ = —
A	Normal 35 mm.	2M, NaNO₃	‖	‖	‖	‖	‖	‖	‖	‖	‖	‖	‖ ‖ = 2 minus
a	Normal 18 mm.	0.25 M, LiCl	0	0	0	0	0	0	0	0	0	0	‖ ‖ ‖ = 3 minus
a	Normal 18 mm.	M, NaNO₃	—	—	—	‖	‖	—	—	—	—	—	‖ ‖ ‖ ‖ = 4 minus
a	Normal 18 mm.	2 M, NaNO₃	‖	‖	‖	‖	‖	‖	‖	‖	‖	‖	
B	Normal 30 mm.	M, MgSO₄, 7H₂O	⊕	⊕	⊕	⊕	⊕	⊕	⊕	⊕	⊕	⊕	Has two oral tentacles rt. sd.
b	Normal 22 mm.	M, MgSO₄, 7H₂O	⊕	⊕	⊕	⊕	⊕	⊕	⊕	⊕	⊕	⊕	⊕ = ⊕
C	Normal 40 mm.	0.2 M, Na₂SO₄, 10H₂O	‖	‖	‖	‖	‖	‖	‖	‖	‖	‖	
c	Normal 35 mm.	0.2 M, Na₂SO₄, 10H₂O	‖	‖	‖	‖	‖	‖	‖	‖	‖	‖	

											Notes
8/4											
D	dial 40 m.	2 M, methyl	—	—	⊕	⊕	0	— g	0	0	0	0	—	—	—	—	—	Shed arily 30 arge and 10 small papillae
d	Normal 25 m.	2 M, methyl	—	—	⊕	⊕	0	0	0	0	0	0					
E	dial 35 m.	2 M, ethyl	—	—	⊕	⊕	0	— g	0	0	0	0					Rt. ra ten absent. the tail short
e	Normal 25 m.	2 M, ethyl	—	—	⊕	⊕	0	—	0	0	0	0					
F	dial 40 m.	0.2 M, lactose	—	—	⊕	⊕	0	—	0	0	0	0					R e on y by allowing several drops
f	Normal 2 m.	0.2 M, lactose	—	—	⊕	⊕	—	— g	0	0	0					
G	Normal 45 m.	M, sucrose	—	—	⊕	⊕	—	0	0	0	0					
g	Normal 30 m.	M, sucrose	—	—	⊕	⊕	0	—	0	0	0	0					an ery defin te (dep. eggs)
H	Normal 45 m.	3 M, glycerine	—	—	⊕	⊕	0	— g	— g	— g	0					
h	Normal 30 m.	3 M, glycerine	—	—	⊕	⊕	0	— g	— g	— g	0					
I	Normal 50 m.	M, sodium acetate											
i	dial 40 m.	M, sodium acetate											
J	dial 55 m.	M, sodium											
j	Normal 30 m.	M, sodium salicylate											
K	Normal 45 m.	0.5 M, lic acid											

TABLE 2—*Continued*

NUMBER AND DATE	SIZE AND CONDITION	STIMULI	Head	Oral tentacles	Dorsal tentacles	Tips oral tentacles	Tips dorsal tentacles	Body	Anterior foot	Middle of foot	Tip of foot	Papillae	REMARKS
k	Normal 23 mm.	0.5 M, oxalic acid	–	–	–	–	–	–	–	–	–	–	
L	Normal 50 mm.	M, tartaric acid	–	–	–	–	–	–	–	–	–	–	
l	Normal 25 mm.	M, tartaric acid	–	–	–	–	–	–	–	–	–	–	Stim. brough indv. nto conv. Particular y head. Conv. = convulsions
M	Normal 45 mm.	0.5 M, citric acid	–	–	–	–	–	–	–	–	–	–	
m	Con. ? 30 mm.	0.5 M, citric acid	–	–	–	–	–	–	–	–	–	–	Many papillae missing
N 8/5	Normal 35 mm.	Sat. sol. R. T. salicylic A.	– g	– g	– g	– g	– g	g	g	g	– g	g	R. T. = room temperature A. = acid
n	Normal 20 mm.	Sat. sol. R. T. salicylic A.	– g	– g	– g	– g	– g	g	g	g	– g	–	Rt. ora tent. los
O	Normal 40 mm.	Sat. sol. R. T. quinine spht.	–	–	–	–	–	–	–	–	g	g	
o	Normal 25 mm.	Sat. sol. R. T. quinine spht.	–	–	–	–	–	–	–	–	g	g	

												Remarks
P	Normal 35 m.	1 per cent HCl	−g	−	−	−g	−g	−	0	−	−	−−	Laying eggs
p	Normal 20 m.	1 per cent HCl	−g	−g	−g	−	−g	−	−g	−	−	−	
Q	Normal 45 mm.	1 per cent HCN	0	0	0	0	0	−	−g	−	−	−	
q	Normal 35 m.	1 per cent HCN	0	0	0	0	0	−	−g	−	−	−	
R	Normal 35 m.	45°C., H₂O	−	−	−	−	−	−	−	−	−	−	With rubber-bulb pipette
r	Normal 22 m.	42°, H₂O	−−	−−	−−	−−	−−	−	−	−−	−−	−−	With rubber-bulb pipette
S	Normal 35 m.	Oiled both	+	+	+	+	++	++	++	++	++	++	Of jellyfish Aquaria vitriao
s	Normal 30 m.	Oiled both	+	+	+	+	++	++	++	++	++	++	He turns *after* the stimulus
T	Normal 35 m.	Oiled both	+	+	+	+	++	++	++	++	+++	+++	Broth of sea-urchin, he sucks the pipette
t	Normal 30 m.	Oiled both	+	+	+	+	++	++	++	++	+++	+++	Broth of sea-urchin; he sucks the pipette
U	Normal 45 m.	Orange flavor	0	0	0	0	0	0	0	0	0	0	With the tip of the foot contracted
u	Normal 40 m.	Orange flavor	0	0	0	0	0	0	0	0	0	0	
V	Normal 45 mm.	Saffrol	0	0	−	−	0	−	−	−	−	−	
v	Normal 30 m.	Saffrol	0	0	−	−	0	−	−	−	−	−	
W	Normal 45 m.	Clear W. oil	⊕	⊕	⊕	⊕	⊕	0	⊕	⊕	⊕	⊕	

TABLE 2—Concluded

NUMBER AND TYPE	SIZE AND CONDITION	STIMULI	Head	Oral tentacles	Dorsal tentacles	Tips oral tentacles	Tips dorsal tentacles	Body	Anterior foot	Middle of foot	Tip of foot	Papillae	REMARKS
													PARTS STIMULATED
w	Normal 35 mm.	Cedar W. oil	⊕	⊕	○	⊕	○	⊕	⊕	⊕	⊕	⊕	
X	Normal 52 mm.	Bergamot O.	⊕	—	○	—	○	⊕	⊕	⊕	⊕	⊕	
x	Normal 42 mm.	Bergamot O.	⊕	—	○	—	○	○	○	○	○	○	
Y	Normal 47 mm.	Origannum	⊕	○	○	○	○	○	○	○	○	○	Sensitive on y on the head
y	Normal 40 mm.	Origannum	⊕	○	○	○	○	○	○	○	○	○	
Z	Normal 50 mm.	Clove oil	—	—	—	—	—	—	—	—	—	—	Sensitive on y when in contact with oi
z	Normal 37 mm.	Clove oil	—	—	—	—	—	—	—	—	—	—	Sets n awareness when drop is suspended

the normal individuals. That is, I placed a number of normal specimens in a narrow trough ca. 60 cm. long, 5 cm. wide, and 6 cm. deep, and tilted it at one end ca. 3°, allowing a gentle stream of sea-water to pass in at the raised end and out in the opposite end through a sieve. Everyone oriented himself toward the current and moved first against it until arriving at the end of the trough. A number of individuals in which the dorsal tentacles had been removed, either partly or totally, either only one or both, behaved in the same manner as the non-mutilated ones, save two of the ones operated on, which seemed to be a little disturbed at first. Otherwise it did not seem to make any difference whether the 'rheotropic' tentacles were present or not. The function of the dorsal tentacles, therefore, does not seem to be 'rheotropic' in Hermissenda. They are tactile organs, however, and may also be slightly gustatory in function.

Type 2—Dendronotus giganteus O'Donoghue

I collected a single specimen of the wonderful species Dendronotus giganteus from between the logs of the floating dock at the station. It measured 140 mm. in length, 60 mm. deep, and the foot was 90 mm. long and 40 mm. wide. (For a complete description, vide O'Donoghue, '21.) This species is much more sensitive to tactile stimuli than is Hermissenda, and is, therefore, not so easily experimented on. The slightest disturbance of the water caused some local contraction of the body. However, the following data were collected, and it may be of interest to study them in comparison with those on Hermissenda (table 3, p. 440).

This table (table 3) shows graphically that the dorsal tentacle, the 'rhinophore,' of Dendronotus is on the whole more sensitive to contact stimuli whether physical or chemical.

Type 3—Melibe leonina Gould

One specimen of this remarkable species was given to me by Dr. Elmer Lund, who found it at the floating dock. The specimen measured 65 mm. in length; the height was 10 mm., and the

TIE 3

NAME OF SPECIES AND DATE	SIZE AND CONDITION	STIMULI	PARTS STIMULATED								REMARKS
			Head	Oral tentacles	Dorsal tentacles	Papillae	Body	Anterior foot	Middle of foot	Tip of foot	
Dendronotus giganteus 8/21	130 mm. Good	Tactile	—	g	—	—	—	⊕	⊕	0	Capillary p pette method
8/4/'21	130 mm. Good	2 M, NH₄Cl	0	0	0	0	0	0	0	0	
8/4/'21	130 mm. Good	2 M NH₄Cl	—	—	—	—	—	—	—	—	By dropping one drop of 2 M, NH₄Cl in ... on given area: The dorsal tentacles withdrew for ½–1 minute
8/4/'21	130 mm. Good	0.25 M, LiCl	0	0	0	0	0	0	0	0	
8/4/'21	130 mm. Good	M, NaC	—	—	—	—	—	—	—	—	
8/4/'21	130 mm. Good	0.5 M, tose	g — g	g — g	— g	g —	g	0	0	0	Movements are accelerated to NaC
8/4/'21	130 mm. Good	M, sucrose	g	—	⁚	—	—	g —	g —	g —	Did no ve as a whole
8/4/'21	130 mm. Good	3 M, g ycerol	0	0	— g	0	0	0	0	0	Did not move as a who c
8/4/'21	130 mm. Good	M sodium ace- te	—	—	—	—	—	0	·	·	M as or hermissenda

width 8 mm.   The hood 30 mm. in diameter in either way; the
foot 35 mm. in length and on the average ca. 5 mm. wide.

As in other species of this genus, the hood is fringed with rows
of cirrhi; in this species it is fringed with two rows, the outer of
which is five times larger than those of the inner row.   But the
cirrhi of the inner row are twice the number of the outer, uniform
in size and arrangement.   That is, there is usually one small
cirrhus at the base of each large, and one between.   The large

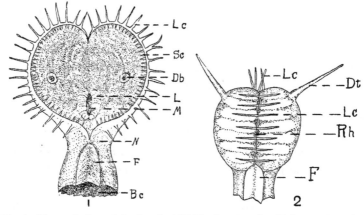

Fig. 1   Ventral view of the hood of Melibe leonina Gould drawn from life by
the author.   Bc., body cavity; Db., base of the stalk of the left dorsal tentacle;
F., foot; L., left lip of the mouth; Lc., large cirrhus; M., mouth; Sc., small cirrhus.

Fig. 2   Ventral view of the hood, contracted, as in the process of swallowing;
note the dorsal tentacles ('rhinophores') are not contracted.   Drawn from life
by the author.   Dt., dorsal tentacle (left); Lc., large cirrhi; F., foot; Rh., rim of
the hood.   (Note that the anterior part of the hood is brought caudad, e.g., to
bring the food near to the mouth so that the lips can take hold of it; compare the
position of the dorsal tentacles with that in figure 1.)

cirrhi terminate ca. 4 mm. on each side from the midventral line
of the caudoventral aspect of the rim of the hood.

This species is highly sensitive to tactile stimuli.   The very
slightest ripple on the water disturbs it.   Figure 1 shows the
position in which the hood is held in life; the large cirrhi (lc.)
are stretched outward and the small cirrhi (sc.) inward.   When
a glass rod is introduced into the water, even with the utmost
care, and at ca. 20 cm. away from the animal, the hood con-

tracts and the organism becomes restive. Complete contraction of the hood is effected by introducing the rod within the area of the hood. When a crustacean, ca. 10 mm. or more (I got it to take two Amphipoda, a Cammarus ca. 15 mm. long and a Caprella ca. 20 mm.) is dropped within the rim of the hood, it closes up the hood firmly (fig. 2) without contracting the tentacles on the back of the hood, opens the mouth (while the hood is yet contracted), and passes the food into the digestive tract. It also ate a strip of cucumarian muscle ca. 60 cu. mm. but its favorite food seems to be Crustacea (Kjerschow Agersborg, '21). Its favorite position is on the surface film with the back inverted. The cirrhi seems to be receptors of tactile stimuli. The animal may try to swallow anything that comes within the rim of the hood, but it does not swallow everything; it actually tries to eject solids which have come within the rim of the hood and which it cares not to eat. In life an inner axial white rod is seen through the wall of the cirrhic cone. This axial rod, as I have shown before ('22, in press), consists of nervous tissue; fine fibers radiate from it to the periphery. The dorsal tentacles are not more sensitive to tactile stimuli than are the cirrhi. But when they are touched with a glass rod they contract within the sheath of their stalk and remain contracted for a short time.

SUMMARY

1. Hermissenda opalescens Cooper responds to tactile stimuli applied to any part of the body: the head, the oral and dorsal tentacles, the body, the various parts of the foot, and the papillae. The dorsal tentacles give the most effective response to a tactile stimulus, such as the end of a glass rod.

2. The head and the dorsal tentacles are most sensitive to acids; but of the two, the latter are more sensitive to acids and salts in solution, the tips giving the most effective response.

3. The oral tentacles are almost as sensitive to stimuli as are the dorsal tentacles, but in addition to a general response of this nature, the oral tentacles also have a selective function in that when they are stimulated by some palatable food, the animal may be made to move in the direction of the stimulus; if the

animal responds to a stimulus negatively, the appliance of it in front of a progressively moving animal may bring it to a halt; the directions of its movements may also be changed.

4. The following table shows the comparative response of the oral and dorsal tentacles of Hermissenda to a piece of gonads of Cucumaria japonica Semper. The difference is practically 100 per cent.

TABLE 4

NUMBER OF INDIVIDUAL	ORAL TENTACLE	DORSAL TENTACLE	TEMPERATURE	REMARKS
1 = 1c	+	0	15°C.	
2 = 1b′	+	0	15°C.	
3 = 1a	+	0	15°C.	
4 = 2a	+	+?	15°C.	Allowing food to touch the right or left ten-
5 = 2b	⊕	⊖	15°C.	tacle, the organism turns or does not re-
6 = 2c	+	0	15°C.	spond, in the direction of the stimulus
7 = 3b	++	⊕?	15°C.	
8 = 3c	++	⊕	15°C.	
9 = 3d	+	0	15°C.	

If this table is compared with the results as given in table 2, Uu-Zz, pp. 437–438, it is seen that the oral tentacles have the power of discrimination between certain substances, such as food and odorous oils, while the dorsal tentacles lack this power for the same substances. There is no evidence that the 'rhinophores' are olfactory in function.

5. The dorsal tentacles of Dendronotus giganteus, like those of Hermissenda, are the most sensitive parts of the body to tactile stimuli (table 3).

6. In Melibe leonina Gould the cirrhi are more sensitive to tactile stimulus than are the dorsal tentacles.

444   H. P. KJERSCHOW AGERSBORG

LITERATURE CITED

AGERSBORG, H. P. KJERSCHOW 1919 Notes on Melibe leonina Gould. Pub. Puget Sound Biol.Sta., vol. 2, no. 49.
1921 Contribution to the knowledge of the nudibranchiate mollusk, Melibe leonina Gould. Amer. Nat., vol. 55, May–June.
1921 a On the status of Chioraera Gould. Nautilus 35: 50–57.
1922 The morphology of the nudibranchiate mollusk Melibe leonina Gould. Jour. Morph. (in press).
ALDER, A., AND HANCOCK, A. 1845 A monograph of the British nudibranchiate Mollusca. The Ray Society, London.
AREY, LESLIE B. 1918 The multiple sensory activities of the so-called rhinophores of nudibranchs. Amer. Jour. Physiol., vol. 46, no. 5.
BERGH, R. 1879 On the nudibranchiate gasteropod Mollusca of the North Pacific Ocean, with special reference to those of Alaska. Proc. Acad. Nat. Sci. Phila., vol. 31, 3d ser. 9.
CLAUS, C., AND GROBBEN, KARL 1910 Lehrbuch der Zoologie. Marburg in Hessen.
COPELAND, MANTON 1918 The olfactory reactions and organs of the marine snails Alectrion obsoleta (Say) and Busycon canaliculatum (Linn.). Jour. Exp. Zool., vol. 25, no. 1.
O'DONOGHUE, CHAS. H. 1921 Nudibranchiate Mollusca from the Vancouver Island region. Trans. Royal Canad. Inst., Toronto, vol. 13, no. 1.
FISCHER, PAUL 1887 Manuel de Conchyliologie et de Paléontologie Conchyliologique, Paris.
HANCOCK, A., AND EMBLETON, D. 1852 On the anatomy of Doris. Phil. Trans. Royal Soc., London, part 1, vol. 142.
JEFFREYS, J. G. 1869 History of naked marine Gastropoda. Brit. Conchol., vol. 5.
LANG, ARNOLD 1900 Lehrbuch der vergleichenden Anatomie der wirbellosen Thiere. Mollusca. Jena.
LANKESTER, E. Ray 1906 A treatise on zoology, part 5, Mollusca. London.
MINNICH, DWIGHT E. 1921 An experimental study of the tarsal chemoreceptors of two nymphalid butterflies. Jour. Exp. Zool., vol. 33, no. 1.
TAPPARONE-CANEFRI 1876 Genere Melibe Rang. Melibe papillosa De Filippi. Memorie della Reale Accademia delle Scienze di Torino, Ser. 2, T. 28, pp. 219–220.

Resumen por el autor, D. E. Minnich.

Un estudio cuantitativo de la sensibilidad tarsal a las
soluciones de sacarosa en la mariposa
Pyrameis atalanta L.

El autor ha demostrado previamente que los tarsos de la
mariposa Pyrameis atalanta L. son sensitivos en contacto con la
estimulación química. Una de las substancias que la mariposa
puede distinguir claramente con los tarsos es una solución IM
de sacarosa. Si se ponen en contacto con esta solución, Pyrameis siempre responde, sin relación alguna con su condicion
nutritiva. El animal puede también responder si los tarsos
tocan el agua destilada, pero solamente después de un periodo
más o menos prolongado de inanición total. Además, la inges-
tión de agua inhibe inmediatamente la respuesta. Es posible,
por consiguiente, mantener a Pyrameis en un estado de 100 por
ciento de reacción a la solución IM de sacarosa, pero solamente
en uno de O por ciento de respuesta al agua sola. Bajo estas
condiciones, la concentración mínima de sacarosa necesaria para
producir una respuesta, esto es, la concentración del umbral de
respuesta a la sacarosa, varía directamente con la condición de la
nutrición. Durante low periodos de inanición de sacarosa el
umbral decrece gradualmente, y puede alcanzar niveles tales
como $M/3200$, $M/6400$ y aún $M/12800$. Pero con la iniciación
de un periodo de dieta a base de sacarosa el umbral se eleva
súbitamente a un nivel generalmente de $M/10$, que se conserva
próximamente constante mientras la dieta continúa. Cuando
se compara con la de otros animales, la sensibilidad de Pyrameis
a la sacarosa está muy desarrollada. Este hecho está sin duda
relacionado con el hecho de que los azúcares forman el alimento
principal de este insecto.

Translation by José F. Nonidez
Cornell Medical College, New York

A QUANTITATIVE STUDY OF TARSAL SENSITIVITY TO SOLUTIONS OF SACCHAROSE, IN THE RED ADMIRAL BUTTERFLY, PYRAMEIS ATALANTA LINN.

DWIGHT E. MINNICH

Department of Animal Biology, University of Minnesota

ONE FIGURE

CONTENTS

INTRODUCTION

It is well known that many insects are extremely sensitive to distance chemical stimulation. The males of certain Lepidoptera, in particular, exhibit at the time of mating a degree of sensitivity probably unique in the animal kingdom. Thus Riley ('94, p. 39) describes an experiment in which a male of Philosamia cynthia Drury was successful in seeking out a female a mile and a half away, and Fabre ('79–'04), in his classic experiments upon several other species of Lepidoptera, reports equally remarkable results. But while we do know something of the degree of olfactory sensitivity possessed by insects, we know absolutely nothing concerning their acuity of taste or contact chemoreception. This is due to the fact that the only organs of taste heretofore described were located on the mouth parts or within the buccal cavity, where experimentation was virtually impossible.

Recently, however, I have shown (Minnich, '21) that the nymphalid butterflies, Vanessa antiopa Linn. and Pyrameis atalanta Linn., possess taste organs or contact chemo-receptors on their ambulatory tarsi. In a more detailed study of

445

Pyrameis alone (Minnich, '22) I have shown further some of the substances which can be differentiated through the tarsal organs. These substances consist of distilled water and several aqueous solutions, including a solution of 1M saccharose. In the course of this investigation two facts were discovered which have made possible a quantitative study of the tarsal sensitivity to saccharose solutions. First, contact of the four ambulatory tarsi with a 1M saccharose solution will always effect an extension of the proboscis, irrespective of the nutritional condition of the animal. From the time of its emergence from the pupa until its death, whether starved or abundantly fed, the butterfly will continue to respond with the utmost constancy to this stimulus. Second, contact of the tarsi with distilled water alone will also effect an extension of the proboscis, but only after a prolonged period of inanition with respect to water. Thus, a butterfly deprived of water for a sufficient number of days, generally four to seven, will finally become 100 per cent responsive to water. This response, however, ceases at once if the animal be allowed to drink water. By the mere administration of water, therefore, Pyrameis can be maintained in a state of 0 per cent responsiveness to water, but of 100 per cent responsiveness to 1M saccharose. Under these conditions, the minimal concentration of saccharose necessary to effect a response is readily determined. It is the purpose of the present paper to report the data obtained in this fashion.

METHODS

All butterflies were hatched in captivity. Upon hatching they were kept without food or water until they became 100 per cent responsive to water. They were then placed on a diet of distilled water which was administered morning and evening. Although some individuals appeared quite weak after the period of total inanition, they generally revived after their first access to water, and for some days the water diet was sufficient to maintain them in a normal condition. Sooner or later, however, the butterflies began to show signs of weakness which gradually increased until further trials became impossible. At this point 1M saccharose was substituted for distilled water, and again the animals generally revived. The diet of saccharose was continued for three days, after which the animals were subjected to a second period of total inanition, followed by a second period of water diet, followed in turn by a second period of 1M saccharose diet. Those individuals which survived the entire experiment were thus carried through six nutritional periods: a first period of total inanition, a first period of water diet, a first period of 1M saccharose diet, a second period of total inanition, a second period of·water diet, and a second period of 1M saccharose diet.

The method employed in making trials was the following. The animal was placed in a holder manipulated by hand, and was held with its four ambulatory tarsi in contact with a flat layer of absorbent cotton contained in a Syracuse watch-glass and saturated with the stimulating substance. Each trial with a given concentration of saccharose was immediately preceded by a trial with distilled water alone. If there was any evidence of response in the preliminary trial the saccharose test was abandoned, thereby avoiding any possibility of misinterpretation. No visible movement of the proboscis during one minute constituted a no response, while extensions of the proboscis during the same interval, whether partial or complete extensions, constituted a response. After each trial with saccharose the feet were carefully rinsed in distilled water and dried by contact with clean filter-paper.

Excepting the first three to five days after hatching and days of responsiveness to water alone, an attempt was made to determine the threshold concentration of saccharose for each animal daily. In general, this was accomplished by means of four determinations, although occasionally as many as six or seven were made. These determinations were made at least one hour apart. Between them, however, tests with other substances, viz., 1M saccharose, 2M NaCl, and M/10 quinine hydrochloride, were being made at minimal intervals of fifteen minutes. As far as I was able to observe, the trials with these various substances in nowise affected the threshold of response to saccharose.

In selecting the concentrations of saccharose for making determinations I followed no fixed, uniform procedure. Instead, I was guided by the immediate reaction of the individual together with its previous behavior. In general, the first day of experimentation was begun with a concentration of M/10 or M/100, while each day thereafter was begun with the concentration equal to or just below the threshold concentration of the previous day. If this concentration produced a response, lesser concentrations were tried; if it produced no response, greater concentrations were tried. This general procedure is brought out clearly in column IV of table 1, where the concentrations used on butterfly no. 13 are given in the order tried, with the reaction indicated in each case.

The quality of saccharose employed in making solutions was U. S. P. Stock solutions of M/10 and M/200 were made up from time to time during the experiments, while the other dilutions employed were prepared fresh daily from these stock solutions.

The present experiments were carried out at the same time and on the same animals as the experiments described in a previous paper (Minnich, '21). For a more detailed account of general methods than the one given here the reader is referred to that paper.

RESULTS

A total of seven butterflies was experimented upon. Three of these died quite early in the experiment, probably as a direct result of starvation. Of the remaining four, two survived to within a few days of the end of the experiment, while two others not only survived the entire experiment, but were still in vigorous condition several days later when they were killed and preserved for further study.

The most complete data were obtained on butterfly no. 13, and, since they are typical of the results obtained on the animals as a whole, they are presented in full in table 1. It will be noted that during the first three days after hatching no trials were made. The first determination was thus made on the 4th day when the threshold concentration was found to be M/200.

TABLE 1

Showing the responses of butterfly no. 13 to distilled water and to various concentrations of saccharose, under different nutritional conditions. Changes in nutritional conditions were always made in the evening after the trials of the day were completed. This is indicated by a repetition of the day number followed by (p.m.). Threshold concentrations are indicated in italics

I CONDITIONS OF NUTRITION	II DAY OF AGE	III RESPONSE TO DISTILLED WATER	IV RESPONSE TO SACCHAROSE SOLUTION
	1	Not tried	Not tried
	2	Not tried	Not tried
	3	Not tried	Not tried
	4	No response	M/10 Complete extension
		No response	M/100 Complete extension
		No response	*M/200 Complete extension*
		No response	M/400 No response
Given neither water nor food	5	Complete extension	Not tried
		Complete extension	Not tried
		Partial extensions	Not tried
		Partial extensions	Not tried
	6	Complete extension	Not tried
		Complete extension	Not tried
		Complete extension	Not tried
		Complete extension	Not tried
	7	Complete extension	Not tried
		Complete extension	Not tried
	7 (p.m.)		
	8	No response	M/200 Complete extension
		No response	*M/400 Partial extensions*
		No response	M/200 Complete extension
		No response	M/400 No response
Offered distilled water morning and evening	9	No response	M/400 Complete extension
		No response	*M/800 Partial extensions*
		No response	M/1600 No response
		No response	M/800 No response
		No response	M/400 Partial extensions
	10	No response	*M/800 Complete extension*
		No response	M/1600 No response

TABLE 1—*Continued*

I CONDITIONS OF NUTRITION	II DAY OF AGE	III RESPONSE TO DISTILLED WATER	IV RESPONSE TO SACCHAROSE SOLUTION
Offered distilled water morning and evening	10	No response No response	*M/800* *Partial extensions* M/400 Partial extensions
	11	No response No response No response No response	M/1600 Complete extension *M/3200* *Complete extension* M/6400 No response M/3200 No response
Offered 1M saccharose morning and evening	11(p.m.) 12	Not tried	Not tried
	13	No response No response No response No response	M/200 No response M/100 No response M/50 No response *M/10 Complete extension*
	14	No response No response No response No response	M/50 No response *M/10 Complete extension* M/50 No response *M/10 Complete extension*
	15	No response No response No response No response	M/50 No response *M/10 Complete extension* M/50 No response *M/10 Complete extension*
Given neither water nor food	15(p.m.) 16	No response No response No response No response	*M/50* *Partial extension* M/100 No response M/50 No response M/10 Complete extension
	17	No response No response No response No response	*M/100 Complete extension* M/200 No response *M/100* *Partial extensions* M/200 No response
	18	Complete extension Partial extensions No response No response	Not tried Not tried M/200 No response M/100 No response
	19	Complete extension Complete extension	Not tried Not tried

TABLE I—*Continued*

I CONDITIONS OF NUTRITION	II DAY OF AGE	III RESPONSE TO DISTILLED WATER	IV RESPONSE TO SACCHAROSE SOLUTION
Given neither water nor food	19	Complete extension No response	Not tried *M/100 Complete extension*
	20	Complete extension Complete extension Complete extension Complete extension	Not tried Not tried Not tried Not tried
Offered distilled water morning and evening	20(p.m.) 21	No response No response No response No response No response	M/1600 No response M/400 No response M/200 No response *M/100 Partial extensions* M/50 Partial extensions
	22	No response No response No response No response No response	M/200 No response M/100 Partial extensions M/50 Complete extension *M/200 Partial extensions* M/400 No response
	23	No response No response No response No response No response	M/400 No response M/200 No response M/100 No response M/50 Complete extension *M/100 Partial extensions*
	24	No response No response No response No response	*M/100 Complete extension* M/200 No response *M/100 Complete extension* M/200 No response
	25	No response No response No response No response No response No response	M/200 Complete extension *M/400 Complete extension* M/1600 No response M/800 No response *M/400 Partial extensions* M/800 No response
	26	No response No response No response	M/800 Complete extension M/3200 Partial extensions *M/6400 Complete extension*

TABLE 1—*Concluded*

I CONDITIONS OF NUTRITION	II DAY OF AGE	III RESPONSE TO DISTILLED WATER	IV RESPONSE TO SACCHAROSE SOLUTION
Offered distilled water morning and evening	26	No response No response No response	M/25,600 No response M/12,800 No response M/6400 No response
Offered 1M sac- charose morn- ing and evening	26(p.m.) 27	No response No response No response No response No response	M/100 No response M/50 No response M/10 *Complete extension* M/50 No response M/10 *Complete extension*
	28(a.m.)	Animal found dead	

The 5th, 6th, and 7th days, no trials were made with saccharose since the animal responded to water. During the 7th day the animal became so weak that further experimentation was impossible. Accordingly, trials were suspended, water was administered, and the butterfly was allowed to recuperate.

On the following morning, viz., the morning of the 8th day, the butterfly had entirely recovered, its responsiveness to water had disappeared, and trials with saccharose were resumed. As the period of water diet continued, the threshold of response to saccharose gradually fell from M/400 on the 8th day to M/3200 on the 11th day. At the close of the 11th day the animal again evinced signs of weakness, and it was necessary to change the diet to 1M saccharose.

Absence from the laboratory prevented any trials on the 12th day. When experimentation was renewed on the morning of the 13th day, however, the butterfly had completely recovered its normal vigor. But the threshold of response had risen from M/3200 to M/10, and at this level it steadily remained throughout the entire period of saccharose diet, viz., the 13th, 14th, and 15th days.

Following the first period of saccharose diet came the second period of total inanition, which lasted from the 16th to the 20th days, inclusive. During the first two days of this period, the

16th and 17th days, the threshold of response to saccharose again fell to M/100. On the 18th, 19th, and 20th days the responsiveness to water gradually reappeared and prevented any conclusive determinations with saccharose.

At the close of the trials on the 20th day the second period of water diet was begun, and on the following morning the responsiveness to water had completely disappeared once more. The threshold concentration of saccharose on this the 21st day was still M/100. The water diet was continued until the close of the 26th day, the threshold concentration of saccharose gradually falling meantime until it reached M/6400. In the last two trials of the 26th day the butterfly had become extremely weak—a fact which may account for the negative results obtained.

The physical condition of the animal on the evening of the 26th day necessitated the administration of saccharose, and accordingly the second period of saccharose diet was begun. On the morning of the 27th day the butterfly appeared greatly revived. But tests with saccharose showed that the threshold of response had again risen to M/10, precisely as it had done at the beginning of the first period of saccharose diet. On the evening of the 27th day the animal seemed quite as vigorous as usual, but on the morning of the 28th day it was found dead.

In figure 1 the threshold concentrations of saccharose for butterfly no. 13 are represented graphically. A glance at the curve shows that during periods of total inanition followed by periods of water diet, i.e., during saccharose inanition, the threshold of response gradually fell, while immediately a period of saccharose diet was begun the threshold abruptly rose. Two facts are thus brought out clearly. First, the minimal effective concentration of saccharose for the tarsi varies directly with the nutritional condition of the animal. During periods of saccharose diet the threshold remained at M/10; during periods of complete inanition and water diet, i.e., saccharose inanition, the threshold gradually fell to M/3200 or M/6400. Second, the minimal effective concentration after prolonged inanition with respect to saccharose may be extremely low. At the close of the first period it was M/3200; at the close of the second, M/6400.

In table 2 I have summarized the results obtained on all seven
butterflies. While the results there presented exhibit rather
wide individual differences, they show the same general features
which are brought out in the behavior of butterfly no. 13.   Thus
the threshold of response in all animals fluctuated directly with
the nutritional condition.   Furthermore, the minimal effective
concentration after prolonged inanition with respect to saccharose
was very low in general.   The extremely low thresholds, M/3200,
M/6400, and M/12,800, which were observed in four of the seven
animals, are particularly noteworthy.

TABLE 2

*Showing the threshold concentrations of saccharose solution, under varying nutri-
tional conditions, in seven butterflies.   An asterisk signifies the death of the
butterfly within twenty hours after the determination indicated*

ANIMAL NUMBER	FIRST PERIOD OF TOTAL INANITION	FIRST PERIOD OF WATER DIET	FIRST PERIOD OF SACCHAROSE DIET	SECOND PERIOD OF TOTAL INANITION	SECOND PERIOD OF WATER DIET	SECOND PERIOD OF SACCHAROSE DIET
11	Not tried	M/100	M/10	M/50	M/12,800	· M/10
12	Not tried	M/200	M/10	M/50	M/400*	
13	M/200	M/3200	M/10	M/100	M/6400	M/10*
22	Not tried	Not tried	M/100	M/400	M/3200	M/100
23	Not tried	M/100*				
24	Not tried	M/12,800*				
25	Not tried	M/400*				

This responsiveness to very dilute solutions of saccharose
affords further evidence of a point previously made by me, viz.,
that the nature of tarsal stimulation is not osmotic, but chemical
(cf. Minnich '21, p. 198).   Throughout the present experiments
all animals were tested four times daily with several solutions
other than those mentioned in this paper.   Among these were
1M saccharose and 2M NaCl.   On the days when the four butterflies mentioned in the foregoing paragraph were responding to
such dilutions of saccharose as M/3200, M/6400, and M/12,800,
three of the four gave no response whatever to 2M NaCl.   Yet
the NaCl solution was vastly the more effective osmotically.
It might be objected that the salt solution produced such a
powerful osmotic effect as to inhibit response, whereas a more

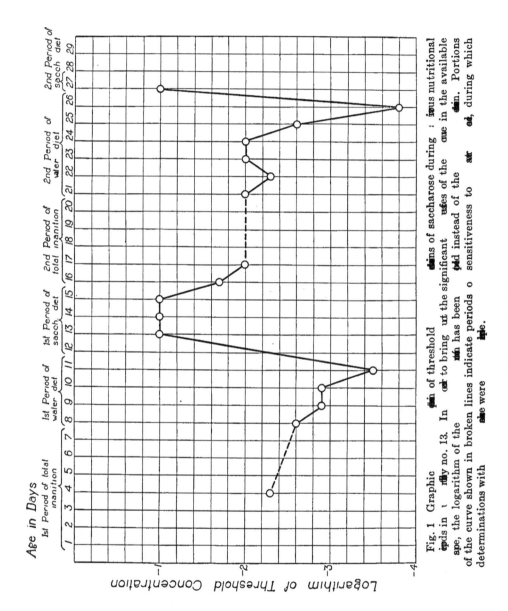

Fig. 1 Graphic ... of threshold ... ins of saccharose during : ... ious nutritional ... ds in ... ... y no. 13. In ... ... to bring ... ut the significant ... ... es of the ... ... in the available ... ge, the logarithm of the ... ... has been ... ... instead of the ... ... during which of the curve shown in broken lines indicate periods o sensitiveness to ... ... determinations with ... ... were ... ...

dilute solution would have produced a response. But this ob jection is at once ruled out by the fact that a 1M solution of saccharose always yielded a 100 per cent response. The tarsal sense organs, therefore, appear to be quite unaffected by osmotic pressure and must be regarded as specific chemoreceptors.

The tarsal sensitivity of the red admiral to saccharose is quite remarkable when compared with the sensitivity of other animals to the same substance. Among the marine invertebrates saccharose appears to have little or no stimulating power apart from the osmotic effect of high concentrations. The same holds true for the lower aquatic vertebrates. According to the table given by Parker ('12, p. 228), Amphioxus, Ammocoetes, Mustelus, and Amiurus are entirely unresponsive to saccharose. Moreover, the weakest solution which the human tongue can detect is but M/50. But as we have seen in two of the seven butterflies tested, nos. 11 and 24, the tarsi were able to discriminate an M/12,800 solution. In other words, the tarsi of these two butterflies were 256 times more sensitive to saccharose than the human tongue.

The sensitivity of Pyrameis to saccharose is thus a highly specialized one. The reason for this is not far to seek. Pyrameis feeds on nectar, exuding sap, and juices of ripe and decaying fruit. In all of these substances sugars are found. In the laboratory I have kept the butterfly alive and in good condition for thirty days on 1M saccharose solution. Sugars thus appear to be the most important food of this insect, and the highly developed sensitivity to saccharose is doubtless directly correlated with this fact. Like the organs of olfaction, therefore, the organs of taste may be very highly developed in certain lepidopterous forms.

CONCLUSIONS

1. In Pyrameis the threshold of response to saccharose solutions varies directly with the nutritional condition of the individual. During periods of total inanition followed by periods of water diet, i.e., during saccharose inanition, it gradually falls. But with the initiation of a period of saccharose diet, it rises abruptly to a level which remains approximately constant throughout the remainder of the period.

2. After prolonged inanition with respect to saccharose the threshold concentration may fall as low as M/3200, M/6400, or even M/12,800 in some individuals. The tarsal sensitivity of Pyrameis to saccharose may thus be as much as 256 times that of the human tongue.

3. The highly developed sensitivity to saccharose is doubtless correlated with the fact that sugars form the chief food of this insect.

BIBLIOGRAPHY

FABRE, J. H. 1879–1904 Souvenirs entomologiques. 7me serie, nos. XXIII, XXIV, XXV. 8me edition. Paris.

MINNICH, D. E. 1921 An experimental study of the tarsal chemoreceptors of two nymphalid butterflies. Jour. Exp. Zoöl., vol. 33, pp. 173-203.
1922 The chemical sensitivity of the tarsi of the red admiral butterfly, Pyrameis atalanta Linn. Jour. Exp. Zoöl., vol. 35, pp. 57-81.

PARKER, G. H. 1912 The relation of smell, taste, and the common chemical sense in vertebrates. Jour. Acad. of Nat. Sciences of Phila., vol. 15, second series, pp. 221-234.

RILEY, C. V. 1894 The senses of insects. Insect Life, vol. 7, pp. 33-41. ·

Resumen por el autor, Stefan Kopeć.

Relación mutua entre el desarrollo del cerebro y los ojos
de los Lepidópteros.

1. Los ojos de la mariposa nocturna Lymantria dispar L., se
desarrollan con completa independencia del cerebro y ganglio
subesofágico. El cerebro solamente ejerce una influencia regula-
dora sobre la dirección de las fibras nerviosas que van desde la
retina del ojo al ganglio óptico. 2. El gérmen de los ojos, ex-
tirpado de la cabeza de una oruga e injertado sobre el abdomen,
se desarrolla normalmente en la nueva posición, a pesar de la
ausencia de conexiones con la cadena nerviosa. 3. La extirpación
de los gérmenes oculares imaginales de las orugas impide el
desarrollo de las capas externas del ganglio óptico del adulto y
produce cambios en la estructura de ciertas capas internas,
suponiendo que tenga lugar la regeneración del gérmen del ojo.
4. El ganglio subesofágico de los ejemplares desprovistos de
cerebro se desarrolla menos que en los individuos normales. La
extirpación del ganglio subesofágico de la larva no ejerce influencia visible sobre la formación del cerebro imaginal.

Translation by José F. Nonidez
Cornell Medical College, New York

MUTUAL RELATIONSHIP IN THE DEVELOPMENT OF THE BRAIN AND EYES OF LEPIDOPTERA[1]

STEFAN KOPEĆ

Government Institute for Agricultural Research, Pulawy, Poland

ONE PLATE (SIX FIGURES)

THE DEVELOPMENT OF IMAGINAL EYES IN BRAINLESS INSECTS

The correlation of the components of the eye in the ontogenetic development of vertebrates has been studied. The experiments undertaken by Spemann on the development of the lens and by Lewis on the development of the cornea and continued by Bell, Le Cron, Dürken, Ekmann, Fischel, King, Wachs, and others proved that while in certain amphibians the lens depends on the optic cup and the cornea on the lens, in others such a dependence cannot be ascertained. The mutual relationship between the development of the nervous system and the onto-genetic formation of the eyes in invertebrates has, so far as I know, not yet been studied. The investigations of Herbst on Crustacea ('96–'16), Carrière ('80) and Hankó ('14) on the Mollusca, since they refer only to the processes of regeneration of the eyes, do not come into consideration here. In my experiments on insects I endeavored to study these relations by examining the development of the eye in moths the caterpillars of which had been deprived of the brain, and the development of the imaginal brain in animals whose caterpillars had been deprived of the eyes.

I removed the whole brain (ganglion supraoesophageale) from several caterpillars of Lymantria dispar L after their last moult.

[1] Paper from the Embryologic-Biological Laboratory, Jagellonian University, Cracow, Poland, presented in the Acad. of Sc., Cracow (cf. Bull. intern. Acad. d. Sc. Cracovie, 1917).

In this manner I also deprived the caterpillars of the larval optic
ganglia (as to method of operation, cf. Kopeć, '18). Some of
the caterpillars underwent pupation and developed to adult
forms of moths. In all the specimens of these moths, or of their
pupae, the eyes were quite normally developed macroscopically
(cf. fig. 1, representing the eyes of brainless moth). In dissections
of eyes which developed without the brain, only a more or less
distinct wrinkling of the surface of the eye may often be noticed.
I consider this deformation to be due only to the mechanical
difficulties experienced by a brainless animal in drawing its
injured head from the skin of the caterpillar or pupa during its
metamorphosis.

Also microscopical researches have demonstrated the normal
structure of 'brainless' eyes (cf. figs. 2. and 3, representing dissections of eye which developed without the brain, with figs. 4 and
5 giving analogous dissections of normal eyes). Both the number
of ommatids and the size and form of all components of the
compound retina were in no wise different from those observed
in normal circumstances. The arrangement of the pigment is
somewhat different in figure 3 of the 'brainless' eye and in figure
5 of a normal one. But also in normal specimens we may observe individual fluctuations in this arrangement of the pigment,
due probably to the time at which they were killed.

On the interior surface of the normal eye I always found the
layer which is called by Berger ('78) 'the nerve-bundle layer'
(Nervenbündelschicht) to be well developed. In this layer I
have discerned two distinct parts: an external layer of single
bundles, consisting of a relatively small number of nerve fibers
coming directly from the retina of the eye (fig. 4, *s.n.b.*) and an
internal one, which I might call the layer of compound bundles
(*c.n.b.*). The compound bundles are made up of single bundles
which are in both layers distinctly separated from one another and
run regularly to the interior of the head, radiating in the direction
of the optic ganglion. The arrangement of the single bundles
does not evidence any changes in the brainless insects (fig. 2,
s.n.b.), while the compound bundles always behave with remarkable irregularity. These latter bundles have an abnormal

arrangement (fig. 2, *c.n.b.*). We get the impression that, in the ontogenetic development of the eye, the brain or the optic ganglion has a determining influence on the direction of these nervous bundles. These anomalies might, however, just as well be attributed to different modes of growing together, which took place during cicatrization of torn larval nervous bundles. The crucial fact here was the behavior of moths deprived of the germs of their imaginal eyes, these germs being subsequently regenerated. However, in these cases the larval nerve-bundles underwent analogous tearing or concrescenses during operation, yet the arrangement of the single as well as compound bundles was quite normal. Consequently, during the formation of the imaginal eye the brain (or the optic ganglion) has an influence on the direction of the nerve-bundles proceeding from the retina to the optic ganglion.

The other layers situated under normal conditions between the layer of nerve-bundles and the imaginal optic ganglion, which are considered by Berger to be a modified part of the optic ganglion, arise through transformation of the larval optic ganglion which was removed together with the larval brain. In view of such anatomic-histological conditions it becomes quite clear that only the nerve-bundle layer described above is present besides the eye sensu stricto in the brainless moth.

The simultaneous removal of brain and of the suboesophageal ganglion has convinced me of a quite similar independence of the evolution of the eye from the second ganglion situated in the head. Stress must be laid on the fact that in no case was I able to detect any processes of regeneration of the removed brain or suboesophageal ganglion. Thus the development of the imaginal eye of moths is quite independent of the presence of the brain and the suboesophageal ganglion.

THE DEVELOPMENT OF IMAGINAL EYES THE GERMS OF WHICH HAVE BEEN TRANSPLANTED ON THE ABDOMEN OF CATERPILLARS

The transplantation of the eyes in vertebrates has been accomplished by Uhlenhuth ('12, '13), who grafted larval eyes of salamanders and newts on other specimens of the same species. He convinced himself that the eyes undergo further evolution after certain transitory processes of atrophy. In order to render the independence of the development of the insectal eye still more striking, I attempted the transplantation of the germ of eyes to the abdomen, i.e., to surroundings which are heterogeneous from an anatomical point of view. I deprived seventy-five caterpillars of the whole lateral part of the head, on which all their ocelli are situated. Johansen ('92) has shown that the material for the imaginal eye is contributed by the hypodermis between the larval eyes, which are to be found on the chitinous plate removed. I then inserted this plate into a large wound made on the fourth abdominal segment of the same caterpillar by the amputation of one of the paired orange-yellow warts. After a few hours the plate had been well fixed in the new position by means of coagulated blood.

Out of forty-one pupae obtained from this series of experiments, fifteen had distinctly demarcated, tiny hill-like convexities in the place where the plate had been grafted. These convexities differ from the integument of the other segment in that they have a more glossy and deeper hue. In thirteen of the moths which emerged from these pupae, a very distinct imaginal eye, in most cases of normal size, was found on the corresponding abdominal segment (fig. 6, *tr. e.*). The eyes were more or less hemispherical, sometimes somewhat wrinkled and divided by furrows. The histological structure of these eyes differs in no way from that of normal eyes. We need only note that the height of the ommatidia was reduced and that there was a proportional shortening of their components. This detail was no doubt connected with the abnormal tension and pressure in the new surroundings of the eye which influence the developing organ. Here also the grafted eyes had a well-developed 'nerve-bundle layer' running abnormally. I was, however, not able to find any nerves radiating from the eye deeper into the interior of the body

THE DEVELOPMENT OF THE BRAIN IN INSECTS DEPRIVED OF THE GERMS OF EYES

In the following series of experiments I removed the above-mentioned plate, which contained the material for the eye of the moth, from one side of young caterpillars after the second or third moult, and without depriving the animals of the brain. I examined its behavior in cases when the removed hypodermis did not undergo regeneration and thus did not form an imaginal eye. At the same time the non-operated side of the head could be used as a standard. It was my intention to ascertain whether the morphological and histological metamorphosis which the larval brain undergoes when metamorphosing into the imaginal organ, is in any way dependent on the development or presence of the eye.

In its principal part the brain of the moth is composed of the large optic ganglia resulting from the metamorphosis of those external parts of the larval brain, which form the optic ganglia of the caterpillar (cf. Johansen, '92, and Bauer, '04). In all the specimens which did not regenerate their eyes the imaginal optic ganglion was more voluminous, though shorter on the operated side than on the normal one. The granular, as well as the molecular layer and the ganglion-cell layer (Körner, Molekular and Ganglienzellenschicht of Berger), was never developed. It follows from the paper of Berger that the tissue which gives rise to these layers is situated in the external parts of the larval brain. This tissue remained intact in the head of the caterpillar operated upon: this I conclude from the fact that I never observed any injuries of the larval brain when, to verify the results of operation, I examined the microscopical sections of several specimens directly after operation. In other words, the absence of this layer of the optic ganglion in specimens which had no regenerated eye can only be interpreted by the absence of some developmental stimuli, derived from the normal eye in course of development or, as I had several times observed, from the regenerated eye. That is to say, when the eye is regenerated all the layers of the imaginal optic ganglion layers are developed quite normally also.

The layer of the external chiasma (äussere Kreuzung) never was found afterward in the operated specimens. The layer of the internal chiasma and the external medullary layer (innere Kreuzung and äusseres Marklager of Berger) were often but slightly developed. The internal medullary (inneres Marklager) was thicker along the longitudinal axis of the brain, thinner along the transverse axis. These anomalies set in a still clearer light the influence exerted by the eye on the development of the imaginal brain or sensu stricto, of its optic ganglion during its metamorphosis. In one of my previous papers I already urged this opinion, but was unwilling to consider it as proved on account of the small amount of experimental material then at my disposal (cf. Kopeć, '13, p. 457). My former conclusion as to the dependence of the formation of the brain on the formation of the imaginal eye in insects is now well founded and it finds perfect confirmations in the important investigations of Herbst ('16) on changes in the structure of the optic ganglion in certain crustacea deprived of the eyes.

In connection with the results discussed here I wish to remark briefly that in the development of both ganglia of the head a certain mutual dependence may be observed. In microscopical sections of brainless specimens we are at once struck by the exceedingly meager development of the suboesophageal nervous ganglia. Its normal development is evidently correlated to some extent with the formation of the brain. The removal of the larval suboesophageal ganglion, on the contrary, has no visible effect on the formation of the imaginal brain.

SUMMARY

1. The eyes of the moths develop in complete independence of the brain and the suboesophageal ganglion. The brain exerts only a regulating influence on the direction of the nerve fibers going from the retina of the eye to the optic ganglion.

2. On the other hand, if the imaginal eye, the germ of which had been removed in the caterpillar, is absent, the external layers of the optic ganglion do not develop at all, and certain internal layers show changes in their structure.

3. The germs of mature eyes grafted from the head of the caterpillar on its abdomen develop normally, notwithstanding the absence of any junction with the nervous chain.

4. In specimens deprived of brain the suboesophageal ganglion develops to a markedly less degree than in normal specimens. The removal of the larval suboesophageal ganglion has no visible effect on the formation of the imaginal brain.

BIBLIOGRAPHY

BAUER, V. 1904 Zur inneren Metamorphose des Centralnervensystems der Insecten. Zool. Jahrb., Abt. f. Anat. u. Ontog, Bd. 20.

BERGER, E. 1878 Untersuchungen uber den Bau des Gehirnes und der Retina der Arthropoden. Arb. d. Zool. Inst. Wien, Bd. 1.

CARRIÈRE, J. 1880 Studien über Regenerationserscheinungen bei Wirbellosen. Würzburg.

HANKÓ, B. 1914 Über das Regenerationsvermögen und die Regeneration verschiedener Organe von Nassa mutabilis. Arch. Entw.-Mech., Bd. 38.

HERBST, C. 1896 Über die Regeneration von antennenähnlichen Organen an Stelle von Augen. I. Mitteilung. Arch. f. Entw. Mech., Bd. 2. 1900 III. Mitteilung. Ibidem, Bd. 9. 1916 VII. Meitteilung. Ibidem, Bd. 42.

JOHANSEN, H. 1892 Die Entwicklung des Imagoauges von Vanessa urticae L. Zool. Jahrb., Abt. f. Anat. u. Ontog., Bd. 6.

KOPEĆ, ST. 1913 Untersuchungen über die Regeneration von Larval-organen und Imaginalscheiben bei Schmetterlingen. Arch. f. Entw. Mech., Bd. 37. 1918 Lokalisationsversuche am zentralen Nervensystem der Raupen und Falter. Zool. Jahrb., Abt. f. allgem. Zool. u. Physiol., Bd. 36.

UHLENHUTH, E. 1912 Die Transplantation des Amphibienauges. Arch. f. Entw. Mech., Bd. 33. 1913 Die synchrone Metamorphose transplantierter Salamanderaugen. Ibidem, Bd. 36.

PLATE

EXPLANATION OF FIGURES

All the ⟩ ⟨ᴡhs re˝ ꭓ to the full-grown stages o the ᴍoth L ᴛᴀ a dispar L.

1 Anterior part of a ᴄᴇ ᴍth ᴍᴇ well developed. ˙ᴘᴅ o brain ⟩ ꭓᴅg the ᴍᴛ ᴀᴇ o arva lᴇ. Eyes

2 Front part of the ᴋᴅ ᴊᴇ of a ᴛh, the caterpillar of ⟨ ᴋh had ⎸ ᴇn ˙ ᴘᴅ of the brain ᴀ the ᴍᴛ ᴄᴇ. ., ᴄᴅ o., ommatidia s.n.b., sing e nerve-bundle ᴊᴛ; . ᴛᴄ, compound nerve- uᴅdle ᴌᴇ.

3 Part of he ᴍᴇ ᴊ, ᴏᴍᴇ magnified; l., ᴅᴅs; cr., crystalline ᴍs r. ⟨ᴍᴇ.

4 Front part o the ᴄᴋᴅ ᴊᴇ of a normᴀ ᴛh. ᴛᴋ ᴇᴋᴅ ad the ᴍᴇ ᴋᴍs as in figure 2. ᴄᴋᴌ ᴍᴅt of the ᴄᴘᴅ nervous bund ᴇ, ˙ ꭓᴅg toward the optic ⟩ ᴀᴍn.

5 Part of the ᴍᴇ ᴊ, ᴏᴍᴇ ꭓᴅᴅ. Tᴌe ᴍᴇ magn ᴀᴍn ᴀᴅ the ᴍᴇ ᴋᴍs as in figure 3.

6 Abdomen of a ᴍoth from the ᴇ rpil aᴋ of which the germ o the imaginal ᴊᴇ hd⎸ ᴇn ᴊᴅd on the fourth ab ᴄᴅᴇ ᴇᴍnt tr. e. ᴊᴍd ᴊ, normally developed.

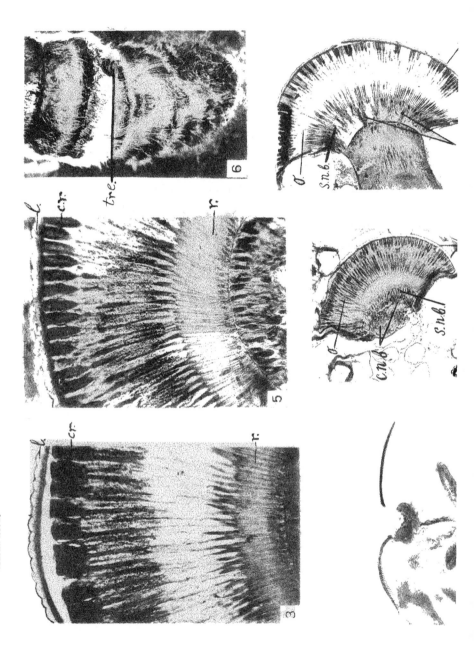

Resumen por el autor, Stefan Kopeć.

Autodiferenciación fisiológica de los gérmenes de las alas injertados en orugas del sexo opuesto.

El gérmen de las alas imaginales injertado en ejemplares de orugas de Lymantria dispar L. del sexo opuesto continúa desarrollándose y las alas diferenciadas a sus expensas presentan su tinte dimórfico propio, en vez de exhibir el del individuo sobre el cual se desarrollan. Los pigmentos de la escama no proceden, por consiguiente, directamente de la sangre desecada sino que son el producto de ciertos cambios químicos, los cuales, según Mayer, tienen lugar en la sangre bajo la influencia de substancias específicas contenidas en las células formadoras de las escamas.

Translation by José F. Nonidez
Cornell Medical College, New York

PHYSIOLOGICAL SELF-DIFFERENTIATION OF THE WING-GERMS GRAFTED ON CATERPILLARS OF THE OPPOSITE SEX[1]

STEFAN KOPEĆ

Government Institute for Agricultural Research, Pulawy, Poland

The development of the pigment in the wings of moths has not yet been sufficiently studied; there is, however, no longer any doubt that the chief ingredient in the formation of the pigments is the blood, or the so-called haemolymph of the animal. How this formation takes place has not yet been ascertained; some authors believe that the pigment is directly caused by the drying up of the haemolymph. On the contrary, according to Mayer ('96, '97), there are certain ferments in the scale-forming cells which render the formation of the pigment possible. On this view the pigment ˉis the product of certain changes which occur in the insect's blood, under the influence of special ferments.

Crampton ('00) is inclined to Mayer's opinion, but I think that his notable experiments are not sufficient to demonstrate it. Crampton united various parts of the pupae of the moth Callosamia promethea by means of paraffin, so that the front part of the body obtained belonged to one sex, the hind part to the other. These bodies developed further until the stage of the adult moth, when each part of the artificially united body showed its specific and dimorphic color, different in front and behind. "We must conclude, therefore," says Crampton, "that the production of the sexually-different ground-colors of the adult moths is determined by some 'ferment' factors which differ in the two sexes, and that the difference in the adult colors is not due to a difference between

---

[1] Paper from the Embryologico-Biological Laboratory, Jagellonian University, Cracow, Poland, presented in the Acad. of Sc. Cracow. (Cf. Bull. Acad. d. Sc. de Cracovie, 1917.)

469

the respective haemal fluids of the two sexes." Crampton's investigation itself, however, does not justify our drawing such a conclusion; at the same time it is quite true, as we shall soon convince ourselves. First, the united parts were relatively large bodies, very thoroughly and abundantly supplied with blood, this being different in each part; it seems to me very probable, therefore, that in the cases given there was no thorough mingling of the two haemolymphs in the artificially joined organism; since it is quite possible that these did not mix at all, Crampton's conclusion does not seem to me to be proved. Secondly, the connection of the chrysalides was possibly made at too late a stage for this supposed influence of one part on the other to be visible. I tried, therefore, to verify Mayer's theory by a somewhat different method. At the same time my experiments were intended to show the morphological and physiological self-differentiation of the insect wing still more distinctly than it has been shown in the experiments on the castration of caterpillars. (See below.)

I removed the germs of the first right wing from male caterpillars of Lymantria dispar L. after their last moult, and in their place I grafted a similar germ from a female caterpillar, and vice versa. It was necessary that both caterpillars should be of about the same age, so that the time of pupation of the one operated upon would coincide most strictly with the time at which the implanted wing of the opposite sex should attain the stage of metamorphosis. Naturally, the fulfillment of this condition was extremely difficult, and on the other hand the metamorphosis of the grafted wing always occurs quite independently of the organism on which it has been implanted. Therefore, if the condition mentioned has not been fulfilled, the implanted wing undergoes metamorphosis at a different time than the caterpillar on which it is grafted and cannot easily come forth and develop subsequently. If we take into consideration also that the delicate germ was often injured during transplantation, it is not astonishing that out of 120 specimens operated upon, the implanted germ developed into the pupal and imaginal wing only in two cases. In these two cases we had to do with male wings developed on female organisms. The grafted wings were somewhat folded and could not be drawn out of the pupal integument by the moth alone. During the extraction of these wings the scales, especially from the upper surface, nearly all remained in the pupal skin, from which they could very easily be removed uninjured. The form and color of the scales were afterward studied exactly under the microscope.

In one case the form of these scales was quite normal, in the other the scales were less deeply dentated than in normal conditions, which seemed to show that they were not completely developed. The color of these wings corresponds to the dimorphic hue which the transplanted wing would have had if it had remained intact in the previous organism; therefore, the grafted wing was in both cases otherwise colored than the normal left wing of its foster-mother. The wings of the male, which had been transplanted on the female, had more or less dark gray or dark brown scales, in contrast to the female wings, which were for the most part white.

I think the results of these experiments certainly speak in favor of Mayer's theory, as it turned out that the cells of the wing germs were able to collect the respective dimorphic pigments in their scales, forming them from the heterogeneous blood of the other sex in the new surroundings. Thus the pigments of the scales are not directly derived from the desiccation of the haemolymph, but must be the outcome of certain chemical processes occurring in the insect blood under the influence of substances formed in the scale-producing cells, and considered by Mayer to be ferments. Owing to these substances, which are different in different parts of the imaginal wing, various pigments are formed in the blood of the insect, and this results in the production of complicated figures on the wings. These substances develop by means of physiological self-differentiation, and their formation is outside the influence of the haemolymph of the other sex. This fact is unexpected, since the investigations made by Dewitz ('09–'16b), Steche ('12 a, b,) and Geyer ('13, '14) have demonstrated that the blood in insects of one and the other sex is not the same, but shows great differences in the chemical qualities as well as in its coloring. It happens that the differences are not important enough to have a decisive influence on the hue of the insect wing. Consequently, the real cause of the color of the wings of dimorphic moths, being distinct for the two sexes, is not the difference of the blood, but the difference of the substances which are present in the cells of the germs of the wings. The physiological self-differentiation of the corresponding cells takes place early, even before the pupation of the caterpillar. In this stage the germ of the yet unformed wing seems to resemble in some degree an exposed but as yet undeveloped photographic plate.

The self-differentiation of the wing of moths, as I have stated it, is sufficient to interpret the known results on the castration of caterpillars, which does not lead to even the slightest changes in the dimorphic coloring of the wings of the adult moth (compare the experiments of Oudemans, '99; Kellogg, '04; Meisenheimer, '07, '09; Kopeć, '08, '11, '13, and Geyer, '13). Prell ('15 a), having castrated caterpillars of the moth Cosmotriche potatoria L., observed, on the contrary, a much larger variability in the direction of a lighter hue in the wings of castrated males than in normal specimens. Prell does not question the negative results of the experiments on the castration of moths hitherto performed, and obtained chiefly on the species Bombyx mori L. and Lymantria dispar L., but he believes that the results ought not to be extended to all forms of moths. It is possible that further investigations made in this direction would lead us to distinguish between those forms of moths which react and those which do not react after castration by showing a change of color of the wings. But the experiments hitherto made by Prell are too small in number to draw any certain conclusion from them. In Prell's investigations we also miss standard experiments, such as might exclude the possibility of the influence of the operation itself, whether the sexual glands were removed from the insect body or not. According to Prell's own words, the castrated caterpillars refused food, and they had also to be far more abundantly sprinkled with water. If we bear in mind that they were subjected to a powerful ether narcosis and that they often lost a large quantity of blood during the operation, we can readily understand that they might have been much weakened by the operation and have responded more readily to all changes of external conditions, such as temperature, moisture, etc., even when these were imperceptible to us.   In contrast to Lymantria dispar, Cosmotriche potatoria belongs to those species of moths whose dimorphic wing-colors undergo distinct changes under the influence of cold; to this difference Prell ascribed his results, which he considers different.   The later investigations of Prell ('15 b) on the castration of various Vanessae, the classical material for the study of the influence of temperature, do not seem to support this opinion.   Castrated Vanessae as well as castrated females of Cosmotriche potatoria undergo no changes after castration, in contrast to males of the latter species.   In this behavior of the Vanessae, Prell ('15 b) sees a proof that in the experiments on males of Cosmotriche the change in hue was not excited by the operation itself, but that we have here to do with the effect of the removal of the sexual glands; for if the lighter hue of the wings of the males of Cosmotriche were to depend on the operation itself, it would have to be admitted, according to Prell, that similar changes would appear also among the Vanessae operated upon, as they are even more sensitive to the influence of external conditions.   I believe, however, that we might just as well suppose a different reaction power of the moth to castration, possibly different in the two sexes or in various forms of moths, as a different behavior of various organisms, in respect to their change of color, affected by debility and supersensitivity resulting from the operation.   In this way the results obtained by Prell may be made to accord with the results of the researches of Oudemans, Kellogg, Meisenheimer, and of my own.

While some authors have seen certain contradictions in the results of different investigations, which, according to them, prove the influence of castration on the dimorphism of the moth, others have drawn the conclusion, from all these experiments, that the germ of the wings is already differentiated early in the larval life, and hence the removal of the gonad or the implantation of glands of the other sex cannot change anything in the coloring of the insect wing. The results of my experiments on the transplantation of the wing-germs confirm the latter conclusion, showing at the same time the principle in virtue of which these animals have quite a different position in regard to the development of their dimorphic secondary characters from that of vertebrates. Steche ('12, a, b) and Geyer ('13), who relied on their experiments on the dimorphic differences of the insect blood, expressed the opinion that the fundamentally different behavior of insects and vertebrates after castration is caused by the fact that the whole body (soma) of the former undergoes sexually dimorphic differentiation from the beginning of life. In the light of my own experiments described in this paper this hypothesis gains a new and important confirmation.

SUMMARY

The germ of the imaginal wings grafted on specimens of caterpillars of Lymantria dispar L. of the opposite sex continues developing, and the differentiated wings have the dimorphic hue proper to them, and not to the specimen on which they develop. The pigments of the scale therefore do not proceed directly from the desiccated blood, but are the product of certain chemical changes which, according to Mayer, occur in the blood under the influence of specific substances contained in the scale-forming cells.

BIBLIOGRAPHY

CRAMPTON, H. E. 1900 An experimental study upon Lepidoptera. Arch. f. Entw. Mech., Bd. 9.
DEWITZ, J. 1909 Die wasserstoffsuperoxydzersetzende Fahigkeit der männlichen und weiblichen Schmetterlingspuppen. Zentrbl. f. Physiol., Bd. 22.
1912 Untersuchungen über Geschlechtsunterschiede. II. Ibidem, Bd. 26.
1916 a Idem., III. Zool. Anz., Bd. 47.
1916 b Idem., IV. Zool. Jahrb., Abt. f. allg. Zool. und Physiol., Bd. 36.
GEYER, K. 1913 Untersuchungen über die chemiche Zusammensetzung der Insectenhaemolymphe und ihre Bedeutung für die geschlechtliche Differenzierung. Zeitschr. f. wiss. Zool., Bd. 105.
KELLOGG, V. L. 1904 Influence of the primary reproductive organs on the secondary sexual characters. Jour. Exp. Zoöl., vol. 1.

Kopeć, St. 1908 Experimentaluntersuchungen uber die Entwicklung der Geschlechtscharactere bei Schmetterlingen. Bull. Acad. Sc., Cracovie.

1911 Untersuchungen uber Kastration und Transplantation bei Schmetterlingen. Arch. f. Entw. Mech., Bd. 33.

1913 Nochmals uber die unabhängigkeit der Ausbildung sekundärer Geschlechtscharactere von den Gonaden bei Lepidopteren. Zool. Anz., Bd. 43.

Mayer, A. G. 1896 The development of the wing scales and their pigment in butterflies and moths. Bull. Mus. of Comp. Zoöl. of Harvard Coll., 29.

1897 On the color and color-patterns of moths and butterflies. Ibidem, 30.

Meisenheimer, J. 1907 Ergebnisse einiger Versuchsreihen uber Exstirpation und Transplantation der Geschlechtsdrusen bei Schmetterlingen. Zool. Anz., Bd. 32.

1909 Experimentelle Studien zur Soma- und Geschlechtsdifferenzierung. Erster Beitrag. Jena: Fischer.

Oudemans, J. Th. 1899 Falter aus kastrierten Raupen, wie sie aussehen und wie sie sich benehmen. Zool. Jahrb., Abt. f. Syst., Bd. 12.

Prell, H. 1915 a Über die Beziehungen zwischen primären und sekundären Sexualcharakteren bei Schmetterlingen. Zool. Jahrb., Abt. f. allg. Zool. und Physiol., Bd. 35, S. 183.

1915 b Ibidem, S. 593.

Steche, O. 1912 a Die 'sekundären' Geschlechtscharaktere der Insekten und das Problem der Vererbung des Geschlechts. Zeitschr. f. ind. Abst. und Vererbungslehre, Bd. 8.

1912 b Beobachtungen über Geschlechtsunterschiede der Haemolymphe von Insektenlarven. Verh. d. Deutch. Zool. Ges.

Resumen por el autor, E. J. Lund.

El control experimental de la polaridad orgánica por medio
de la corriente eléctrica.

II. La polaridad eléctrica normal de Obelia.   Una prueba
de su existencia.

En el tallo de la colonia de Obelia existe una diferencia definida
de potencial eléctrico, tal que la región apical de crecimiento
es electronegativa con respecto a la región media o más basal.
Esta polaridad eléctrica está asociada con el tejido viviente del
cenosarco y no se origina en ninguna otra estructura del tallo,
porque no existe diferencia de potencial en: a) los tallos que se
abandonan para que mueran y se maceran en agua de mar; b)
los tallow desprovistos mecánicamente del tejido vivo, y c) los
tallos en los cuales el tejido vivo muere mediante la acción del
cloroformo.

La magnitud del descenso de potencial varía en trozos de
tallos de diferentes colonias.   También varía a lo largo del tallo
de la misma colonia, siendo mayor en la región apical en vías de
crecimiento activo.   La conclusión general derivada de los
experimentos es que, puesto que las diferencias normalmente
inherentes del potencial eléctrico tienen lugar en el tallo de
Obelia y están asociadas con el crecimiento apical, antonces
debiera ser posible inhibir o modificar el proceso del desarrollo
por una aplicación apropiada de una E.M.F. de orígen externo.
En un trabajo precedente el autor ha demostrado la posibilidad
de esta modificación.

Translation by José F. Nonidez
Cornell Medical College, New York

EXPERIMENTAL CONTROL OF ORGANIC POLARITY BY THE ELECTRIC CURRENT

II. THE NORMAL ELECTRICAL POLÁRITY OF OBELIA. A PROOF OF ITS EXISTENCE

E. J. LUND

The Puget Sound Marine Biological Laboratory and the Department of Animal Biology, University of Minnesota

TWO FIGURES

In the previous paper (Lund, '21) it was shown that by passing an electric current of proper density lengthwise through an isolated internode from the stem of Obelia, it was possible to inhibit polyp formation on the end turned toward the cathode, while under these same conditions of current density a normal polyp formed on the end of the internode which was turned toward the anode. It was also shown that in order to inhibit regeneration of polyps on apical internodes, it was necessary to use a higher-current density than that which was necessary for inhibition of polyp formation in internodes from basal levels of the stem. The final general conclusion drawn from the experiments was, that since it was possible in this way to inhibit selectively polyp formation, the electromotive force applied from an external source probably acted upon some kind of system in the regenerating internode which was electrical (ionic) in its nature and which in all probability was closely associated with the mechanism which determines the polarity of the regenerating tissue. This conclusion did not rest alone upon the evidence from the experiments reported in the preceding paper, for several investigators have reported more or less convincing evidence that differences of electrical potential occur in hydroid stems of Tubularia, Pennaria, and Campanularia (see, e.g., Mathews, '03). More attention seems to have been paid to these normal continuously existing potential differences in plants than in animals (Buff, '54; Kunkel, '81; Elfving, '82; Burdon-Sanderson, '82, and especially Müller-Hettlingen, '83).

In the present paper evidence will be presented, which it is believed amounts to a proof of the existence of a normal difference of electrical potential in the living tissue of the stem of Obelia, the occurrence of which was suggested by the experiments in the previous paper.

EXPERIMENTAL

The galvanometer which was used to detect the currents obtained by leading off from different levels of the stem, was a Leeds Northrup instrument of somewhat more than one thousand megohm sensitivity.[1] Under the conditions of the procedure in the experiments, it was difficult or impossible to make non-polarizable Zn-$ZnSO_4$ electrodes which maintained a perfect iso-electric condition. Consequently, the simple arrangement of a copper wire inserted in a bent tube filled with sea-water, as shown in figure 1, was used as a readily washable electrode. The two electrodes used gave a sufficiently constant difference of potential for the purpose of the present experiments. The same pair of electrodes were used in all the experiments.[2]

Suppose now that the electrodes are placed at a definite distance apart and a piece of stem is placed upon them such that the ends rest in the drops of sea-water as shown in figure 1. The galvanometer will show a certain deflection due to the P.D. between the electrodes. If, now, the stem is reversed in its position and the galvanometer deflection is identical with the first, then obviously the current is the same. But, if the deflection is less, then either one of two possibilities exists: 1) An inherent P.D. exists in the stem such that it opposes the P.D. between the electrodes or, 2) the ohmic resistance is greater in the latter position than in the first position. If the first of these possibilities is the correct interpretation, then obviously the deflection with the stem in the first position is greater than would occur if no P.D. inherent in the stem existed between the ends of the stem. The correct measure of the current due to the

---

[1] The instrument was kindly loaned by Prof. H. L. Brakel, of the Department of Physics of the University of Washington.

[2] It is apparently impossible to reproduce a Cu-$CuSO_4$ electrode.

P.D. inherent in the stem is therefore represented by one-half the difference between the deflections. By proper calibration of the galvanometer, it therefore would be possible to determine the value of the current due to the P.D. between the ends of the stem, and by direct measurement of the resistance of the stem the P.D. inherent in the stem would of course be known.

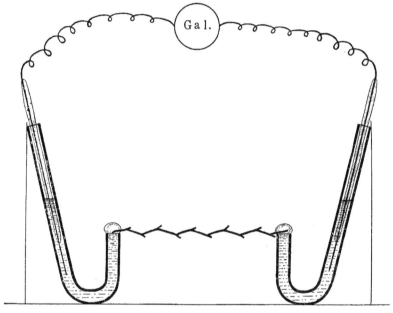

Figure 1

The second interpretation of the difference in deflection, when the stem occupies the two positions on the electrodes, would be that the current can pass more readily in one than in the opposite direction through the stem. The properties of the stem would therefore partially resemble the behavior of a rectifier toward an alternating current.[3] That this possibility should be taken

[3] According to F. Elfving (Bot. Zeit., Bd. 40, 1882, S. 257–263), Kunkel (Arbeit d. bot. Inst. Wurzburg, Bd. 2, Heft 2) found that the resistance of the shoot (seedlings) when a weak 'current' was passed from base to apex through the shoot was less than the resistance of the same shoot when the 'current' was reversed. No experiments on the root are mentioned, but Elfving briefly states that an electric

into consideration will be clear from such experiments as those reported by Bayliss ('11) on certain colloidal salts when uniform distribution in a solvent is prevented by means of a membrane permeable to the anion, but impermeable to the cation. However, it seems that in most if not all such cases a P.D. exists between the outside and inside of the membrane, so that in either case we should have to account at least in part for the difference in deflection by an inherent P.D. in the stem.

In order to make clear the peculiarities of the records given in the tables, a sample series of tests upon an apical piece of the main stem of a colony is given as follows:

TEST	APICAL END OF STEM PLACED ON		DIFFERENCES BETWEEN SUCCESSIVE DEFLECTIONS IN MILLIMETERS
	Left electrode—Deflection in millimeters on left scale	Right electrode—Deflection in millimeters on left scale	
1	21.0 ←——	—————— 22.5	−1.5
2	20.0 ←——	—————— 26.5	−6.5
3	17.5 ←——	— — ——— 24.0	−6.5
4	17.0 ←——	—————— 23.0	−6.0
5	15 0 ←——	—————— 22.5	−7.0
6	14 0 ←——	—————— 20.0	−6 0

The numbers represent the magnitude of the galvanometer deflections in millimeters on the scale. The direction of the deflection was the same in all the tests given in this paper. The arrows indicate the sequence of the tests; thus the first test with apical end of the piece on the right electrode gave a deflection of 22.5 mm. The position of the piece on the electrodes was then reversed. The deflection was now 21 mm. in the same direction on the scale. Returning the piece to its first orientation gave 26.5 mm., and again placing the apical end on the left electrode gave 20 mm. The tests were repeated in this way until six pairs of readings were obtained in serial order. The duration

current, sufficiently weak to be practically harmless when passed through the root of a seedling from base to tip, produced marked injury when passed through the root in the opposite direction. Recent experiments in this laboratory by Mr. Emmett Rowles on the growth of roots through which an electric current was passing seemed to indicate a difference in the effect on growth when the current passed in opposite directions.

of six pairs of tests was approximately ten minutes.   It will be noted that the magnitude of the deflections decreased as time went on, but the differences between the deflections were about the same after the first test.   The decrease in deflection is caused by the increase in resistance of the stem due to evaporation of the trace of water on its surface after removal from sea-water. A proof that this is the correct explanation will be given later when the tests on dead stems are considered.   The stems were manipulated by means of a wet camel's-hair brush.   In a series of tests on the same piece the distance between the electrodes was necessarily always the same.   In the above tests it was 24 mm.

In the last column of the table above are given the differences between the successive galvanometer deflections.   The − sign indicates that the apical end is electronegative to the basal end and, as in the tables which follow, the + sign indicates that the apical end of the piece is electropositive to the basal end.[4]

Since, for lack of space, it is impossible to publish the actual readings of the galvanometer deflections, the differences only between successive galvanometer deflections are given in the tables to follow.   One important thing to be noted is, that since the deflections typically decreased in magnitude in successive tests on the same stem, then if, as sometimes happened, the difference of potential in a stem was small and at the same time the fall in total deflection due to drying was rapid, the differences between such successive deflections would have a + sign, and therefore apparently indicate that the apical end was electropositive to the basal end.   Such a condition sometimes occurred in the tests in the tables given below.   Therefore, the occasional occurrence of a + sign opposite a small (0.5 to 1.5 mm.) difference in deflection in some of the following tables does not necessarily indicate that the apical end was electropositive. Only a consistent occurrence of the + sign in successive large (2 to 10 mm.) differences between deflections can be taken as evidence that the apical end is actually electropositive.

The ohmic resistance of equal lengths of stem from different colonies varies.   This is also true of pieces of equal length from

---

· [4] This refers to the current inside the stem.

different levels in the same stem. The resistance of pieces of living stems 25 mm. in length varied, in a considerable number of determinations, from about 180,000 to 280,000 ohms, depending upon the diameter of the stem and certain other conditions. The resistance of the galvanometer was 1000 ohms and that of the copper—sea-water electrodes was 5300 ohms. Constancy of the deflection when the electrodes are constant will therefore largely depend upon the constancy of the resistance of the stem.

In table 1 A are given five series of tests upon the main stems of five different colonies designated by the numbers at the top of the table. The apical three-fourths of the main stem was isolated from its branches in each case. Immediately after removal of the branches, the tests were made in the manner described above. It will be seen by comparing the differences between deflections that a definite and marked difference of electrical potential occurs. The average differences between the deflections in the last five tests in A were 6.1, 8.1, 3.8, 9.6, 8.6 mm. There is a perfect uniformity in direction of the P.D. inherent in the stem, such that the apical end is electronegative to the basal end. These pieces of stem were now placed in finger-bowls in fresh sea-water and allowed to regenerate for twenty-four hours. The tests were then repeated with the results given below in table 1 B. Comparison of the differences between deflections will show that these have decreased markedly during the preceding twenty-four hours. The differences are, however, quite definite and generally in the same direction as in the previous tests.

Two more experiments were carried out in which a total of ten stems were used. The period of regeneration between the tests was forty-eight hours. A marked decrease in the P.D. occurred in nine of the stems, but the original direction of the P.D. was apparently retained. One stem showed no P.D. between its ends at the end of the forty-eight-hour period.

The following tentative explanation of this decrease in P.D. is offered at the present time. From the above experiments and others to follow it is easily shown that the apical end which is the actively growing region of the stem is electronegative to the older basal end. Now, after isolation of the piece of the stem, the regeneration process sets in at the basal as well as the apical end and also at the cut ends of the branches. Usually

TABLE 1

Electrical difference of potential between ends of apical three-fourths of the main stems of Obelia colonies; the latter are designated by the numbers at the top of the table. Numbers represent differences between successive galvanometer deflections on the scale. A gives the tests on the main stem made immediately after removing branches. B, tests made twenty-four hours after removing branches, i.e., at the end of twenty-four hours' regeneration. The — sign indicates that the apical end is electronegative to the basal end. The + sign may or may not indicate the opposite condition. See text for explanation

	DIFFERENCE BETWEEN SUCCESSIVE GALVANOMETER DEFLECTIONS				
	1	2	ȣ	4	5
A	− 0.5	−7.0	−2.5	− 7.0	0 0
	− 7 0	−6.0	0.0	− .	− 4.
	− 8.0	−7.5	−3.5	− .	−10.
	− 9.5	−8.5	−3.5	− .	−10.
	− 7.5	−9.5	−4.0	−1 .	− 8.
	+ 1.0	−9.0	−8 0	− .	− 6.
	− 6.5			− .	− 8.
Average......	− 6.1	−8.1	−3 8	− 9.6	− 8.6
B	−14.0	0 0	+3.5	+ 2.5	+ 4.0
	− 8.5	0 0	+1.5	+ .	0
	− 3 0	− .	+ .	+	− 1.
	− 3 0	+ .	−	−	− 0.
	− 2 0	+	− .	−	+ 1.
	− 2.0	−	− .	− .	− 2.
	+ 1.0	+ .	.	− .	− 1.
	− 1.5	− .	− .	− .	− 1.
		− .		− 2.5	− 2.
		− .			− 0.
		− .			
		−			
Average	− 1.5	−2.7	−1.4	− 2.1	− 1.6

polyps appear at all these places. This means of course that after twenty-four hours of regeneration of the basal end the tissue has taken on the rôle and properties of a growing tissue similar to that at the apical end, and consequently its electrical potential should approach that of the apical end, with the result that the potential difference between the basal and apical ends should tend to disappear, which it does.

PROOF THAT THE DIFFERENCE OF ELECTRICAL POTENTIAL IS ASSOCIATED WITH THE LIVING TISSUE IN THE STEM

Differences in electrical potential such as described above for Obelia and originally for Tubularia, Pennaria, and Campanularia by Mathews ('03) might readily be conceived to originate elsewhere than in the living tissues of the stem. It is therefore necessary to remove completely any doubt on this point before we proceed with the analysis of the problem. Three different methods of testing the question will be given.

Dead stems. Table 2 shows the results of tests on stems isolated from colonies left to die and macerate in sea-water in the laboratory for several weeks. It is clear from a comparison of the differences between deflections that no detectible P.D. exists between the ends of the stem.

Living stems from which the tissue has been removed mechanically. The most striking proof of an inherent P.D. in the coenosarc of the stem is given in table 3. Stems isolated from five different actively growing colonies were tested immediately. The results are given in table 3 A. All the tests show that a marked P.D. exists between the ends of the stem and that the apical end of the stem is electronegative to the basal end. After the tests in A were completed, the coenosarc was mechanically removed by rolling the round end of a glass rod along the stem. The perisarc tube is elastic and returns perfectly to its original shape, automatically becoming washed free from cells and finally filled with sea-water instead of the living tissue. In this way an ideal control for a crucial test is provided. Table 3 B gives the results of the tests on the perisarc tubes filled with sea-water. No potential difference is evident. But, as was explained above in connection with table 1, the deflections in successive tests on the same perisarc tube decreased. This is due to an increase in ohmic resistance of the stem perisarc, caused by evaporation. The results in table 1 should be judged in the light of the results brought out in tables 2 and 3, and the experiments given below.

TABLE 2

Showing absence of an electrical difference of potential in stems from two colonies which had been allowed to die in sea-water. A gives the tests on the main stem immediately after removing branches. B, tests on same stems after washing in fresh sea-water

		DIFFERENCES BETWEEN GALVANOMETER DEFLECTIONS	
		1	2
A		−1.0 +. +. −. −. −. +. +. −. +. −.	+1.5 +1. +0. 0. −0. −1. −0. −0. −0. +0. +0.
Average.................		−0.05	+0.1
B		−1.5 −. −. −. −. −. −. −. −.	−0.5 0. −0. 0. 0. −0. 0.
Average.................		−0.7	−0.2

Stems killed in chloroform. The apical halves of the main stem of five different colonies were isolated and the apical internode removed from each one. The pieces were tested immediately. In table 4 A are given the results. It will be noted that all but stem no. 3 show a definite P.D. in the same direction as that in the previous experiments. The question naturally arises, why did not stem no. 3 show a more definite and larger P.D.? Some light on this question will be given later.

TABLE 3

Showing that the electrical difference of potential in a living stem disappears after the living tissue (coenosarc) in the stem has been removed mechanically and sea-water substituted for it in the perisarc tube. A, tests on living stem with coenosarc. B, tests on same stems after mechanical removal of coenosarc. Numbers at top of table refer to main stems of five different colonies. The − sign indicates that the apical end is electronegative to the basal end

		DIFFERENCES BETWEEN SUCCESSIVE GALVANOMETER DEFLECTIONS			
	1	2	3	4	5
A	−1.5	−5.0	−5 0	− 5.0	− 2.5
	−7.0	−8 0	−5.5	− 7.5	− 4.5
	−5.0	−8 0	−5 0	− 7.0	− 6.5
	−5 0	−8.0	−5.0	− 8.5	− 8.5
	−5.5	−9 0	−5.5	−10 0	−11.0
	−6 0	−9.5	−6.5	−10 5	−11.0
Average......	−5.0	−8.0	−5.4	− 8 0	− 7.3
B	+4.	−3 0	+1.5	0 0	0.0
	−1.	−.	−1.0	0.0	0.0
	0.	.	0 0	0 0	0.0
	0	−	0 0	− 0.5	0.0
	−0	− .	−1.0	0 0	− 0.5
	0 0	− .	−0 5	− 0.5	+ 0.5
		−			
		−			
Average	+0 4	−1 5	−0.1	− 0.1	0.0

After the tests in A were made, all the pieces were treated with chloroform-saturated water for thirty minutes, then quickly rinsed in fresh sea-water and tested again. The tests after this chloroform treatment are given in B. It is clear that the P.D. has practically entirely disappeared. The differences in stems nos. 1 and 4 are small but definite. Retreatment of stems 1 and 4 obliterated all trace of P.D. The average differences between the deflections as given in the table are somewhat misleading, for they are the averages of all the tests rather than the averages of the later tests in each series, which, as explained above, are the most significant, e.g., stem no. 3 A. In general, differences between galvanometer deflections of 1 mm. on the scale proved to be significant whenever the electrical resistance' of the stem

TABLE 4

Showing that treatment of the living stem with chloroform-saturated sea-water causes the disappearance of the normal difference of potential in the stem. A, tests made on living stems immediately after removal of branches. B, tests on the same stems after treatment in chloroform-saturated sea-water

	DIFFERENCES BETWEEN SUCCESSIVE GALVANOMETER DEFLECTIONS				
	1	2	3	4	5
A	+2.0	−1.0	+2.5	+0 5	−3.0
	+1.0	−1.5	+1.5	0.0	−7.0
	0.0	−2 0	1.0	−2.0	−8.5
	0.0	−3.0	— .	−2.0	−8.0
	−4.0	−5 5	— .	−2.5	−7.5
	−6.5	−2 0	— .	−3.5	−6.0
	−5.0	−4 5	— .		
	−5.0				
Average......	−2.2	−2.9	−0.3	−1.6	−6.3
B	+1.5	0 0	+3.0	+1.5	0.0
	+1.0	+1.0	+1.0	0.0	0.0
	0.0	−0.5	+1.5	0 0	−0.5
	0.0	0.0	−0.5	−1.5	−0.5
	−0.5	0.0	0.0	−1.5	−1.0
	−1.0		0 0		−0.5
	−1.5				
Average......	−0.07	−0.1	+0.9	−0.3	−0.4

remained approximately constant. The limit of error is therefore to be judged to some extent from the conditions of the experiment and nature of the results.

All the stems were now placed in fresh sea-water. No regeneration occurred, showing that loss of the normal P.D. in the stem was associated with death of the living tissue. A few experiments were carried out in which various concentrations of ether were used. The results indicate that a definite decrease

and removal of the P.D. can be brought about by this anaesthetic.
The complete investigation of the effect of anaesthetics upon the
P.D. will be left for later consideration. The purpose of the
previous experiments is simply to furnish a conclusive demonstration of two facts: 1) A normal difference of electrical potential exists along the stem of Obelia. 2) This potential difference
is associated with the living condition of the tissue of the stem.

TABLE 5

*A represents a series of tests on apical pieces of the main stems of five different colo-
nies. B represents a series of tests on the corresponding basal pieces of the stems
of the same colonies. The lengths of apical and basal pieces of the same stem are
equal. Note that the potential differences in the apical and basal pieces are opposite in direction*

	DIFFERENCES BETWEEN SUCCESSIVE GALVANOMETER DEFLECTIONS				
	1	2	3	4	5
A	−3.0 −3.5 −3.0 −2.5	−2.5 −1.5 −1.5	−2.0 −3.5 −4.5	−7.5 −5.5 −5.5	−9.0 −8.5 −7.5
Average......	−3.0	−1.8	−3 3	−6.8	−8.3
B	+5.0 +5.5 +5.0	+2.5 +1.5 +1.0	+3.0 +2.0 +2.0	+4.5 +3.5 0 0 +0.5	+6.5 +6.5 +5.5
Average......	+5.1	+1.8	+2.3	+2.1	+6.1

THE DIRECTION AND MAGNITUDE OF THE FALL OF ELECTRICAL POTENTIAL ALONG THE STEM

Because of the nature of the method used in the preceding and
following experiments, it should be clear that the results are not
what would be desired in a quantitative investigation of this
problem. But, nevertheless, as will be indicated in the following
two experiments, some idea may be gained of the magnitude and
direction of the fall of electrical potential along the stem by
comparing the magnitude of the differences in deflection produced
by apical and basal halves or quarters of the same stem.

During the work represented in part by the experiments given above, the impression was gained that the amount of the P.D. in comparable pieces from different stems was distinctly different. In some pieces of stem very little, if any, difference could be detected, for example, table 4 A, stem 3, while in other pieces it would be relatively large. Therefore it was decided to compare in a preliminary way the magnitude and direction of the differences between deflections when pieces from the apical and more basal regions of the same stem were used. Three experiments were carried out, in each of which five main stems of actively growing colonies were used. In the first experiment two pieces from the same stem were compared. The total length of each of the stems

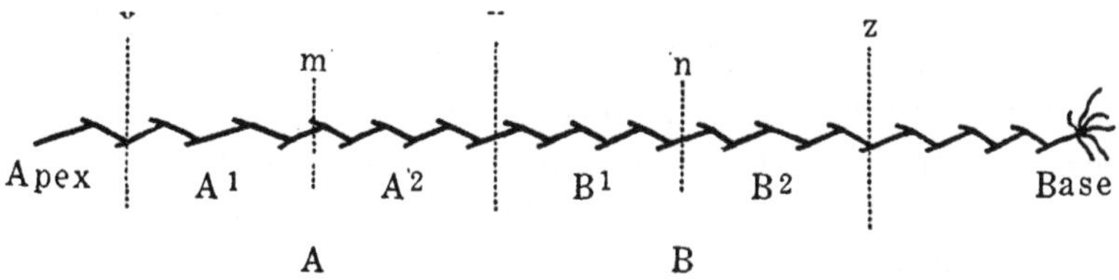

Figure 2

numbered 1 to 5 in table 5 were, respectively, 110, 80, 100, 110 115 mm. Three cuts were made, as shown in figure 2, x, y, and z. The lengths of the pieces included between x and z in stems 1 to 5 were, respectively, 75. 70, 64 and 66 mm. The cut at y was made exactly in the middle so as to give two pieces A and B of equal length. The tests on the A pieces are given in table 5 A and the tests on the corresponding B pieces in B of the same table. There is a distinct P.D. in the A pieces such that the apical ends are all electronegative to the basal end of the same piece. While in the B pieces the direction of the P.D. is reversed. The magnitude of the P.D. varies but is quite distinct in every case. A second experiment similar to the first except that the cuts y and z, figure 2, were made proportionately nearer the apical end of the stem, again showed that the apical ends of the A pieces were electronegative and that in three of the B pieces the basal end was electronegative to the apical end. The other two B pieces showed no significant P.D.

This peculiar opposed direction of fall of potential in the apical and basal pieces of the same stem suggested a further test by the procedure in the following experiment. Five actively growing colonies, the main stems of which are numbered 1 to 5 in table 6, were used. The total length of the stems were, respectively, 100, 85, 75, 115, and 95 mm. The piece from each stem included between the cuts x and z in figure 2 was tested first. The lengths of the pieces from stems 1 to 5 were, respectively, 64, 57, 57, 83 and 70 mm   The results are given in table 6 W. It will be noticed that while the apical ends in four of the stems are electronegative to the basal ends, still the differences are small. In the piece from stem no. 1 no P.D. could be detected. Each one of the five pieces was now cut at y, figure 2, into two pieces of exactly equal length and then tested immediately. The results from the apical pieces are given in table 6 A, while the corresponding tests on the B pieces are given in the same table in B. It will be observed that the P.D. in every one of the A pieces is quite large, the apical end again being electronegative to the basal end. All the B pieces show a marked P.D. in the opposite direction; this difference is, however, not as large in the B pieces as in the corresponding A pieces.

Each one of the A and B pieces was now cut into two equal parts at m and n, figure 2, and tested at once. The pieces from the same stem are numbered A^1, A^2, B^1, B^2, as in figure 2. It will be seen that the A^1 pieces show an unmistakable electronegative condition of the apical end   Pieces A^2 of stems 1, 4 and 5 show the same condition as the A pieces, while the A^2 pieces of stems nos. 2 and 3 are doubtful. The B^1 pieces also show individual differences, while the B^2 pieces of stems nos. 1, 3, 4, and 5 show a quite distinct P.D. in the opposite direction. Piece B^2 of stem no. 2 is doubtful.

From the experiments above it appears that the apical or growing end of the stem shows the most distinct difference of potential. The middle pieces vary more or less, depending upon the individual stem, while the basal pieces tend to have the fall of potential reversed.

· TABLE 6

W represents a series of tests upon the 'whole' (see text) stem of five different colonies. A represents corresponding tests on apical half and B a corresponding series of tests upon the basal half of the 'whole' stem after cutting the latter into equal halves. Note the small P.D. in W and the opposite direction of the potential in the A and B pieces. A¹ and A² represent tests upon apical and basal halves, respectively, of the A pieces. B¹ and B² represent tests upon apical and basal halves, respectively, of the B pieces. Note the definite difference of potential in the apical pieces A¹ and the marked tendency to reversed direction of fall of potential on the basal pieces B²

	DIFFERENCES BETWEEN SUCCESSIVE GALVANOMETER DEFLECTIONS				
	1	2	3	4	5
W	0 0	−1 0	− 0.5	− 0.5	− 1.0
	0 0	−0 5	− 1 0	− 1.5	− 1.5
	0 0	−0 5	− 1.0	− 2 5	− 2 0
		−1.5			
Average......	0 0	−0 8	− 0 8	− 1.5	− 1.5
A	− 3 5	−6.5	− 3.0	− 6 0	− 8.5
	− 4 0	−5 0	− 2.5	− 7 0	− 8.5
	− 5.0	−4.5	− 4 5	− 7.0	− 5.5
Average......	− 4.2	−5.3	− 3.3	− 6 6	− 7.5
B	+ 0.5	+2.5	+ 2 0	+ 2 5	+ 2.0
	+ 2.0	+5.5	+ 2 0	· + 4.5	+ 3.0
	+ 3.0	+7.0	+ 2 0	+ 4.0	+ 3.0
	+ 1.5				
Average	+ 1.7	+5.0	+ 2.0	+ 3.6	+ 2.6
A¹	− 2 0	−6 0	− 8.0	− 5 0	−10 0
	− 9.5	−5 0	−10 0	− 5 0	− 9.5
	− 6.0	−3 0	−12 5	− 6 5	− 9 0
Average	− 5 8	−4 6	−10 2	− 5 3	− 9 5
A²	− 4 0	+3 0	− 0 5	− 6 5	− 8.5
	− 5.0	−0.5	− 2 5	− 6 5	− 3.5
	− 6.0	+3 5	0.0	−11.0	− 6.5
Average... ..	− 5 0	+2 0	− 1 0	− 8.0	− 6.1
B¹	−10 0	−6.5	+ 2 0	+ 0 5	− 5.5
	− 4.5	−1.0	+ 4.5	+ 2 5	− 1.5
	− 5 5	+1.0	+ 3 0	+ 1 5	− 2.5
	− 5.0		+ 0.5		
Average	− 6.2	−2.1	+ 2 5	+ 1.5	− 3 0
B²	− 3 0	−2.5	+ 3.5	+ 2.5	+ 2.0
	+ 4.0	−9.5	+ 1.5	+ 3 0	+ 3.5
	+ 3.0	+3.0	+ 4.0	+ 4 0	+ 4.5
	+ 2.5				
Average. ...	+ 1.6	.	+ 3.0	+ 3.1	+ 3.3

All that the writer desires to show by these experiments is: 1) that the magnitude of the fall of electrical potential differs along the length of the stem and, 2) that under the conditions of these experiments, the direction of the fall of potential may be different in more basal regions than in the apical part of the same stem. This latter result appears somewhat surprising, in view of the general morphological similarity of the different levels of the stem. It appears that the small potential difference (or absence of P.D., table 6 W, 1) shown by the whole pieces between cuts x and z was due to the fact that the apical and basal ends of these pieces were both electronegative to the middle region of the piece. This result suggests a similarity to the results obtained by Müller-Hettlingen ('83) on growing seedlings.

CONCLUSIONS

It is evident that the observations reported in this paper appear complementary to the results reported in the first paper on the effects of an electric current on the regeneration process in the internodes of Obelia. Now, if differences of electrical potential are inseparably associated with structural polarity and apical growth, then it follows that the stem or certain regions of the stem of Obelia—and also very probably each branch—serves as a conductor of a continuous electric current of perhaps varying intensity, which passes within the tissues of the stem and completes the circuit in the surrounding sea-water. If the basal region is electronegative to the middle region of the stem, then of course the direction of the current would be the opposite in the lower parts of the stem to that in the apical region of the stem. In any case, if these constant bioelectric currents are inseparably associated with apical or basal growth, it is logical to expect that if they be appropriately opposed or augmented by the application of an external source of E.M.F., then the normal growth processes should show the effects of such interference.

If this reasoning is correct, then an appropriate application of an external electromotive force to regenerating and growing tissues should be a unique instrument for the investigation of the causes and conditions of structural polarity and symmetries which are everywhere a common fundamental characteristic of the individual in organic nature, whether it be a cell, group of cells, or whole organism.

There is another significant point to which I wish to call attention. If cells, cell groups, and individuals, e.g., polyps in a structurally and physiologically correlated system like Obelia, possess an inherent mechanism for establishing directed electromotive forces within themselves in relation to their structural or functional polarities, then it will be clear that here we may have a normal electrical mechanism which may serve as a determining factor in the process of orientation of cells with respect to one another during embryogeny and regeneration. I do not wish to carry this line of reasoning farther in this paper than to suggest the possibilities. In following papers will be presented experimental evidence relative to this question.

Because of the nature of the experimental method employed in the experiments above, it was necessary to manipulate the stems and pieces in air. This obviously limited the degree of accuracy of the results A further detailed and quantitative study of the phenomena will be undertaken when appropriate and sufficiently accurate methods for measurement have been devised.

SUMMARY

1. A definite difference of electrical potential occurs in the stem of the colony of Obelia. This confirms the conclusion arrived at by means of a different method in the preceding paper, that an electrical polarity is one primary associated condition for the development of morphological polarity.

2. The electrical polarity in the stem of Obelia is associated with the living tissue of the coenosarc and does not originate in any other structure of the stem for no potential difference occurs in the following: a) Stems left to die and macerate in sea-water; b) stems from which the living tissue has been removed mechanically, and, c) stems in which the living tissue has been killed by chloroform. ·

3. The potential difference in the stem is not associated directly or simply with the mechanical injury due to cutting, and there-

fore is not identical with the current of injury in, for example, muscle, nerve, and other tissues.

4. The magnitude of the fall of potential varies in pieces of stems from different colonies. It also varies along the length of the stem of the same colony, being greatest in the apical actively growing region.

5. A limited number of experiments indicated that the direction of the fall of potential in basal regions of the stem may be opposite to that in the apical region.

6. The difference of electrical potential between apical and basal ends of pieces of stem after twenty-four or forty-eight hours of regeneration was less than the potential difference in the same pieces immediately after isolation. A tentative explanation of this result is given in the text.

7. The general conclusion from the experiments is, that since normally inherent differences of electrical potential occur in the stem of Obelia and are associated with apical growth, then it should be possible to inhibit or modify developmental processes by appropriate application of an E.M.F. of external origin. This was shown to be possible in the previous paper.

LITERATURE CITED

BAYLISS, W. M. 1911 IIJ. The osmotic pressure of electrolytically dissociated colloids. Proc. Roy. Soc. Lond., vol. 84 B, pp. 229–253.

BUFF, H. 1854 Ueber die Electricitatserregung durch lebende Pflanzen. Annalen d. Chem. u. Pharm., Bd. 89, S. 76–89.

BURDON-SANDERSON 1882 On the electromotive properties of the leaf of Dionea in excited and unexcited states. Phil. Trans., vol. 173, pt. I, pp. 1–56.

ELFVING, F. 1882 Ueber eine Wirkung des galvanischen Stromes auf wachsende Wurzeln. Bot. Zeit., Bd. 40, S. 257- 264.

KUNKEL, A. J. 1881 Electrische Untersuchungen an pflanzlichen und tierischen Gebilden. Pflug. Arch., Bd. 25, S. 342–379.

LUND, E. J. 1921 I. Effects of the electric current on regenerating internodes of Obelia commissuralis.

MATHEWS, A. P. 1903 Electrical polarity in the hydroids. Am. Jour. Physiol., vol. 8, pp. 294–299.

MÜLLER-HETTLINGEN, J. 1883 Ueber galvanische Erscheinungen an keimenden Samen. Pflug. Arch., Bd. 31, S. 193-214.

SUBJECT AND AUTHOR INDEX

495

9 780483 806016